HANDBOOK OF SWITCHGEARS

HANDBOOK OF SWITCHGEARS

BHEL
Bharat Heavy Electricals Limited

McGraw-Hill
New York Chicago San Francisco Lisbon London Madrid
Mexico City Milan New Delhi San Juan Seoul
Singapore Sydney Toronto

The McGraw·Hill Companies

Cataloging-in-Publication Data is on file with the Library of Congress.

Copyright © 2007 by The McGraw-Hill Companies, Inc. All rights reserved. Printed in the United States of America. Except as permitted under the United States Copyright Act of 1976, no part of this publication may be reproduced or distributed in any form or by any means, or stored in a data base or retrieval system, without the prior written permission of the publisher.

1 2 3 4 5 6 7 8 9 0   BKM BKM   0 1 3 2 1 0 9 8 7 6

ISBN-13: 978-0-07-147696-6
ISBN-10: 0-07-147696-2

This book was first published in India in 2005 by Tata McGraw-Hill.

The sponsoring editor for this book was Stephen S. Chapman and the production supervisor was Richard C. Ruzycka. The art director for the cover was Handel Low.

This book was printed on acid-free paper.

McGraw-Hill books are available at special quantity discounts to use as premiums and sales promotions, or for use in corporate training programs. For more information, please write to the Director of Special Sales, McGraw-Hill Professional, Two Penn Plaza, New York, NY 10121-2298. Or contact your local bookstore.

Information contained in this work has been obtained by The McGraw-Hill Companies, Inc. ("McGraw-Hill") from sources believed to be reliable. However, neither McGraw-Hill nor its authors guarantee the accuracy or completeness of any information published herein, and neither McGraw-Hill nor its authors shall be responsible for any errors, omissions, or damages arising out of use of this information. This work is published with the understanding that McGraw-Hill and its authors are supplying information but are not attempting to render engineering or other professional services. If such services are required, the assistance of an appropriate professional should be sought.

FOREWORD

It gives me immense pleasure to be writing the Foreword to this book.

Bharat Heavy Electrical Limited (BHEL) is the largest engineering and manufacturing enterprise of its kind in India, and one of the leading international companies in the field of power equipment manufacturing. BHEL started manufacturing switchgears in 1961 and has since made significant strides in the field. In this book, efforts have been made to reflect and put forward the vast experience and expertise gained by BHEL over these 44 years. The book fulfils the long felt need of practicing power engineers, who had otherwise to refer to several articles to solve their various queries on switchgear design, application, and protection. It makes clear that the choice and use of switchgear is far from standard, and that a thorough knowledge of the matter is a must.

I personally find this book, brought out by one of the leading organisations in the switchgear manufacturing industry in India, informative, exhaustive, and practical in its approach.

The book covers not only the general concepts but also provides consolidated information on switchgears in the context of Indian conditions. This approach is beneficial for the utilities practising engineers. Starting from the basics, it enables even engineering students to understand the subject. The topics of each chapter have been skillfully selected and cover all aspects of switchgears.

I sincerely hope that this book will serve as a useful reference, not only to practicing switchgear engineers but also to institutions and academicians.

I wish all success to this publication.

Prof. Dr. René Peter Paul Smeets
KEMA High-Power Laboratories
The Netherlands

PREFACE

Switchgear is an important link in any power system network, including transmission and distribution systems. These days, increased emphasis is being given to designing the best possible switchgear and associated equipment system. This book aims to provide a holistic view of all the aspects of switchgear, and highlights the role it plays alongwith its associated systems, in the control and regulation of power flow in power system network.

Bharat Heavy Electricals Ltd (BHEL) is a pioneer in the switchgear industry having more than 40 years' experience with various circuit breaker technologies. This book is the collective effort of 22 experts from BHEL, a leading switchgear manufacturer in India.

Chapter 1 explains comprehensively the technological aspects of medium and high voltage circuit breakers ranging from 3.3 kV to 800 kV class and also the comparison of various switching technologies. The arcing phenomenon has been discussed in detail to explain and establish the superiority of the technology. Different types of operating mechanisms, viz. spring, pneumatic, hydraulic, electrical and magnetic have been elaborated along with their relative advantages and disadvantages. It is rare to find all these details in one place in a consolidated manner and as such serves as very useful information for practising engineers and learners.

The requirements of conventional low voltage switchgear and emerging trends for various applications have been briefly brought out in Chapter 2.

Chapter 3 deals elaborately with medium voltage switchgears rated from 3.3 kV to 33 kV class and provides information about various forms of construction in metal enclosed, metal clad and porcelain clad design. To bring more clarity on constructional aspects, the definition of various terms have been taken from relevant IEC and IS. Present and future trends in medium voltage switchgears have also been explained.

High voltage circuit breaker ranging from 66 kV to 800 kV class have been covered in Chapter 4. This elaborates on the historical background, present status and future outlook of technological developments. The chapter also provides insight into various interrupting media like air, oil and SF6 gas. The latest trends in the design of SF6 gas circuit breakers and operating mechanisms are discussed.

Since the information available on GIS is mostly with reference to global conditions, efforts have been made in Chapter 5 to provide details relevant to the Indian environment with reference to material availability, operating conditions at site and conditions in Indian industry.

Chapter 6 is entirely devoted to establish usefulness of auto-reclosers and sectionalisers. The case study on auto-reclosers clearly highlights the great potential this product has for rural electrification and power distribution to important load centres in India.

The theory, design and constructional aspects of instruments transformer along with application and testing requirements have been fully covered in Chapter 7. The details of instruments transformer testing and analysis of testing results along with standardisation and cost considerations have been provided based on BHEL experience and expertise gathered during long association with switchgears.

Chapter 8 narrates various types of surge arresters, their characteristics, construction and application aspects. The details of experimental study in the use of combinations of surge protection devices forms very useful information for all practicing engineers.

The basics of various relay technologies have been dealt with in Chapter 9. The cost comparison of numerical relay with electromechanical and static relay and their advantages are covered in this chapter. This information is useful for learners in protection field.

Chapter 10 strives to explain various aspects of application of medium voltage switchgear and highlights the considerations for selection of suitable circuit breakers for different applications. The details given are based on BHEL's rich experience in the field of switchgears.

Energy metering, as covered in Chapter 11, emphasises the features of modern static meters and state-of-the-art technology which BHEL has adopted in many prestigious power projects.

Chapter 12 traces a variety of control and interlocking schemes used in medium voltage switchgear which are mostly application specific. These schemes are based on experience gathered by BHEL over a period of time.

Protection schemes for medium voltage switchgears have been dealt with in Chapter 13 and cover all protection aspects. The details are expected to provide an insight to beginners and students of protection philosophy.

Chapter 14 deals with generator protection for large generators rated 500 MW and above. The details regarding classification of tripping and grouping of protection is the result of extensive experience and expertise gained by BHEL in the manufacturing of large generators and their protection systems.

Information on EHV transmission system control and protection is provided in Chapter 15, fully based on the latest practices in manufacturing of control and relay panels for switchyard projects up to 400 kV class.

Chapter 16 covers the basics of integrated protection and control system for sub-station and is based on experience gained during the development of numerical protective relays and SCADA system at corporate R&D unit of BHEL.

Chapter 17 captures the erection and commission requirements of switchgear. All aspects regarding requirements and precautions for erection and commissioning of indoor and outdoor switchgear for satisfactory performance have been covered in this chapter.

The operation and maintenance aspects of switchgear under various conditions have been covered in Chapter 18. It provides useful information about the dos and donts to be followed during site work, inspection and on maintenance procedures.

The performance of a switchgear under normal/abnormal conditions is verified by subjecting it to different type of tests as per the guidelines laid down in national/international standards. Chapter 19 briefly covers the significance and procedures used for different tests with as many practical cases as possible.

Chapter 20 is a compilation of details of all major standard organisations and various applicable standards for switchgear products. This serves as readily available information for practising engineers.

The details regarding statutory requirements and electricity rules covered under Chapter 21 provides necessary guidelines to switchgear designers and site engineers to ensure safety of personnel and property.

The book reflects the expertise gained by BHEL in the field of switchgears and offers a wide and thorough coverage. The work here is informative as it follows a practical approach and highlights all salient features of switchgear. I am confident that the book will serve as a useful reference for all practising engineers and students.

P.T. Deo
Executive Director
BHEL, Bhopal

EDITORIAL COMMITTEE

Shri R K Shukla, Chief (Switchgear Engineering), has eight years' experience in transformer and tap changer design, plus 25 years in switchgear design and development. He holds a BE in Electrical from Ravishankar Shukla Technology, Raipur.

Shri Indrapal, Additional General Manager, has 30 years experience in LV and MV switchgears. He holds a BSc Engg (Electrical) from Aligarh University.

Shri S G Deshpande, Sr. Dy. General Manager, has 32 years' experience in design of control and protection for power system application. He holds a BE Electrical from Devi Ahilya University, Indore.

CONTRIBUTORS

H S Jain, Addl. General Manager, has 30 years' experience in development of switchgear and gas insulated power equipment at Corp. R&D. He has a BE Electrical from Jiwaji University, Gwalior, and a Ph D from IIT, Bombay.

M P Soni, Addl. General Manager, has worked for 28 years in R&D of power system controls and protection at Corp. R&D. He holds a BE Electrical and a Masters in Control Systems from Jabalpur University, and a Ph D from IIT, Mumbai.

N N Shenoy, Addl. General Manager, has worked on design of tap changer for 10 years and on switchgear application for 20 years. He holds a BE Electrical from Kerala University.

S K Sahasrabudhe, Sr. Dy. General Manager, has 15 years' experience in design of control and protection for power system application and 11 years in switchgear application, He has a BE Electrical from Pune University and an M Tech in Power Systems from Nagpur University.

D K Dikshit, Sr. Dy. General Manager, has worked for 25 years in design and development of medium voltage switchgear. He holds a BE Electrical from Bhopal University.

A K Murgai, Sr. Dy. General Manager, has worked for 25 years in design of control and protection for power system application. He holds a B Tech Electrical from IIT, Delhi.

Santosh Kumar, Dy. General Manager, has 24 years' experience in design and development of medium voltage switchgear. He has a B Sc in Mechanical Engineering from Ranchi University, and an M Tech from Barkatullah University, Bhopal.

M K Shrivastava, Dy. General Manager, has 24 years' experience in design of HT Motor Controls, AVR & EHG, DC HSCBs and medium voltage switchgear application. He holds a BE Electrical from lIT, Roorkee, and an MBA in Marketing and Finance from Barkatullah University, Bhopal.

Neelam Bhogal, Dy. General Manager, has 24 years' experience in switchgear design and application. She holds a BE Electrical from Bhopal University and a PGDOM from IGNOU.

Shubra Chaturvedi, Sr. Manager, has worked for 15 years on design of motor crane control and for five years in the design of outdoor switchgear and autorecloser sectionalizer. She holds a BE in Electronics and Telecommunication from Devi Ahilya University, Indore.

P K Upadhyay, Manager, has 12 years' experience in design of control and protection for power system application. He has a BE in Electronics and Telecommunication from Barkatullah University, Bhopal.

Sanjay Dhavle, Manager, has worked for 12 years in switchgear application. He holds a BE Electrical from Rani Durgawati University, Jabalpur.

Vivek Johri, Dy. Manager, has 18 years' experience in switchgear design and application. He has a BE Electrical from Rani Durgawati University, Jabalpur.

H R Patel, Dy. Manager, has worked for 13 years in switchgear design and application. He has a BE Electrical from Guru Ghasidas University, Bilaspur.

M Mohan Rao, Sr. Engineer, has nine years' experience in R&D activities related with GIS at Corp. R&D. He has a B Tech from Venkateshwara University, Tirupati, and a Masters in High Voltage Engineering from IIS, Bangalore.

M A Khan, Sr. Engineer, has four years' experience in power plant control systems and five years in switchgear design and application. He holds a B Tech Electrical from HP University, Shimla, and an M Tech in Power Systems from Kurukshetra University.

Divya Bhat, Sr. Engineer, has six years' experience in switchgear design and a BE Electrical from Bhopal University.

Manisha Singh, Sr. Engineer, has six years' experience in switchgear design and application and holds a BE Electrical from Devi Ahilya University, Indore.

Murtaza Hussain, Sr. Engineer, has five years' experience in switchgear design and application. He holds a BE Electrical from Ravishankar Shukla University, Raipur, and an M Tech in Control Systems and Infrastructure from IIT, Delhi.

Yellaiah Babu B, Sr. Dy. General Manager, has 29 years' experience in design of EHV switchgear. He holds a BE Electrical from Osmania University, Hyderabad, and an M Tech from IIT, Chennai.

INTRODUCTION

Switchgear is a generic term and includes the entire range of switching devices and their combination with associated control, measuring, protecting and regulating equipment. The assemblies of such devices and equipment with associated inter-connections, accessories, enclosures and supporting structures, intended, in principle, for use in connection with the generation, transmission, distribution and conversion of electric energy also form a switchgear.

We all are familiar with low voltage switches and re-wireable fuses in our homes. A switch is used for opening and closing an electric circuit while a fuse is used for over-current and short-circuit protection. Every electric circuit needs a switching device and a protective device. The switching and protective devices have been developed in various forms. Thus switchgear can be taken as a general term covering a wide range of equipment concerned with the switching, protection and control of various electrical equipments.

Similarly, switching and current interrupting devices play a significant role in the modern electrical network, right from generating stations, transmission sub-stations at different voltages, distribution sub-stations and load centres. The switching device here is called a circuit breaker. The circuit breaker, along with associated devices for protection, metering and control regulation, is called a switchgear.

A switchgear has to perform the functions of carrying, making and breaking the normal load current like a switch. In addition, it has to perform the function of clearing the fault current for which sensing devices like current transformers, potential transformers and various types of relays, depending on the application, are employed. There also has to be provision for metering, controlling and data, wherein innumerable devices are used for achieving the switchgear function. Thus a switchgear can include a circuit breaker, current transformers, potential transformers, protective relays, measuring instruments, switches, fuses, MCBs, surge arrestors, isolators and various associated equipments.

Switchgears are necessary at every switching point in the power system for the reasons mentioned above. There are several voltage levels and fault levels between the generating stations and final load points. Hence in the various applications, the requirements of switchgears vary depending on the location, ratings and local requirements. Besides the supply network, switchgear is required in industrial works, industrial projects, domestic and commercial buildings.

BHEL has very wide experience in the field of design, development, manufacturing, type testing, installations, commissioning, operation and maintenance of various types of switchgear equipments. During the last four decades, the switchgear industry has seen the transition of four switching technologies, viz. those involving air, oil, gas and vacuum. Finally, the emergence of environment-friendly vacuum technology in medium voltage switchgears and of SF6 gas technology in high voltage and gas insulated sub-station (GIS) promises bring stability.

This book on switchgear from BHEL covers the industry's experience of the last four decades in the field of medium and high voltage switchgears, GIS, associated equipments, and control and protection schemes for generation, transmission and distribution applications.

The technological aspects of medium and high voltage circuit breakers ranging from 3.3 kV to 800 kV class are dealt with in Chapter 1. The technological backgrounds and advantages of various switching technologies, viz. those involving air, oil, gas and vacuum are elaborately discussed. Arc energy and arc quenching play a very important role in an interrupting media. The arcing phenomenon is dealt with in detail to understand and establish the superiority of a technology. Different types of mechanisms viz. spring, pneumatic, hydraulic, electrical and magnetic are being used for opening and closing of the circuit breakers. The type of mechanism in a circuit breaker depends upon the energy and speed required for arc interruption. Being simple and economical, the spring mechanism has been popular in both medium and high voltage switchgears. Magnetic actuators are now being introduced in medium voltage circuit breakers up to 12 kV, 40 kA due to their inherent advantages of long life, suitability for a large number of operations, maintenance-free operation and smaller number of components.

Switchgears for low voltage applications are rated up to 1000 V ac and 1500 V dc. Oil circuit breakers (OCBs), air circuit breakers (ACBs), switch fuse units (SFUs), off-load isolators, HRC fuses, earth leakage circuit breakers (ELCBs), miniature circuit breakers (MCBs) and moulded caste circuit breakers (MCCBs) are the commonly used devices in low voltage switchgear. The requirements of conventional switchgears and new trends for various applications like incomers, sub-incomers and feeder protections are briefly discussed in Chapter 2.

Switchgears for medium voltage applications are rated from 3.3 kV to 33 kV class. Medium voltage switchgears are mainly used for the distribution of electrical energy connected to various electrical networks. Chapter 3 deals elaborately with the historical background of minimum oil circuit breakers, bulk oil circuit breakers, air magnetic switchgears, SF6 gas insulated switchgears, and the latest trends in vacuum switchgear and gas insulated switchgears. This chapter also provides information about various forms of construction in metal enclosed, metal clad and porcelain clad design. The design aspects, related to current rating, temperature rise limits, short-circuit current requirements, vibrational stresses and insulation aspects are covered elaborately.

High voltage circuit breakers ranging from 66 kV to 800 kV class are covered in Chapter 4. This chapter deals with the historical background, present status of technological developments and the future outlook of high voltage switchgears. This chapter also provides insight into various interrupting media like air, oil and SF6 gas which have been successfully used in the development of high voltage circuit breakers. The latest trends in the design of SF6 gas circuit breakers and operating mechanisms are also discussed. System requirements covering important duties like short-circuit (terminal fault), short line fault, transformer magnetising current, energisation and breaking of long transmission lines, breaking of capacitive current and out-of-phase switching are also elaborately discussed in this chapter.

The requirement of gas insulated metal enclosed systems is specified in IEC 517. Sub-stations and switchyard equipments like circuit breakers, disconnectors, earth switches, busbars, CTs and PTs are metal encapsulated and pressurised with SF6 gas. Chapter 5 details the design and manufacturing aspects of various components/assemblies of GIS system used in both the medium and high voltage class of switchgears. The use of GIS equipments reduces requirement for space and improves reliability, thus increasing their popularity for application in metro cities and hydro stations where the costs of space and construction are very high.

Auto reclosures and sectionalisers are used in the distribution networks of medium voltage switchgears upto 33 kV class. These equipments are useful for the fast automatic restoration of supply following transient faults in the system due to frequent lightning surges and in areas where power lines run through forests and trees. These equipments are extensively used in the continents of America and Australia due to their distinct advantages over conventional switchgears. Although this product has not yet gained popularity in India, it has great potential for rural electrification and power distribution to important load centres in India. Chapter 6 is entirely devoted to this topic in view of its potential importance. This chapter deals with the technical specifications, construction, and technological aspects of auto re-closures and sectionalisers, along with case studies for various applications.

Instrument transformers (current and voltage) are used in switchgear installations for the measurement of electrical parameters for protection and metering purposes. Chapter 7 provides an overview of instrument transformers while covering theories of technical parameters that define terminologies, design and constructional aspects, important design considerations, application aspects, and testing requirements. Due to the existence of a large number of key system parameters, numerous verieties of CTs and VTs are being used. This chapter also provides suggestions for the standardisation and cost reduction of instrument transformers. With the emergence of numerical technology in the relaying and metering field, current and voltage sensors are replacing conventional CTs and VTs. This chapter also briefly describes current sensors (Rogowski coil) and voltage sensors.

Surge suppressors/arrestors are very important components of switchgears and sub-station installations. These are used to protect the sub-station equipments from temporary over-voltages, switching impulses, and lightning impulses, and to a certain extent, very fast transient over-voltages. Historically. spark gaps with air insulation were used as surge suppressors. Lightning arrestors, surge capacitors, surge suppressing reactors, and thyrite resistors with series gaps were used in the past for this purpose. Innovation in this field has resulted in the advent of two commonly used types, viz. the metal oxide-based (ZNO) type and C-R type of surge arrestors/suppressors. Chapter 8 deals with the various types of surge arrestors, their characteristics, and construction and application aspects.

Protective relays are vital parts of the switchgear equipment. The function of a protective relay is to initiate a signal to circuit breakers for disconnecting the element of the power system when it develops a fault. There have been rapid developments in relaying technology during the last two decades. The most important advancement has been due to the advent of computer technology which has helped in the development of numerical relays. Chapter 9 details various relay technologies like electromechanical, static, microprocessor-based, micro-controller-based and numerical technologies. Although all the technologies are presently in use, numerical relay technology is gaining popularity over the others due to its communication capabilities which facilitate integrated control, protection and metering features.

When a switchgear has to be applied in an electrical power system, certain considerations need to be kept in mind regarding the location of the equipment whether indoors or outdoors, system parameters (system earthing, frequency, insulation level, ratings, ambient conditions), etc. These are detailed in Chapter 10, which discusses the application of medium voltage switchgears. The requirement of switchgears for various applications in a generating station, transmission substation and distribution system are also detailed in this chapter.

The metering of electrical energy is a very important aspect of the switchgears system. Metering features are provided at all the electrical installations like generation, transmission and distribution. Chapter 11 deals with various types of electrical energy meters in use, as also their safety requirements, typical connection diagrams and communication features. The chapter also highlights the features of online energy monitoring system.

Control and inter-locking schemes constitute a very important aspect of medium voltage switchgears. The switching operation involves a variety of control and inter-locking schemes. Chapter 12 deals with the details of control and inter-locking schemes for medium voltage switchgear. This chapter covers safety schemes using position limit switches, voltage selection schemes, tripping schemes, trip circuit supervision schemes, alarm schemes, synchronising schemes and automatic supply transfer schemes. Although the functional requirements of each type of scheme are generally the same, there are variations due to auxiliary supply distribution philosophy, system/bus configurations, the number of breaker trip coils, and alarm and tripping logics. The chapter describes each scheme with certain typical examples.

Chapter 13 deals with five principles of power system protection, while describing the possible faults, their detection and methods of deriving fault parameters. It also describes in detail the most commonly used protection schemes in medium voltage switchgears.

The generator plays a key role in the electrical power system. The composition of a generator, generator transformer and unit auxiliary transformer is considered as a unit and the protection associated with it is called unit protection. Chapter 14 describes various features of unit protection. Fault conditions of the unit are identified and the protection system is designed to avoid damages to the equipments as well as to achieve maximum utilisation of the generator capacity. The classification of tripping is described in detail. A typical generator protection scheme is also described.

In Chapter 15, an attempt has been made to describe the control and protection philosophy normally adopted in the EHV transmission system. The experiences gained while executing the various EHV transmission jobs have also been listed. The chapter describes the control and protection requirements for various circuits in the EHV transmission system. Emphasis has been laid on inter-tripping schemes in transmission lines. The present trend of integrated control and protection is also touched upon.

With technological advancements in discrete electronic and electromechanical devices for power system protection, metering and supervisory controls are being replaced by integrated protection and control systems. Chapter 16 provides an overview of integrated protection and control systems. A three-level architecture, with metering and protection at the highest level, control and sub-station monitoring at the middle level, and data analysis and archiving at the lowest level, is described in this chapter.

The erection and commissioning of switchgear is an expert job. The creation of switchgear includes placing the equipment on foundation, assembly of loose supplied parts, mechanical and electrical inter-connections and connection of control and power circuits. The switchgear equipments are commissioned after the erection work is completed. Commissioning is generally done to carry out pre-defined checks and tests. Testing is necessary to ensure that the installation work meets the specified requirements and safety norms. Chapter 17 elaborately discusses the erection and commissioning requirements of switchgears.

A switchgear is designed and manufactured to ensure the desired operation and maintenance of its parts for providing reliable service throughout its life. The major functions of switchgears are protection, control and facilitating the maintenance of the electrical network including the switchgear itself. Chapter 18 deals with the operation and maintenance aspects of the switchgears under various conditions. This chapter also provides information about the dos and donts during site work, requirement of spares during maintenance, inspection and maintenance procedures.

The performance of a switchgear is verified by subjecting it to various tests. The guidelines for these tests are laid down in the form of national/international standards. The test requirements of switchgears vary with their applications and site conditions. Different tests required to ensure the satisfactory performance of the equipments at the site are chosen from the tests specified in standards like IEC62271-100, IEC60298 and IEC60694. Chapter 19 briefly covers the significance of and procedures used for different tests. The switchgear design criteria for meeting test requirements have also been explained wherever necessary.

Reference standards are documents issued by national and international standards like BSI, IEC, ISO, DIN, etc. These standards provide references for the designing, manufacturing, testing, installation, operation and maintenance of any product. Chapter 20 provides an overview of various standards related to switchgear products.

Every country lays down certain statutory and regulatory requirements with regard to electrical installations covering various safety aspects. The Indian Electricity Rules, 1956, have been issued by the Central Electricity Board, Government of India. Some of the rules pertaining to the construction, installation, protection, operation and maintenance of switchgear products are discussed in Chapter 21. Brief explanatory notes covering various aspects that need to be considered during the design, manufacture and installation of switchgear equipments against each applicable rule, are also discussed in this chapter.

R.K. Shukla

CONTENTS

Foreword v
Preface vii
Acknowledgements xi
Editorial Committee xiii
Contributors xv
Introduction xix

1. **Circuit Breaker Technologies**   1.1–1.39
   1.1 Arc Phenomenon: Introduction   *1.1*
       1.1.1 Role of Arc in Current Interruption   *1.2*
       1.1.2 Arc Interruption Theories   *1.2*
       1.1.3 Arcing Process   *1.3*
       1.1.4 High-pressure Arcs: General Characteristics   *1.4*
       1.1.5 Vacuum Arc   *1.11*
   1.2 Circuit Breaker Technology   *1.13*
       1.2.1 Bulk Oil Circuit Breakers (BOCBs)   *1.13*
       1.2.2 Minimum Oil Circuit Breakers (MOCBs)   *1.16*
       1.2.3 Air Circuit Breakers (ACBs)   *1.17*
       1.2.4 Air Blast Circuit Breakers (ABCBs)   *1.21*
       1.2.5 SF6 Gas Circuit Breakers   *1.22*
       1.2.6 Vacuum Circuit Breakers (VCBs)   *1.24*
       1.2.7 Gas Insulated Switchgear (GIS)   *1.27*
   1.3 Circuit Breaker Operating Mechanisms   *1.29*
       1.3.1 Circuit Breaker Opening Requirements   *1.31*
       1.3.2 Circuit Breaker Closing Requirements   *1.32*
       1.3.3 Types of Circuit Breaker Mechanisms   *1.32*
       1.3.4 Motor-driven Mechanisms   *1.38*
       Reference   *1.39*

2. **Low Voltage Switchgear**   2.1–2.7
   2.1 Conventional Incomer   *2.1*
   2.2 Conventional Sub-Incomer   *2.3*
   2.3 Conventional Feeder Protection   *2.3*
       2.3.1 Motor Control   *2.4*
       2.3.2 Other Industrial Load Control   *2.4*
       2.3.3 Lighting/Domestic Load Control   *2.4*

xxvi Contents

    2.4 Emerging Trends *2.5*
        2.4.1 Present Trends in Incomer *2.5*
        2.4.2 Present Trends in Sub-incomer *2.6*
        2.4.3 Present Trends in Feeder Protection *2.7*

3. Medium Voltage Switchgear   3.1–3.30
    3.1 Historical Background *3.1*
    3.2 Present Status *3.5*
    3.3 Future Trends *3.6*
    3.4 Constructional Forms *3.7*
        3.4.1 Metal-enclosed Switchgears *3.7*
        3.4.2 Live Tank Switchgears (Outdoor Porcelain-clad Construction) *3.14*
    3.5 Design Parameters *3.15*
        3.5.1 Normal Current Rating (Thermal Aspects of Design) *3.15*
        3.5.2 Short-circuit Current *3.20*
        3.5.3 Insulation Aspects of Design *3.24*
    3.6 Vacuum Switchgear *3.25*
        3.6.1 Vacuum as an Interrupting Medium *3.26*
        3.6.2 Demands Made on Circuit Breakers *3.27*
        References *3.30*

4. High Voltage Circuit Breaker   4.1–4.21
    4.1 Classification of Circuit Breakers *4.2*
        4.1.1 Classification by Voltage *4.2*
        4.1.2 Classification by Location *4.3*
        4.1.3 Classification by External Design *4.3*
        4.1.4 Classification by Interrupting Media *4.4*
    4.2 Development of Single-Pressure Puffer Breakers *4.6*
        4.2.1 First Generation SF6 Circuit Breakers *4.6*
        4.2.2 Second Generation SF6 Circuit Breakers *4.6*
        4.2.3 Third Generation SF6 Circuit Breakers *4.8*
        4.2.4 Design Features of Puffer Circuit Breaker *4.8*
    4.3 Operating Mechanisms for Circuit Breakers *4.10*
        4.3.1 Spring Mechanism *4.11*
        4.3.2 Pneumatic Mechanism *4.11*
        4.3.3 Hydraulic Mechanism *4.12*
        4.3.4 SF6 Gas Dynamic Mechanism *4.12*
        4.3.5 Motor-driven Mechanisms *4.13*
        4.3.6 Comparison of Operating Mechanisms *4.13*
    4.4 System Requirements of Circuit Breakers *4.14*
        4.4.1 Terminal Faults *4.14*
        4.4.2 Short Line Faults (SLF) *4.16*
        4.4.3 Transformer Magnetizing and Reactor Currents *4.17*
        4.4.4 Energisation of Long Transmission Lines *4.18*
        4.4.5 Switching of Unloaded Transmission Lines and Capacitor Banks *4.19*
        4.4.6 Out of Phase Switching *4.20*

5. Gas Insulated Sub-station/Switchgear (GIS)   5.1–5.26
    5.1 Design *5.2*
        5.1.1 Single Line Diagram *5.2*

	5.2	Bay Information *5.3*	
	5.3	Classification of Gimes *5.3*	
	5.4	Design Consideration *5.6*	
	5.5	Components/Modules *5.7*	
		5.5.1 Busbar *5.7*	
	5.6	Disconnectors (Isolators) *5.13*	
	5.7	Circuit Breaker *5.14*	
	5.8	Current Transformer *5.15*	
	5.9	Earth Switch *5.15*	
	5.10	Accessories *5.16*	
	5.11	Control Panel *5.17*	
	5.12	Constructional Aspects of GIS *5.18*	
		5.12.1 Enclosure *5.18*	
		5.12.2 Seals and Gaskets *5.19*	
		5.12.3 Gas Circuit *5.20*	
		5.12.4 Expansion Joints *5.20*	
		5.12.5 Current Transfer and Plug-in-Joints *5.22*	
		5.12.6 Support Structure *5.22*	
	5.13	Installation and Maintenance of GIS *5.23*	
	5.14	Lifecycle Cost of GIS *5.24*	
	5.15	Conclusion *5.25*	
		References *5.26*	

6. Auto-recloser and Sectionaliser — *6.1–6.14*

    6.1 Technical Specifications for an Auto-recloser *6.3*
    6.2 Control Transformer *6.5*
    6.3 Control Unit for Auto-recloser for 12 kV System *6.5*
    6.4 Working of Auto-recloser *6.6*
    6.5 Need for Auto-recloser *6.7*
        6.5.1 High Lightning Discharges *6.7*
        6.5.2 Large Concentration of Customers *6.8*
        6.5.3 Remote Site/Difficult Access *6.8*
        6.5.4 Bush Fire-prone Areas *6.8*
    6.6 Automatic Sectionaliser *6.9*
        6.6.1 Control Relay Unit for Sectionaliser *6.10*
    6.7 User's Benefits *6.11*
    6.8 A Case Study of Auto-recloser at KPTCL *6.13*

7. Current Transformers and Voltage Transformers — *7.1–7.28*

    7.1 Theory of Instrument Transformers *7.2*
        7.1.1 Theory of Current Transformers *7.2*
        7.1.2 Theory of Voltage Transformers *7.5*
    7.2 Technical Parameters of Instrument Transformers *7.6*
    7.3 Design and Constructional Aspects of Instrument Transformers *7.8*
        7.3.1 Types of Instrument Transformers *7.8*
        7.3.2 Types of Voltage Transformers *7.10*
        7.3.3 Design of Instrument Transformers *7.11*
        7.3.4 Some Important Design Considerations *7.12*
        7.3.5 Application of Instrument Transformers in Various Protection Schemes *7.12*

xxviii Contents

    7.4 Instrument Transformer Testing and Analysis of Test Results *7.16*
        7.4.1 Type Testing of Instrument Transformers *7.16*
        7.4.2 Applicability of Test Certificates *7.17*
        7.4.3 Acceptance of Class PS CT Using Magnetisation Curves *7.18*
    7.5 Standardisation and Cost Considerations of CTs and VTs *7.19*
        7.5.1 Recommendations for Standardisation and Cost Reduction *7.20*
        7.5.2 A Typical Instrument Transformer Standardisation Model *7.22*
    7.6 Modern Current Sensors (Rogowski Coil) *7.24*
        7.6.1 Theory of Rogowski Coil *7.24*
        7.6.2 Design Considerations of Rogowski Coil and Integrator *7.25*
    7.7 Voltage Sensors *7.26*
    7.8 Field Performance of Instrument Transformers *7.27*

8. Surge Protection   8.1–8.20
    8.1 Surge Protection Devices *8.2*
        8.1.1 Metal Oxide Type Surge Arrester *8.4*
        8.1.2 Capacitance Resistance (C-R) Type Surge Suppressor *8.7*
        8.1.3 Application of Surge Protection Devices *8.9*
    8.2 Over-Voltages in the Electrical System *8.10*
        8.2.1 Over-voltages in Sub-stations *8.10*
    8.3 Experimental Study in Use of Combinations of Surge Protection Devices *8.11*
    8.4 Surge Protection Devices for Gas Insulated Sub-stations *8.13*
        8.4.1 Over-voltages in GIS *8.13*
        8.4.2 Why Gas Insulated Metal Oxide Arrester (GISA) for GIS? *8.13*
        8.4.3 Location of Arrester *8.15*
        8.4.4 Performance Evaluation *8.16*
    8.5 Performance Monitoring of Surge Protection Devices *8.19*
        8.5.1 Surge Counters *8.19*
        8.5.2 Measurement of the Leakage Current *8.19*
        8.5.3 Measurement of the Resistive Component of Leakage Current *8.19*
        References *8.20*

9. Protective Relays   9.1–9.23
    9.1 Electromechanical Relays *9.2*
        9.1.1 Operating Principle of Electromechanical Relays *9.2*
        9.1.2 Attracted Armature Relays *9.3*
        9.1.3 Moving Coil Relays *9.5*
        9.1.4 Induction Relays *9.5*
        9.1.5 Thermal Relays *9.8*
        9.1.6 Motor Operated Relays *9.8*
        9.1.7 Mechanical Relays *9.8*
    9.2 Static Relays *9.9*
        9.2.1 Development of Static Relays *9.9*
        9.2.2 Relay Circuits Using Analogue Techniques *9.10*
        9.2.3 Relay Circuits Using Digital Techniques *9.11*
        9.2.4 Microprocessors *9.11*
        9.2.5 Housing of Static Relays *9.11*

        9.2.6   Quality Control and Testing   *9.12*
        9.2.7   Component Testing   *9.13*
    9.3   Numerical Relays   *9.13*
        9.3.1   Relay Hardware   *9.14*
        9.3.2   Relay Software   *9.15*
        9.3.3   Multiple Protection Characteristics   *9.16*
        9.3.4   Adaptive Protection Characteristics   *9.16*
        9.3.5   Data Storage   *9.16*
        9.3.6   Instrumentation   *9.17*
        9.3.7   Self-check Feature   *9.18*
        9.3.8   Communication   *9.18*
        9.3.9   Additional Functionality   *9.21*
        9.3.10 Size of Numerical Relays   *9.21*
        9.3.11 Cost of Numerical Relays   *9.21*
        9.3.12 HV and Environment Testing   *9.22*
    9.4   Conclusion   *9.23*

10. **Application of Medium Voltage Switchgear** 10.1–10.8
    10.1   Various Considerations   *10.1*
        10.1.1   Location Considerations   *10.1*
        10.1.2   Rating Considerations   *10.2*
        10.1.3   Ambient Considerations   *10.3*
        10.1.4   System Earthing Considerations   *10.3*
        10.1.5   Seismic Considerations   *10.3*
        10.1.6   Over-voltage Considerations   *10.4*
    10.2   Application of Switchgears in Power System   *10.4*
        10.2.1   Generator Circuit Breakers   *10.4*
        10.2.2   Switchgears for Power Plant Auxiliaries   *10.5*
        10.2.3   Switchgears for Transmission Sub-stations   *10.6*
        10.2.4   Switchgears for Distribution Applications   *10.6*
        10.2.5   Switchgears for Industrial Applications   *10.6*
        10.2.6   Switchgears for Rural Applications: (Auto-reclosers and Sectionalisers)   *10.7*
        10.2.7   Circuit Breakers for Earthing Applications   *10.7*
    10.3   Protection Requirements   *10.8*

11. **Energy Metering** 11.1–11.9
    11.1   Need for Metering   *11.1*
    11.2   Types of Electrical Energy Meters   *11.1*
    11.3   Features of Modern Meters   *11.2*
    11.4   Mechanical Requirements   *11.3*
    11.5   Auxiliary Supply Requirements   *11.3*
    11.6   Accuracy of Meter and Limits of Error   *11.3*
    11.7   Balanced/Unbalanced Load Meters   *11.3*
    11.8   Communication Feature   *11.4*
    11.9   "Online Energy Monitoring System"   *11.4*
        11.9.1   System Topology   *11.7*
        11.9.2   System Features   *11.7*
        11.9.3   Remote Access   *11.8*

11.10 Storage and Processing of Meter Readings   *11.9*
11.11 Main and Check Energy Meters   *11.9*
11.12 Grid Frequency Management   *11.9*
11.13 Future Trends   *11.9*

12. Control and Interlocking Schemes for Medium Voltage Switchgear   12.1–12.25
12.1 Safety Schemes Using Position Limit Switches   *12.2*
    12.1.1 Test and Service Position Limit Switches   *12.3*
    12.1.2 Test and Service Position Limit Switch Interlocks   *12.4*
    12.1.3 Breaker Moving Portion Auxiliary Switch Interlocks   *12.4*
    12.1.4 Breaker Fixed Portion Auxiliary Switch Interlocks   *12.4*
    12.1.5 Earthing Truck (Test to Service) Limit Switch   *12.4*
12.2 Voltage Selection Schemes   *12.5*
12.3 Tripping Schemes   *12.6*
    12.3.1 Shunt Tripping Scheme   *12.6*
    12.3.2 Series Tripping Schemes   *12.7*
    12.3.3 Capacitor Tripping Scheme   *12.9*
12.4 Trip Circuit Supervision Schemes   *12.9*
12.5 Alarm Schemes   *12.10*
    12.5.1 Alarm Cancellation Scheme   *12.11*
    12.5.2 Alarm Annunciation Scheme   *12.11*
    12.5.3 DC Fail Annunciation   *12.13*
12.6 Synchronising Schemes   *12.13*
    12.6.1 Manual Synchronisation   *12.14*
    12.6.2 Check Synchronising Feature   *12.14*
    12.6.3 Auto-synchronising   *12.17*
12.7 Automatic Supply Transfer Schemes   *12.18*
    12.7.1 Automatic Bus Transfer Scheme   *12.18*
    12.7.2 Auto-changeover Scheme   *12.23*

13. Protection Schemes for Medium Voltage Switchgear   13.1–13.17
13.1 5-S Principles   *13.1*
13.2 Faults   *13.2*
13.3 Fault Detection   *13.2*
13.4 Derivation of Relaying Quantities   *13.2*
13.5 Motor Protection   *13.4*
    13.5.1 Types of Faults   *13.4*
13.6 Transformer Protection   *13.7*
    13.6.1 Philosophy of Protection   *13.7*
    13.6.2 Inter-tripping   *13.9*
13.7 Feeder Protection   *13.9*
    13.7.1 Over-current and Earth Fault Protection   *13.10*
    13.7.2 Sensitive Earth Fault Protection   *13.11*
    13.7.3 Auto-reclosing   *13.11*
    13.7.4 Protection of Parallel Feeders   *13.14*
    13.7.5 Back-up Protection   *13.14*
13.8 Capacitor Protection   *13.14*
    13.8.1 Over-current Protection   *13.16*
    13.8.2 Over-voltage Protection   *13.16*

14. Generator Protection   14.1–14.7
    13.8.3 Unbalance Protection *13.16*
    13.8.4 Under-voltage Protection *13.17*

14. Generator Protection   14.1–14.7
  14.1 Classification of Tripping *14.2*
    14.1.1 Class 'A' Tripping *14.2*
    14.1.2 Class 'B' Tripping *14.3*
    14.1.3 Class 'C' Tripping *14.3*
  14.2 Grouping of Protections in Class A1, A2, B And C Tripping *14.3*
    14.2.1 Class 'A1' Tripping *14.3*
    14.2.2 Class 'A2' Tripping *14.4*
    14.2.3 Class 'B' Tripping *14.4*
    14.2.4 Class 'C' Tripping *14.5*
  14.3 Protection Philosophy *14.5*
    14.3.1 Typical Groupings of Protections for a Large Unit (500 MW) *14.5*
  14.4 Less Severe Fault Conditions *14.6*
  14.5 Modern Trends in Generator Protection *14.7*
    References 14.7

15. EHV Transmission System: Control and Protection   15.1–15.21
  15.1 Typical Circuits in an EHV Transmission System *15.2*
  15.2 Types of EHV Systems *15.2*
    15.2.1 One-and-a-half Breaker System *15.2*
    15.2.2 Two Main and a Transfer Bus System *15.2*
  15.3 Control and Protection Functions *15.5*
  15.4 Auxiliary Supply Distribution Philosophy *15.6*
  15.5 Breaker Related Protection *15.7*
  15.6 Transmission Line Protection *15.8*
    15.6.1 Distance Protection *15.8*
    15.6.2 Inter-tripping Schemes *15.8*
    15.6.3 Programmable Scheme Logic (PSL) *15.14*
    15.6.4 Other Line Protection Functions *15.15*
    15.6.5 Pilot Wire Differential Protection *15.15*
    15.6.6 Tee Differential Protection *15.16*
    15.6.7 Auto-reclosing Requirements in One-and-a-half Breaker System *15.16*
  15.7 Transformer Protection *15.17*
  15.8 Reactor Protection *15.17*
  15.9 Bus Bar Protection *15.17*
  15.10 Islanding Scheme *15.18*
  15.11 Integrated Control and Protection *15.18*

16. Integrated Protection and Control Systems for Sub-station   16.1–16.6
  16.1 Protection and Metering System *16.2*
  16.2 Control and Sub-station Monitoring *16.5*
  16.3 Data Analysis and Archiving Level *16.6*
    References 16.6

17. Erection and Commissioning of Switchgears   17.1–17.18
  17.1 Dispatch *17.2*
    17.1.1 Outdoor Switchgears *17.2*
    17.1.2 Indoor Switchgears *17.3*

17.2 Inspection on Receipt at Site *17.4*
17.3 Storage and Handling *17.4*
 17.3.1 Precautions to be Followed for Indoor Switchgears *17.5*
 17.3.2 Precautions to be Followed for Outdoor Switchgears *17.5*
17.4 Installation *17.6*
 17.4.1 Outdoor Switchgears *17.6*
 17.4.2 Indoor Switchgears *17.6*
17.5 Erection *17.7*
 17.5.1 Outdoor Switchgears *17.7*
 17.5.2 Indoor Switchgears *17.7*
17.6 Assembly *17.9*
 17.6.1 Outdoor Switchgears *17.9*
 17.6.2 Indoor Switchgear *17.9*
17.7 Small Wiring and Connection of LT Cables *17.10*
17.8 Filling with Insulation Medium *17.10*
17.9 Earthing *17.10*
17.10 Final Inspection *17.11*
 17.10.1 Indoor Switchgear *17.11*
 17.10.2 Common for Both Indoor and Outdoor Switchgear *17.11*
17.11 Testing and Commissioning *17.12*
 17.11.1 Visual Inspection Check *17.12*
 17.11.2 Mechanical Check *17.13*
 17.11.3 Electrical Check *17.13*
 17.11.4 Operational Check *17.14*
 17.11.5 Insulation Resistance Check *17.14*
 17.11.6 High Voltage Check *17.15*
17.12 Final Commissioning/Load Testing *17.15*
17.13 Dos & Donts *17.17*

18. Operation and Maintenance of Switchgear 18.1–18.17

18.1 Operation of Switchgear *18.2*
 18.1.1 Dos and Donts for Switchgear Operation *18.3*
18.2 Switchgear Maintenance *18.3*
 18.2.1 Maintenance Requirements *18.4*
 18.2.2 Inspection Schedule *18.4*
 18.2.3 Preparation Before Maintenance *18.5*
 18.2.4 Maintenance Procedure *18.5*
 18.2.5 Switchgear Spares *18.6*
 18.2.6 Lubrication of Operating Mechanism *18.8*
 18.2.7 Troubleshooting Guide *18.8*
 18.2.8 Service Feedback *18.17*

19 Testing of Switchgear 19.1–19.19

19.1 Routine Tests *19.2*
 19.1.1 Dielectric Test on the Main Circuit *19.2*
 19.1.2 Dielectric Test on Auxiliary and Control Circuits *19.2*
 19.1.3 Measurement of the Resistance of the Main Circuit *19.3*

        19.1.4 Tightness Test *19.3*
        19.1.5 Design and Visual Checks *19.3*
        19.1.6 Mechanical Operation Tests *19.3*
   19.2 Type Tests *19.4*
        19.2.1 Dielectric Tests *19.5*
        19.2.2 Radio Interference Test *19.7*
        19.2.3 Measurement of the Resistance of Main Circuit *19.8*
        19.2.4 Temperature Rise Test *19.8*
        19.2.5 Short Time Current Withstand Test *19.10*
        19.2.6 Verification of the Degree of Protection Test *19.11*
        19.2.7 Electromagnetic Compatibility (EMC) Test *19.12*
        19.2.8 Mechanical Operation Tests at Ambient Temperature *19.12*
        19.2.9 Short-circuit Duties Test *19.13*
        19.2.10 Capacitive Current Switching Tests *19.15*
        19.2.11 Environmental Test *19.16*
        19.2.12 Critical Current Test *19.18*
        19.2.13 Short Line Fault Test *19.18*
        19.2.14 Out-of-Phase Making and Breaking Tests *19.19*
   19.3 Special Tests *19.19*

20. Reference Standards 20.1–20.5
   20.1 Standard Organisations *20.2*
        20.1.1 Indian Standards *20.2*
        20.1.2 International Standards *20.2*
        20.1.3 National Standards of Other Countries *20.2*
   20.2 Standards Pertaining to Switchgear *20.2*
        20.2.1 Indian Standards for Switchgear *20.3*
        20.2.2 International Standards for Switchgear *20.3*
   20.3 Standards Pertaining to Associated Equipment, Material, Processes and Guidelines Required for Switchgear *20.5*

21. Statutory Requirements and Electricity Rules 21.1–21.8
   21.1 Indian Electricity Rules, 1956 (Clause No. 29) *21.1*
   21.2 Indian Electricity Rules, 1956 (Clause No. 35) *21.2*
   21.3 Indian Electricity Rules, 1956 (Clause No. 48) *21.2*
   21.4 Indian Electricity Rules, 1956 (Clause No. 51) *21.3*
   21.5 Indian Electricity Rules, 1956 (Clause No. 64) *21.4*
   21.6 Indian Electricity Rules, 1956 (Clause No. 64A) *21.5*
        21.6.1 Interlocks *21.5*
        21.6.2 Protection *21.6*
   21.7 Indian Electricity Rules, 1956 (Clause No. 67) *21.6*
   21.8 Indian Electricity Rules, 1956 (Clause No. 120) *21.7*

Index I.1–I.7

HANDBOOK OF SWITCHGEARS

Chapter 1

CIRCUIT BREAKER TECHNOLOGIES

D.K. Dikshit

This chapter covers various technological aspects related to circuit breakers. The arc phenomenon which is common to all circuit breakers, has been discussed in the chapter, covering the arc phenomenon in gases and vacuum media. Circuit breaker technologies, which are based on the medium of arc interruption have also been covered. This includes the interruption of arc in various media such as oil, air, sulphur hexafluoride (SF6) and vacuum.

The technology of gas insulated switchgear (GIS) is briefly covered in this chapter. Various types of mechanisms, i.e. spring operated mechanism, pneumatic, hydraulic and magnetically actuated mechanisms that are used in circuit breakers are also described in this chapter.

1.1 ARC PHENOMENON: INTRODUCTION

Arc is an essential part of any circuit breaker operation where contact is physically parted. The study of arc and its behaviour is necessary for switchgear designers. This chapter explores the various types of arcs, their properties and behaviour so as to facilitate an understanding of the complexities of electrical circuit breaking. The basic design of circuit breaker interrupter essentially depends upon how the extinction of arc is managed.

The heart of an ac circuit breaker is a switching element having a variable resistance. This variable resistance is a high-pressure arc burning in the corresponding gas in various circuit breaker mediums such as air, oil and SF6. Whereas in vacuum circuit breakers, it is a vacuum arc burning in electrode vapor. In certain other switching devices, which include high-voltage dc converters, is either a low-pressure mercury arc or a 'discharge' in the crystal lattice of a solid-state device.

The physics of discharges in circuit breakers is understood only qualitatively. Many of the processes occurring during this phenomenon are too complex to be subjected to analytical treatment. Since our knowledge of gas properties, particularly at high temperatures, is not very accurate, there are several uncertainties. Nevertheless, progress in circuit breaker-arc physics has been considerably rapid in the last 10–15 years, owing mainly to advances in aerodynamics, and computational and experimental techniques.

The study of arc phenomena, though very complex, is of immense importance for understanding the design and operational characteristics of circuit breakers. In this chapter, an attempt has been made to discuss some of the physical processes, which are vital to an understanding of the phenomena of electric arcs.

1.1.1 Role of Arc in Current Interruption

When two current-carrying contacts open, an arc bridges the contact gap and prevents an abrupt interruption of the current. This arc is useful in a way as it provides a low resistance path for the current after contact separation, thereby preventing current chopping and associated abnormal switching over-voltages in the system. In the case of alternating current (ac), arc is momentarily extinguished at every current zero. To make the interruption complete and successful, re-ignition of the arc between the contacts has to be prevented after a current zero.

It is thus evident that the arc plays an important role in the process of current interruption and therefore must not be regarded as an undesirable phenomenon. It must also be realised that, in the absence of the arc, the current flow would be interrupted instantaneously, and due to the rate of collapse of the associated magnetic field, very high voltage would be induced which would severely stress the insulation of the system. On the other hand, the arc provides a gradual, but quick, transition from the current-carrying to the current-breaking states of the contacts. It thus permits the disconnection to take place at zero current without inducing the potentials of dangerous values. The function of an arc-control device in a circuit breaker is therefore clearly to employ the beneficent action of the arc as efficiently as possible.

1.1.2 Arc Interruption Theories

In the early 1930s, Slepian put forward the concept of a 'race' between the growing dielectric strength of an arc gap after arc extinction and the re-striking voltage of the circuit. However, it was realised later that the two quantities were not independent of each other, which led Cassie in 1939 to put forward the first energy-balance theory of arc extinction, and Slepian and Browne in 1941 to suggest that turbulence in the gas flow was the means by which the rapid dissolution of the arc was obtained. At this time, a considerable amount of work, both experimental and theoretical, commenced around the world, on the manner in which the resistance of the arc changed during the current zero period.

In 1939, Cassie proposed a model of arc in which the arc was assumed to have a cylindrical column with a uniform temperature and current density, so that its area varies to accommodate changes in the current. The power dissipation was assumed to be proportional to the column cross-section. This model was intended to represent an air-blast arc and was represented by the following differential equation:

$$Rd/dt\ (1/R) = 1/\theta\ \{(v/v_0)^2 - 1\}$$

where R is the arc resistance, v, the arc voltage at any instant, v_0, the arc voltage in steady state, and θ the arc time constant (i.e. the ratio of energy stored per unit volume to the energy loss rate per unit volume).

In the steady state, Cassie's equation leads to a constant voltage characteristic $v = v_0$, which is qualitatively typical of the heavy circuit regime of circuit breaker arcs.

A few years later, in 1943, Mayr proposed a somewhat improved model, in which the arc was assumed to be of a fixed diameter but of varying temperature and conductivity, and the power loss occurred from the surface of the arc only. This model was described by the differential equation:

$$Rd/dt\ (1/R) = 1/\theta\ \{(v_i/w_0)^2 - 1\}$$

where i is the arc current at any instant and w_0 is the energy loss from periphery of arc at steady state.

It has been found that Cassie's model best describes the period before current zero whereas Mayr's model represents better the post-arc regime. In 1948, Browne developed a model of the arc that combined both these models. Many other models have also been proposed since then, but none of them relates the terms of the equation numerically with the physical properties of the interrupting medium. An exception, however, is the theory evolved by Butler and Whittaker in 1972.

The fact that energy is stored in the arc column means that the conductance will fall to zero only some time after the current zero and some 'post-zero' current will flow. If immediately after current zero, the rate of rise of recovery voltage is greater than a critical value, the decaying arc channel is re-established by ohmic heating. In this period, which is controlled by the energy balance, thermal failure can therefore occur. After a successful thermal interruption, the re-striking voltage can reach such a high peak value that the gap fails through dielectric breakdown. This is called dielectric failure in the peak regime of re-striking voltage. A typical feature of this failure mode is that the voltage decay occurs so fast that it cannot be resolved on an oscillogram while the voltage in the thermal failure mode decreases and approaches the arc voltage during a time interval lasting for several micro-seconds. The nature of voltage variation with time under the two failure modes is illustrated in Fig. 1.1.

1.1.3 Arcing Process

Arcs in ac circuit breakers occur in two ways. When the current-carrying contacts are being separated, arcing is possible even when the circuit e.m.f. is considerably below the minimum cold electrode breakdown voltage, because of the ions neutralising the electronic space charge and thus allowing large currents to flow at relatively low voltage gradients. This way of occurrence of an arc is common to both dc and ac circuit breakers. The another way of occurrence of arc happens only in ac circuit breakers. In this case, the arc is extinguished every time the current passes through zero and can re-strike only if the transient recovery voltage across the electrodes,

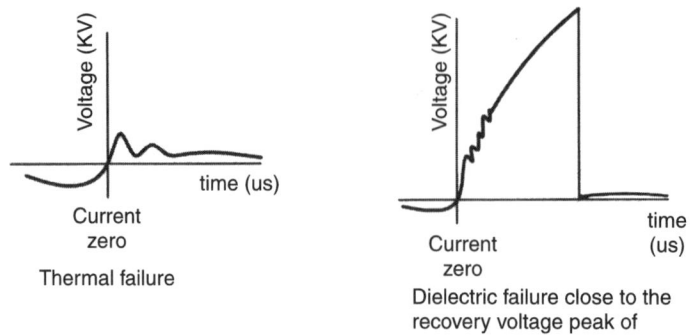

Fig. 1.1: The nature of voltage variation with time under two failure modes

already separated and continuing to separate, reaches a sufficiently high value causing breakdown. The function of an ac circuit breaker is to prevent re-striking of the arc, which depends upon the following important factors:

- The nature and pressure of the medium of arc;
- The external ionising and de-ionising agents present;
- The voltage across the electrodes and its variation with time;
- The material and configuration of the electrodes; and
- The nature and configuration of the arcing chamber.

To the physicist, the arc appears in many forms and is a subject of great scientific interest. To the switchgear designer, the arc all too often appears to have only one objective, viz. self-preservation. The power arcs try to evade de-ionisation systems by striking in new and unexpected directions, and by re-igniting after they appear to have been successfully interrupted, if the design has any weakness. Thus while the physicist would be interested in studying the arc plasma properties, including particle densities, temperatures and distributions, and electrical conductivity, the switchgear designer would be more concerned with the insulating and dielectric recovery properties of the arcing medium so that he can use them to interrupt as high a current and voltage as possible.

Arcs used in circuit breakers can be broadly classified into two categories, viz. high-pressure arcs, with ambient pressures of one atmosphere and above, and vacuum arcs, with ambient pressures below 10^{-4} Torr. Since neither mercury arcs nor semiconductor junctions have been used in normal circuit breakers, these types of discharges are not discussed in this chapter.

1.1.4 High-pressure Arcs: General Characteristics

The arc-interrupting medium in high-pressure circuit breakers is a flowing gas, commonly either air or SF6 in blast-type circuit breakers, and hydrogen in oil circuit breakers. Table 1.1 shows the actual composition of the gases involved in different circuit breakers, as also the nominal composition used for theoretical studies and the pressure ranges involved. From the table, it is clear that the gases involved most often are air, nitrogen, hydrogen, and sulphur hexafluoride.

Circuit Breaker Type	Approx. Actual Composition		Gas Composition Assumed by Theoreticians		Approx. Pressure Range
Air	Air	100%	Air or N_2	100%	1 atm
Air blast	Air	100%	Air or N_2	100%	5–100 atm
Oil	H_2	70%	H_2	100%	1–100 atm
	C_2H_2	20%			
	CH_4	6%			
	Others	4%			
SF6	SF6	100%	SF6	100%	2.5–15 atm

Table 1.1 Pressure and Composition of Gases in Circuit Breakers

The high-pressure arc may be taken to consist of three distinct regions: (1) cathode region, where electrons and metal vapour are emitted from the metal electrode into the plasma, either by field or by thermionic emission, (2) an arc column, where the current is carried by moving electrodes and ions, and (3) an anode region, where electrons from the discharge enter the electrode, and where metal vapour may also enter the discharge.

1.1.4.1 Electrode Regions

The electrode regions are usually considered to be relatively unimportant, as far as arc extinction is concerned, in most high-pressure circuit breakers. However, it is useful to study some properties of these regions. In front of the cathode, the arc usually constricts into a bright spot with a very high current density, which may vary from 10^7 to 10^{12} A/sq.m. depending on the material and the experimental condition. The cathode voltage drop is usually quite low, around 10 V. In front of the anode, the appearance of the contraction is usually not quite as pronounced. However, the anode voltage drop is also comparable to the cathode voltage drop.

The constriction of the arc near the electrodes causes vapour jets to form. These jets can affect the performance of arc chutes and arcing horns in arc-chute circuit breakers, and can significantly affect the ionisation and thermal conditions of the arc column in blast type circuit breakers.

1.1.4.2 Plasma Region (Arc Column)

The arc column in the high-pressure arc is a cylindrical region in which ionised gas gives almost the exact equality of positive and negative charge density, so that very high current densities may be supported with relatively low electric fields.

1.1.4.3 Ionisation Processes

Ionisation in arcs takes place when any process occurs in which sufficient energy can be supplied to neutral particles to detach one or more electrons. This energy may be supplied by the direct impact of electrons, positive ions, neutral atoms or photons. All of it need not be supplied in one encounter; as a particle may be excited by one collision and ionised by the next.

In circuit-breaker arcs, two major processes produce the ionisation. During most of the current-conduction period, the ionisation that takes place is almost wholly thermal ionisation, owing to the thermal agitation of the very hot gas particles, but ionisation by direct electron impact in the electric field of the discharge can also occur very close to current zero. It is generally believed that the amount of ionisation produced in arcs by other processes is very small.

1.1.4.3.1 Thermal Ionisation An ideal gas at normal temperature is a good electrical insulator, because it consists of molecules most of which are in the ground state and hence electrically neutral. It can be made to conduct only when some means are employed to create free electrons and ions. Although some free electrons and ions are present in the gas, which are created by ultraviolet radiation, cosmic rays, radioactivity of the earth, etc., these are, as such insufficient to sustain conduction through the gas. In order to understand how the gas becomes conducting, it must be remembered that a gas consists of molecules which move about with considerable velocities and which are not all the same but are grouped about a mean value in a Maxwellian distribution. The mean kinetic energy of the molecules, $1/2\, mv^2$ is equal to $3/2\, kT$, where T is the absolute temperature and k, the Boltzmann's constant. It is common knowledge that air molecules at room temperature (300° K) move randomly at a speed of approximately 500 m/second and collide very frequently at 10^{10} times/sec. Nevertheless the kinetic energy of these molecules (the average of which is proportional to the absolute temperature) is too low to enable even the fastest particles to excite those with which they collide.

If the gas is heated up, some of molecules will breakdown at the most severe collisions and dissociate into their atoms. Energies of 5.1 eV and 9.1 eV respectively are needed to dissociate an O_2 and N_2 molecule.

As the temperature is increased further, some molecules and atoms are deprived of some valance electrons and the ionised gas called plasma becomes a conductor. This process accounts for much of the conductivity in low-voltage arcs. The degree of ionisation depends on the pressure, the temperature and the ionisation potential of the gas or vapour. The ionisation potential is the work done in removing an electron from an atom or molecule and is measured in electron volts; more than one electron may be removed per particle in highly ionised plasmas. The degree of ionisation X in a thermally-ionised gas was first derived by Saha in 1920 on the basis of thermodynamic reasoning and is given by the following equation:

$$\{X^2/(1 - X^2)\}\, p = 3.16 \times 10^{-7}\, T^{5/2} \exp(-eV_i/kT)$$

where p is the total pressure in atmosphere, T, the gas temperature in degrees Kelvin, e, the electron charge in coulombs, V_i, the ionisation potential of the gas atoms in electron volts and k, the Boltzmann's constant. In many practical cases, the degree of ionisation is sufficiently small to justify the simplifying substitution of unity for $(1 - x^2)$. The ionisation potentials are as follows:

$$\text{Cu (vapour)} = 7.7 \text{ eV} \qquad N_2 = 15.5 \text{ eV}$$
$$O_2 = 12.2 \text{ eV} \qquad F_2 = 17.8 \text{ eV}$$
$$H_2 = 15.4 \text{ eV} \qquad S = 10.3 \text{ eV}$$

An important characteristic of the arc plasma when considering the arc extinction process is its thermal conductivity. Figure 1.2a compares the thermal conductivities of SF6 and Nitrogen (N_2). The peaks in the curves correspond to the molecular dissociation temperatures which are around 2000 K for SF6 and 6500 K for N_2. The result of this is that the arc plasma in SF6 is surrounded by a cylinder of cool gas with a high thermal conductivity and a very low electrical conductivity. In a Nitrogen arc the outer surface is at a much higher temperature and correspondingly more conductive.

This is illustrated in Fig. 1.2b which shows temperature profiles for both gases. The result of the cool gas around the arc plasma and the small diameter of the arc core in the SF6 arc gives it a very short thermal time constant, which is important when considering the extinction process.

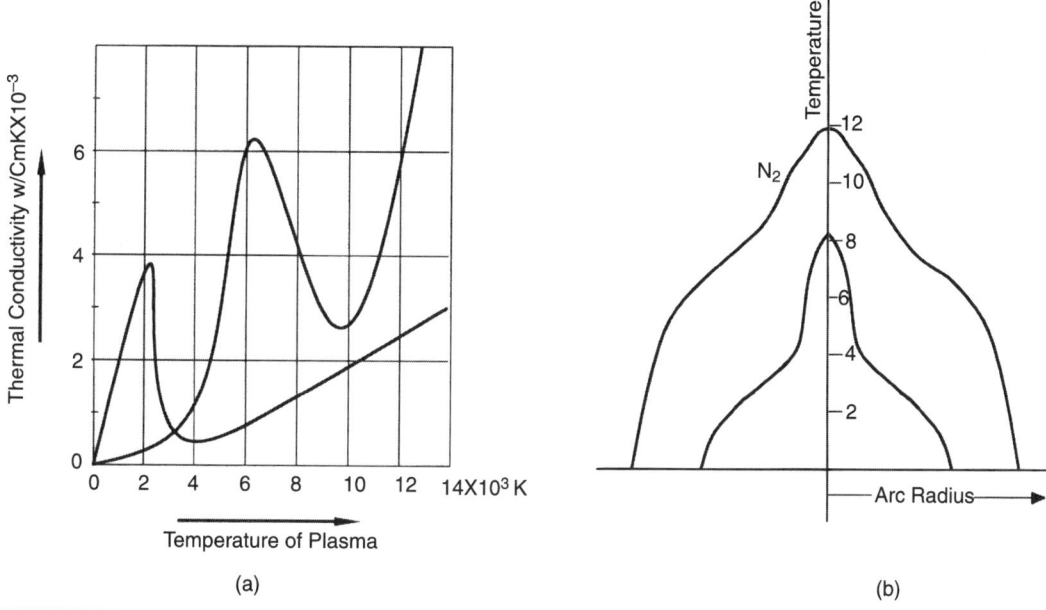

Fig. 1.2: (a) Thermal conductivity of SF and N, (b) SF and N arc temperature profiles

1.1.4.3.2 Ionisation by Electron Collision: Ionisation by electron collision occurs in a gas by electrons that are directly accelerated by high electric gradients in the arc. In this case, the electron velocities are not only much higher than they would be if they were in thermal equilibrium with the surrounding gas at arc temperatures, but are no longer random in direction and are directed by the electric field. The probability of ionisation by collision is, of course, zero when the impacting electron has less energy than the ionisation potential of the gas.

1.1.4.4 De-ionisation, Re-combination and Attachment Processes

If all sources of ionisation are removed from an ionized gas, it rapidly assumes a neutral state by ion diffusion to the walls and by volume re-combination of the positive and negative ions (Fig. 1.3). In de-ionisation by diffusion, the negative ions, or electrons, and positive ions move to the walls under the influence of concentration gradients, and thus completing the process of re-combination (Fig. 1.4). The probability of electron attachment varies widely with different gases and is markedly affected by the presence of certain gases and vapour as impurities. The gases with this property of attaching an electron are termed as electronegative gases. This electron affinity of a gas is expressed in electron volts, and varies from zero for noble gases to a maximum of 3.9 eV for fluorine. Some values are as follows:

$$F = 3.9 \text{ eV} \quad H = 0.7 \text{ eV}$$
$$Cl = 3.8 \text{ eV} \quad O_2 = 1 \text{ eV}$$

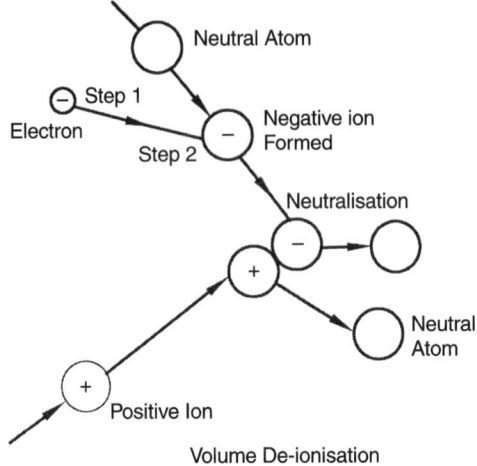

Fig. 1.3: Re-combination of +ve and –ve ions

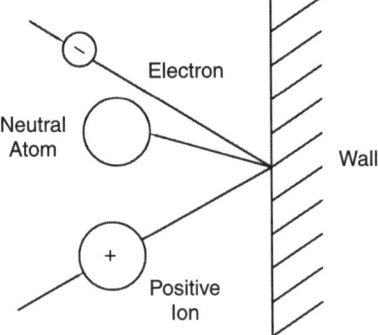

Fig. 1.4: De-ionisation at Surfaces

The mean time of electron attachment is the average time for which an electron will exist free in a gas at normal temperature and pressure (NTP) before being absorbed by a neutral particle to form a negative ion. Table 1.2 shows the value of mean time of electron attachment in seconds for different gases.

The table shows that an electron in pure CO would remain free on an average for about 7×10^{-4} sec, while in F or SF6 it would remain free for only about 4×10^{-9} sec. The noble gases do not form negative ions by attachment of electrons nor do pure nitrogen and hydrogen. Obviously, an electric field would speed up the electrons so as to render attachment improbable. High temperatures also reduce the probability of attachment so that in a high-pressure arc column, few negative ions are likely to form an attachment. When oppositely charged particles re-combine to form

Gas	Attachment time in seconds
CO	7.2×10^{-4}
NH_3	3.35×10^{-4}
Air	6.3×10^{-7}
O_2	1.94×10^{-7}
Cl_2	4.67×10^{-9}
F	4×10^{-9}
SF6	4×10^{-9}

Table 1.2 Mean time of electron attachment

neutral particles, they emit energy, which is equal in amount to the original ionisation energy of the particles. The ionisation and re-combination processes go on simultaneously in an arc while in an equilibrium situation, the rate of ionisation exactly balances the rate of re-combination.

With decreasing current, re-combination of additional charged particles takes place to adjust the arc condition to new values of current. Re-combination occurs more rapidly at low temperatures, i.e. when the arc is cooled by external means, than at high temperatures. In vacuum interrupters, where the mean free path is many cm long, the re-combination takes place mainly on solid surfaces. In high-pressure circuit breakers, however, this process is not important, because the mean free path, of the order of 10^{-5} cm and below is so short that ions in the critical region of the discharge channel almost never meet solid surfaces.

1.1.4.5 Arc Column Characteristics

In the high temperature arc column, the particles are in thermal equilibrium. They are all in rapid random motion, but in the absence of an electric field, no net movement occurs. When an electric field is applied to the gas, the charged particles gain a drift velocity superimposed on their random thermal motion. The drift velocity is proportional to the voltage gradient, with the coefficient of proportionality being termed as the particle mobility. The heavier the particle, the lower the mobility, since the particle loses its directed velocity at each collision of the random walks, and has to be re-accelerated in the direction of the field again. Similarly the higher the pressure, the lower the mobility, since collisions occur more frequently.

The contribution to the total current by the charged particles is directly proportional to their mobility and hence to ion current. This ion current is only a very small fraction, about 0.1 per cent of the total current, and is usually neglected. When an electronegative gas is present, the electrons become attached to form negative ions as the temperature falls and thus the conductivity of the arc column reduces very rapidly. This attachment is of great importance in SF6 circuit breakers during and after the current–zero periods. Thus if the nature of the gas is known, it is possible to calculate many of the electrical properties of the plasma as a function of temperature, since the ionisation density can be calculated from the temperature by the Saha equation. The calculated electrical conductivity of air at different pressures as a function of temperature is shown in Fig. 1.5.

The relationship between arc current and arc voltage is known as the arc characteristic. The static characteristics apply when changes of current impressed on the arc take place slowly, while

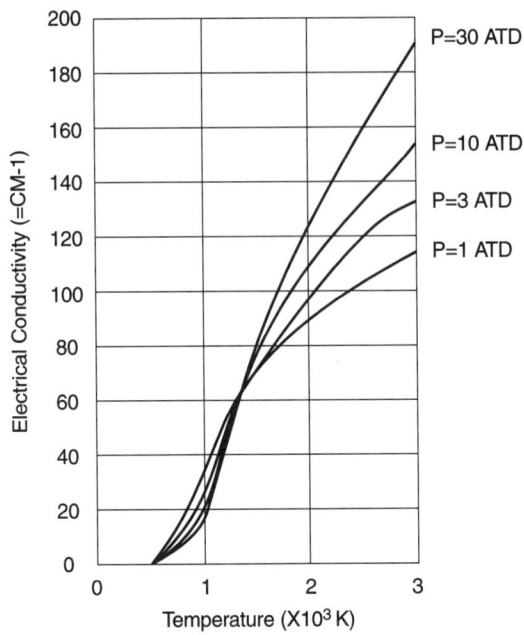

Fig. 1.5: Electrical conductivity of air at different temperatures

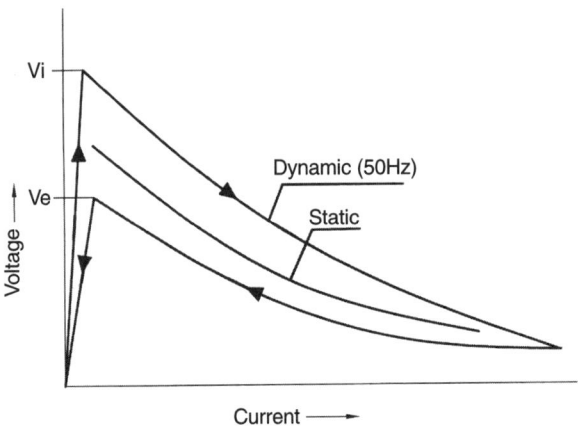

Fig. 1.6: Typical static and 50 Hz characteristics

the dynamic characteristics apply when these changes take place rapidly. Figure 1.6 shows typical static and 50 Hz characteristics. With an increase in the arc current, the temperature rises and the process of ionisation becomes more active, thereby increasing the conductivity of the medium and resulting in a decrease in the arc voltage. The initial breakdown of the gap between the electrodes requires a high ignition voltage Vi at zero current.

If the current through the arc changes suddenly, in step-function form, the arc initially acts as a linear resistor, with the voltage change being proportional to the change in current. Since the conductance of the arc depends on the stored energy, it cannot change instantly but exponentially approaches the level determined by the static characteristic for the new value of current. In air-blast arcs, the arc time constants are short (10 to 100 micro-seconds) because of a very large ratio of power loss to energy stored, whereas in freely burning atmospheric pressure arcs, the voltage may take a milli-second or more to return to the value given by the static characteristic. Ve represents the extinction voltage.

1.1.4.6 Heat Loss from Arc Plasma

The heat loss from an arc column can take place through three mechanisms: conduction, convection and radiation. In breakers with plain break arc in oil, arc in chutes or tubes or narrow slots, nearly all the heat loss is due to conduction. If, however, a gas flow is present either due to the action of the arc itself (as in an ordinary arc in air) or to an external blast, the loss occurs due to a conduction-convection process. At pressures close to 1 atm, radiation is not a significant factor in heat dissipation from the arc. At these pressures, the arc can also be considered as optically thin, i.e. the radiation from the inner portion of the arc is not absorbed to any significant degree before it leaves the plasma. At higher pressure, however, the radiation may become a very important factor in energy dissipation and the plasma also becomes optically thick. In order to understand the quantitative effects of radiation, all known emission processes must be considered.

1.1.4.7 Arc Extinction

The final extinction of the arc requires a rapid increase of the dielectric strength in the medium between the circuit breaker contacts, which is achieved either by the de-ionisation of the arc path or by the replacement of the ionised gas by cool fresh gas. The various de-ionisation processes include high pressure, cooling by conduction, forced convection and turbulence.

At a given temperature, a pressure increase results in a corresponding increase in the density and reduction in mean free path of the particles constituting the discharge. This increases the collision rate and the voltage required to maintain the arc. Another effect of the increased density of particles is a higher rate of de-ionisation due to the re-combination of oppositely charged particles.

The voltage required to maintain ionisation increases with a decrease in temperature because cooling effectively increases the resistance. Cooling by conduction of arc heat to adjacent parts, e.g. baffles, splitters and insulating walls, or by the use of a gas such as hydrogen that has high diffusion rate, heat absorption rate and thermal conductivity, is therefore effective in arc extinction.

Efficient cooling is also achieved by forced heat convection, which may be brought about by a gas blast directed along the discharge as is done in air blast and SF6 circuit breakers. Such a blast also compels the arc to shrink in diameter. Under certain conditions, a flow of gas creates turbulence near the surface of an arc, and it is believed that this effect may be very important in arc extinction.

1.1.5 Vacuum Arc

General Characteristics: The vacuum arc differs from the general class of low and high pressure arcs in the sense that in vacuum arc, the electrons, ions, and neutral atoms are all derived from the electrodes by the arc itself. The base gas pressure here is so low that it plays no significant role in the conduction process. In this sense, the vacuum arc is therefore really a metal vapour discharge.

An ac vacuum arc between two simple disc-shaped electrodes burns in either a diffuse or a constricted mode, depending largely on the magnitude of the current and the size and spacing of electrodes. With copper electrodes of dimensions typically found in a vacuum interrupter, the arc would generally burn in a diffuse mode at currents up to about 10 kA and in a constricted form, above this current.

1.1.5.1 Cathode Region

The vapour necessary to sustain vacuum arc comes mainly from the cathode spots. Each spot carries a mean current dependent on the cathode material, which is about 100 A for copper. The current density at the spots is estimated to be 10^{10}–10^{11} A/m^2, depending on the cathode material.

In the absence of magnetic fields, a single cathode spot moves randomly on the cathode surface with a small velocity. At higher currents, when many spots exist in parallel, the self-magnetic fields are more intense, and the motion of the spots due to mutual repulsion may predominate over their random motion and corresponds to the retrograde motion of the spots in an external magnetic field.

The arcing voltage at low currents (< 300 A) which is confined to the cathode region, depends wholly on the cathode material and is related to the physical properties of material. The value is generally in the range 8–30 V depending on the material. At high currents (a few kiloamperes), however, there can be appreciable plasma and anode voltage drops.

1.1.5.2 Plasma Region

The region between the electrodes of a diffuse vacuum arc is roughly conical in shape, the apex of the cone being at the cathode spot and the base on the anode as shown in Fig. 1.7a. The general picture of vacuum arc is illustrated diagrammatically in Fig. 1.7b. The vapour, which has a high density at the cathode spot, expands into the vacuum and perhaps at a distance of 10 mm from the cathode, the density will be representative of a pressure of 10^{-3} torr. Thus, an electron traversing the inter-electrode gap experiences conditions of high pressure near the cathode where the mean free path is perhaps 10^{-7} m, to conditions of low pressure in the plasma where it is of the order of tens of millimetres.

At low currents, the voltage drop in the plasma region of a low-current arc is negligible. The voltage gradient is less than 0.01 V mm^{-1}. At high currents, however, the gradient may be up to a few volts per millimetre.

1.1.5.3 Current-zero Processes

Successful current interruption by a vacuum arc depends upon the rapid condensation of the metal vapour into the anode and shield near and at current zero. As the current falls, the number of cathode spots falls until we have only one spot at currents between 100 A and zero. The estimated values of metal vapour density, which are in good agreement with the measurements, are so low throughout the gap that the gap is substantially an insulator. This prevents re-ignition of the vacuum arc after the current zero.

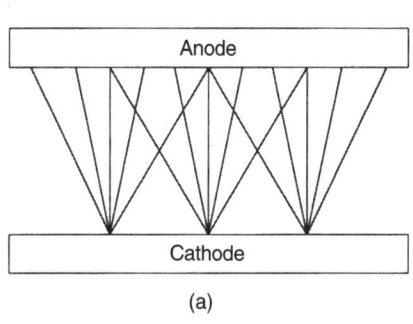 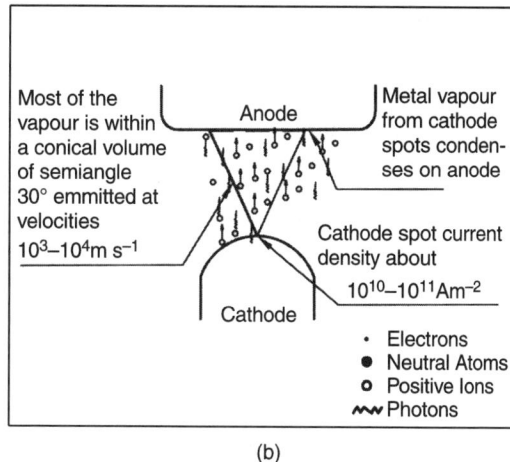

Fig. 1.7: Vacuum arc diagram

1.2 CIRCUIT BREAKER TECHNOLOGY

High voltage circuit breaker technology has changed radically in the past 15 years. Most utility systems are having mix population of bulk oil, minimum oil, vacuum, air blast, SF6 two-pressure, and SF6 single-pressure circuit breakers.

The SF6 single-pressure circuit breaker has become the current state of the art technology at transmission voltages (72.5 kV and above). However, SF6 gas has been identified as a greenhouse gas, and safety regulations are being introduced in many countries in order to prevent its release into the atmosphere.

Vacuum circuit breaker has emerged as the dominant technology in the medium voltage range due to its superior features such as long contact life, lack of maintenance requirement, low operating energy requirement and high reliability.

This chapter briefly discusses the trends in the technology of circuit breakers such as bulk oil circuit breakers (BOCBs), minimum oil circuit breakers (MOCBs), air blast circuit breakers (ABCBs), SF6 circuit breakers (SF6) and vacuum circuit breakers (VCBs).

1.2.1 Bulk Oil Circuit Breakers (BOCBs)

The circuit breakers in which the contacts part in oil and wherein the oil also provides the main insulation between the live part and earthed metal tank are called bulk oil circuit breakers. These types of breakers are designed in all voltage ranges from 1000 V to 330 kV.

Although the earliest switching devices employed air as the extinguishing medium, it was not long before a better dielectric was needed, and insulating oil soon appeared on the switchgear scene. Initially it was used in small quantities just around the contacts. As larger powers and voltages were required, the oil was also used as an insulating medium in an earthed metal tank and the bulk oil circuit breaker was born.

Initially, the contacts were separated in the oil bath, and the movement of the oil, the cooling effect of the hydrogen created from the oil by the arc, and the distance which separated the contacts, were the factors that cooled the arc path and created the dielectric strength needed to interrupt the relatively small fault currents in the early days of power distribution. As fault currents increased, the increase in pressure due to the vaporisation of the oil increased the de-ionising properties of the hydrogen bubble surrounding the arc path, and increased its dielectric strength, thus preserving the breaking capacity of the device. Obviously care had to be taken to ensure that the mechanical strength of the container was sufficient to withstand the higher pressures as circuit breaker ratings increased, and the proportioning of the air space above the oil became a critical design parameter.

Further research work led to the development of the side-vented 'explosion pot' or the 'cross jet pot', which forms the basis of arc control device used in oil circuit breakers.

Figure 1.8 is a cross-section drawing of a typical side-vented arc control device. The principle of operation of this device is that the pressure developed by the vaporisation and dissociation of the oil is retained in the pot by withdrawing the moving contact through a stack of insulating plates having a minimum radial clearance around the contact. Thus there is practically no release of this pressure until the moving contact uncovers one of the side vents, created by cutting a slot in one of the plates.

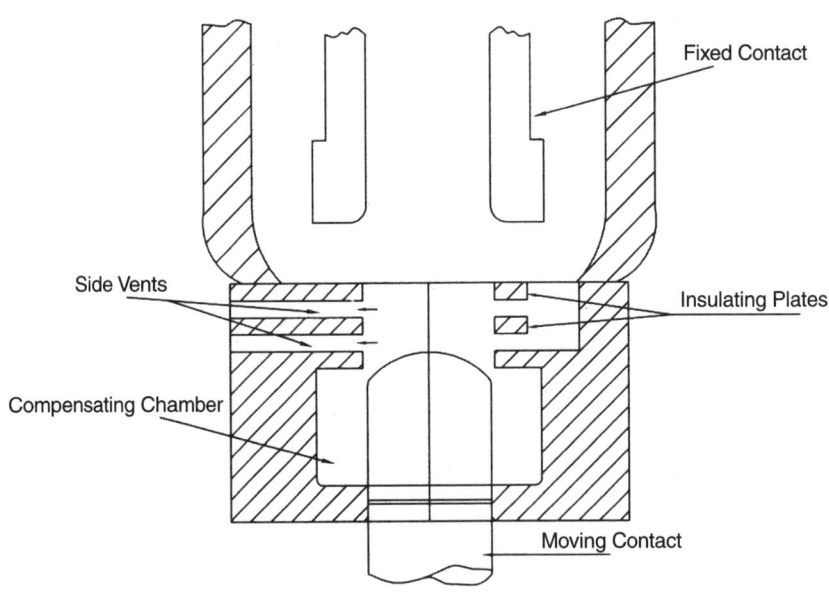

Fig. 1.8: Side-vented arc control device

The compressed hydrogen gas can then escape across the arc path, thus exerting a powerful cooling action on the ionised column. When the current zero is reached, the post-arc resistance increases rapidly due to this cooling action and clearance occurs. At currents below the maximum rating of the circuit breaker, the cooling action is less vigorous, but the degree of ionisation is also less, which is why clearance is still achieved. The design of an effective, wide range, arc control device requires a careful balance of the vent areas, vent spacing and contact speed to ensure consistent performance over the full range of currents. These circuit breakers face problems when clearing low currents, such as the load current of the circuit breaker. One of the refinements made in the design to improve the interruption of small currents was the introduction of a supplementary oil chamber below the side vents. This is known as a compensating chamber and provides a fresh source of oil to be vaporised in order to feed more clean gas back across the arc path when clearing low currents. As with all types of circuit breakers which use the power of the arc to provide the means for its own extinction, there exists a 'critical zone' in the breaking current range.

Bulk oil circuit breakers usually have a metal top dome, through which six bushings pass to connect into the system (Fig. 1.9a and 1.9b), and the arc control devices, generally of the form illustrated in the Fig., are attached to the bottom ends of these bushings. The moving contacts are then arranged on a bridge, which, in turn, is fastened to an insulated drive rod. A removable tank containing the insulating oil is bolted to the dome. The earlier designs had two sets of contacts

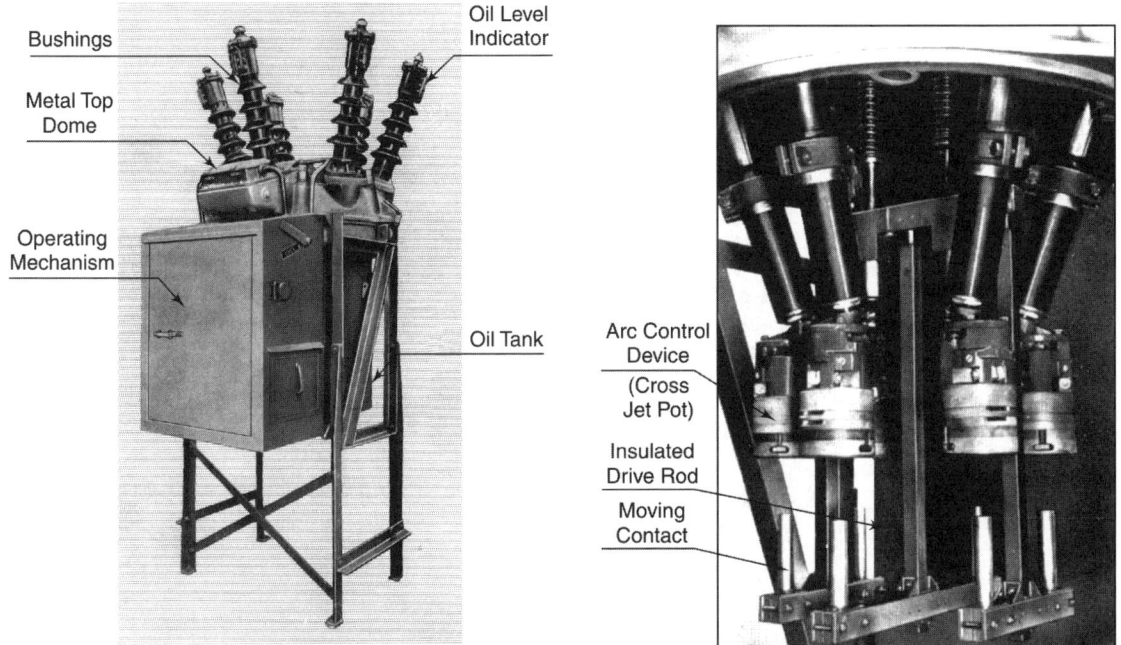

Fig. 1.9: (a) View of bulk oil circuit breaker, (b) Breaker with oil tank lowered

and arc control devices in series in each phase, and this type of oil circuit breaker was called 'double break'. However, considerations of the voltage division between these two breaks during the recovery period led some manufacturers to create single break designs.

The development of bulk oil circuit breakers was limited to the 330 kV class. The limitations were mainly due to the use of a very large amount of oil (app. 50,000 l for a 330 kV circuit breaker), the requirement of a very high speed of contact separation and the use of a large energy mechanism. After bulk oil devices, minimum oil circuit breakers and air blast circuit breakers were introduced.

1.2.2 Minimum Oil Circuit Breakers (MOCBs)

These types of circuit breakers utilise oil as the interrupting media. However, unlike bulk oil circuit breakers, these designs place the interrupting units in insulating chambers at live potential. This feature of the design of MOCBs reduces the requirement of oil, and these breakers are therefore known as minimum oil circuit breakers. These designs are available in voltages ranging from 1000 V to 765 kV using the multi-break technique. A typical view of 36 kV MOCB indicating main parts is shown in Fig. 1.10a.

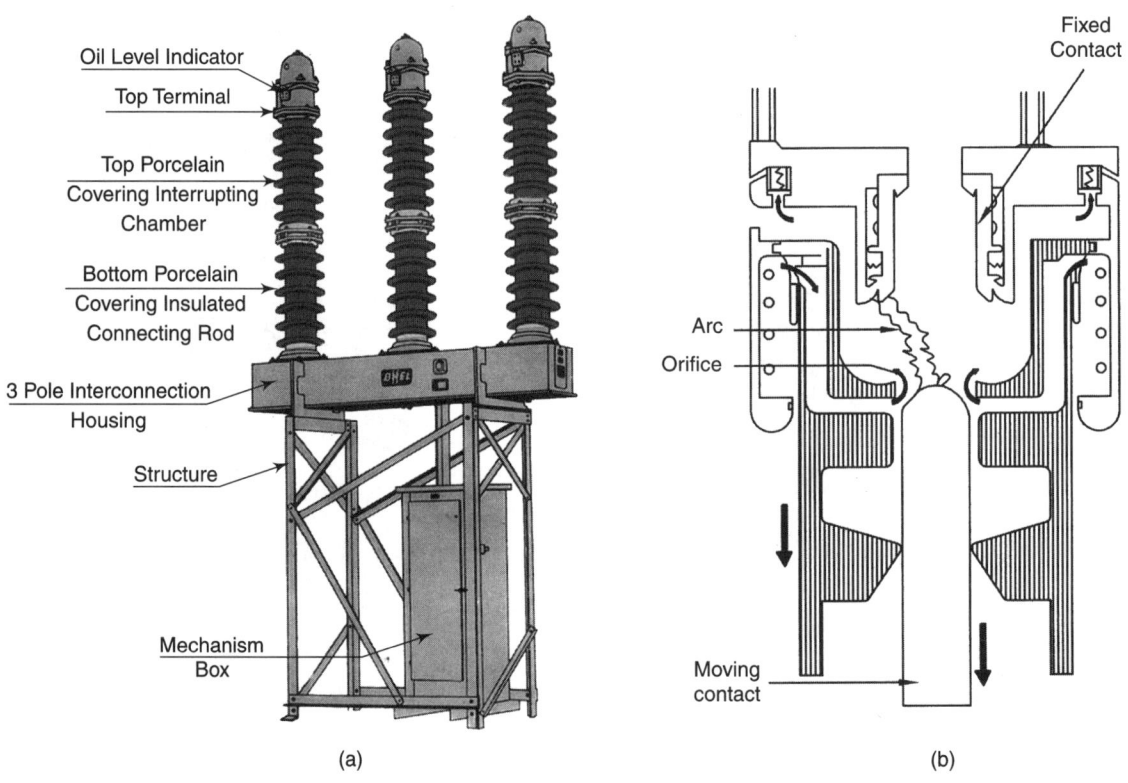

Fig. 1.10: (a) Typical view of 36 kV MOCB, (b) Cross-section of interrupting chamber

This type of breaker is widely used in transmission and distribution network. A modern innovation in this field is the achievement of current interrupting capacity up to 50 kA with the interrupter heads hermetically sealed and lightly pressurised with dry nitrogen. In this type of construction, moisture, which has the most detrimental effect on oil, is kept out which facilitates long service life. Unit ratings upto 145 kV with single interrupter and 245 kV with two interrupters per pole have been developed for transmission application.

In an oil circuit breaker, the arc drawn across the contacts is contained inside the interrupting pot and thus the hydrogen bubble, formed by the vaporised oil (gas) is also contained inside the chamber. As the contacts continue to move and when the moving contact rod separates itself from the orifice at the bottom of the chamber, an exit similar to a nozzle becomes available for exhausting the hydrogen that is trapped inside the interrupting chamber (see Fig. 1.10b).

It is well-known that minimum oil circuit breakers are sensitive to high TRV peak and prone to re-strikes during switching capacitor banks. However, this problem has been effectively countered by pressurising the interrupter heads with dry nitrogen.

These types of breakers were found to be more suitable for applications in countries with very low ambient temperatures like Canada, Russia and Finland, where the gas blast circuit breakers become more sensitive to icing and liquification problems.

Due to their cross blast designs, minimum oil HV circuit breakers suffer from few major inherent problems such as sensitivity to high peak voltage and the effects of pre-arcing. Usually the out-of-phase switching test is the most difficult, test on MOCBs, which generally dictates the number of interrupters to be used in series for a particular voltage class. The energy released due to pre-arcing is very high and high strength insulating materials are required to enable cross jet pots and jet plates to withstand high-pressure surges. The number of breaks per pole also depends upon the ability of the arc interrupting chamber to withstand the energy released due to pre-arcing, which is a function of current in the arc and voltage per break.

Presently this technology is considered outdated due to the advent of VCB technology in medium voltage and SF6 technology in the high voltage range.

1.2.3 Air Circuit Breakers (ACBs)

A circuit breaker in which the contacts open and close in air at atmospheric pressure is defined as an air circuit breaker. The oil circuit breaker developments described above led to almost complete replacement of the earlier designs of air circuit breakers in medium voltage distribution systems by oil circuit breakers in many countries. However, in some countries, notably France and Italy, the air circuit breaker was developed and used for systems up to 15 kV, but in general, its use is restricted to low voltage applications or high security installations where the risk of an oil fire or oil contamination of the environment is too high to be tolerated.

Countries following the American practice used air circuit breakers almost exclusively for systems up to 15 kV until the advent of the new vacuum and SF6 technologies. Figure 1.11a shows main parts of an air circuit breaker trolley.

The principles of arc interruption followed in an air circuit breaker are rather different from those in any other type of circuit breaker. While the objective of both types of circuit breakers is

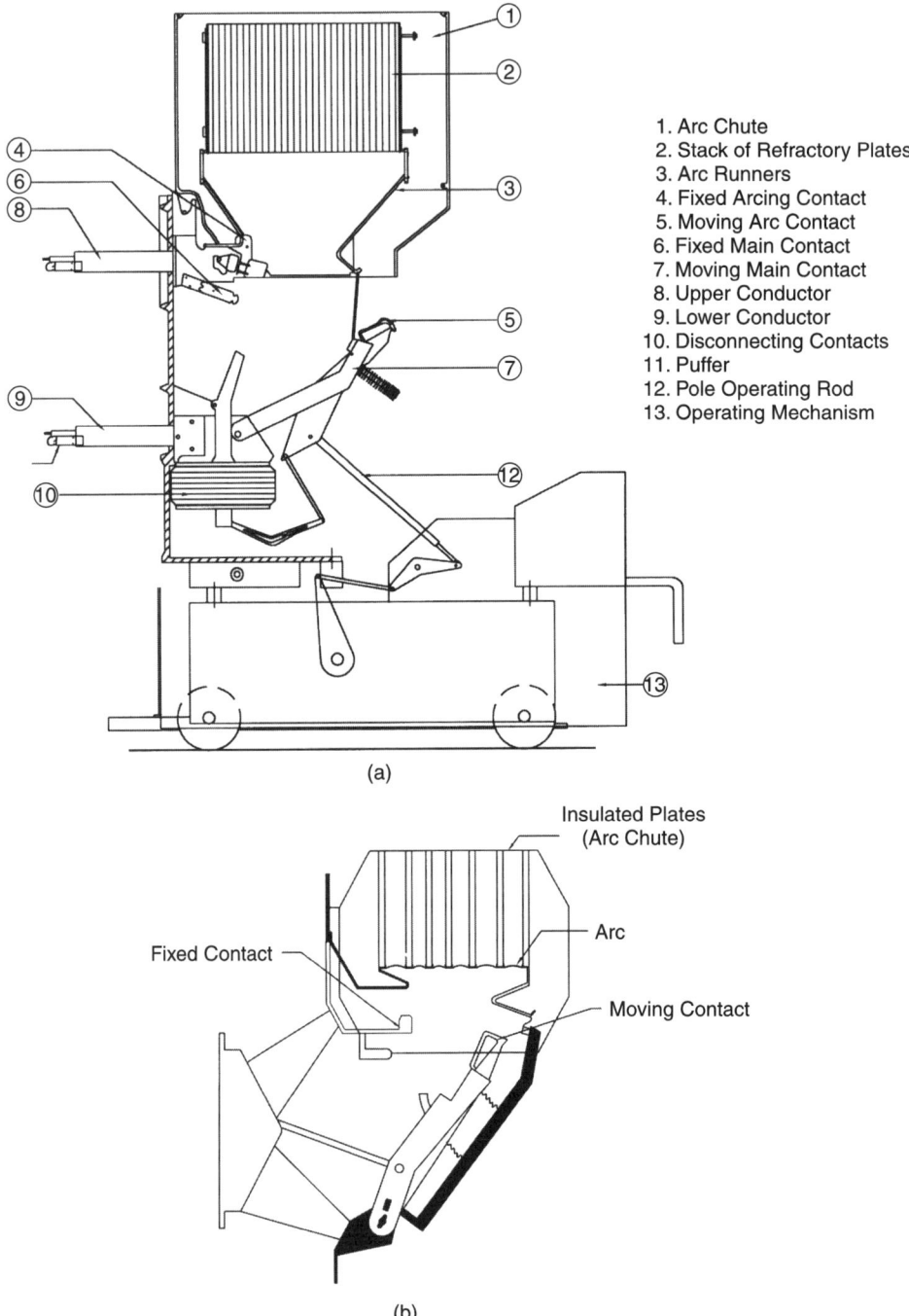

Fig. 1.11: (a) Main Parts of an air circuit breacker trolley, (b) Cooling arc in an insulated plate arc chute

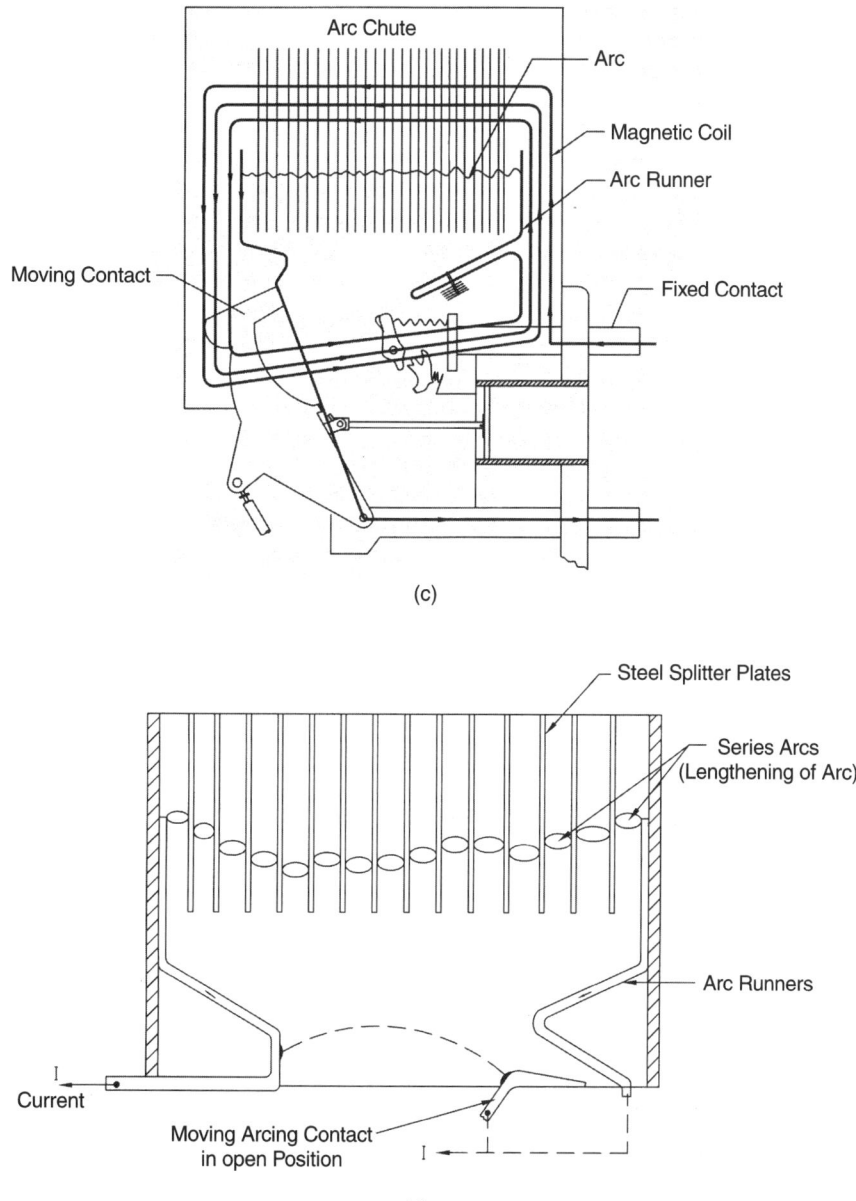

Fig. 1.11: (c) Arc chute with magnetic coil, (d) Arc chute with metal arc splitter plates

the same, i.e. to prevent the resumption of arcing after current zero by creating a situation wherein the contact gap will withstand the system recovery voltage, the air circuit breaker does this by creating an arc voltage in excess of the supply voltage.

This can be done in three ways:

- Intense cooling of the arc plasma, so that the voltage gradient is very high;
- Lengthening the arc path to increase the arc voltage; and
- Splitting up the arc into a number of series arcs.

The first objective is usually achieved by forcing the arc into contact with as large an area as possible of insulating material. All air circuit breakers are fitted with a chamber surrounding the contact and arc zone, usually called the 'arc chute', because the arc is driven into and through it. If the inside is suitably shaped, and the arc can be made to conform to the shape, the arc chute walls can help achieve cooling. This type of arc chute needs to be made from some kind of refractory material, with asbestos compounds once being the favoured choice. However, because of increasing awareness of the health hazards associated with the use of asbestos, other materials such as high temperature plastics reinforced with glass fibre and ceramics are being used for this purpose. Figure 1.11b shows cooling of arc in an insulated plate arc chute.

The second objective is achieved concurrently with the first, if the walls of the arc chute are shaped in such a way that the arc is not only forced into close proximity with them, but also driven into a serpentine channel. The lengthening of the arc and the simultaneous increase in the voltage drop per unit length soon lead to a high arc voltage, and a high arc resistance. This changes the system power factor so that the instantaneous value of the supply voltage as the current approaches zero is much below its peak value. Finally, the high value of resistance represented by the arc influences the damping of the TRV oscillation such that the amplitude factor is reduced almost to unity.

The preceding paragraphs explain what is required, and how to achieve it, but a heavy current arc is an intractable thing and one of its main tendencies is to keep itself as short as possible since that helps to maintain its ionisation level, and hence its current-carrying capability. Thus, while it is not difficult to design convoluted walls to arc chutes, it is also not easy to persuade the arc to enter them! The usual arrangement of an air circuit breaker is to arrange the arc chute above the contacts, and enable the contacts and the connections leading to them to form a tight loop so that a magnetic field is produced within the contact gap acting upon the arc in such a manner as to drive it up into the chute. In this design, increase in fault current results into increase in magnetic field which elongates the arc and thus helps in arc extinction process.

The use of the magnetic circuit complicates the contact structure, as it is not advisable to retain the coils in circuit continuously, and arrangements have to be made to insert them into the circuit during the opening of the circuit breaker. This is usually achieved by splitting the arc runner, a metal strip that runs up from each contact to guide the arc into the required path so that the coil is automatically inserted into the current circuit as the arc runs up towards the chute. Refer Fig. 1.11c showing an arc chute with magnetic coil.

The heavier the fault current, the more effective will be the air circuit breaker, until it either becomes so effective that the arc runs straight through the arc chute and re-establishes itself as a nice short arc outside the chute, or the pressure inside the chute below the plates, created by the arc, becomes so great that the arc is prevented from rising into the plates. Either case obviously constitutes the upper limit of breaking capacity for a given arc chute, which needs to be at a respectable margin above the designed rating.

When it is interrupting maximum fault current, an air circuit breaker, rated typically at 40 kA and 12 kV, would emit an impressive fireball from its arc chute, if a large cooling baffle is not fitted to contain it. The result is that air circuit breakers of heavy ratings constitute large, heavy and expensive pieces of plant and machinery. They do, however, have advantages for certain applications, particularly where the system switching parameters are severe, as the high arc resistance makes the air circuit breaker virtually independent of the circuit in which it is connected.

However, the third technique can facilitate a reduction in energy generation by using metal arc splitter plates refer Fig. 1.11d in the arc chute, which not only increases the voltage gradient, but also helps to introduce a large number of anode and cathode voltage drops. A combination technique has been designed which uses an assembly of small arc chutes which are individually of the 'cooling and lengthening' type, but which are collectively arranged such that the initial arc is split into a number of series arcs, each in its own mini arc chute. This is a more complicated design but one which is suited to higher distribution voltages. It also has another advantage in that it eliminates the need for providing extra components for the creation of a magnetic field in order to drive the arc into the chute.

Thus the art of designing an effective air circuit breaker requires the correct combination of chute shape, magnetic field development and air puffer to facilitate a uniform arc extinction performance over the whole current range. Several parameters have to be balanced for this, including some that have not yet been mentioned such as arc chute height, length and outlet area. The mechanical forces developed in a heavy-duty air circuit breaker are also considerable.

A final point concerning the characteristics of the air circuit breaker is that the interrupting technique described above is the only one that does not really depend on the presence of a current zero. If the arc voltage can be made to exceed the supply voltage, the arc will be extinguished, irrespective of whether the current is ac or dc, which makes it suitable for the control and protection of dc circuits.

Although these circuit breakers are considered obsolete for medium voltage application, they continue to be preferred choice for high current rating in low voltage application.

1.2.4 Air Blast Circuit Breakers (ABCBs)

This type of circuit breaker has been used earlier for open terminal HV application, for system voltages of 245 kV, and 400 kV up to 765 kV, especially where faster breaker operation was required. In the later design, a two-cycle operation was achieved and even faster operating systems with a total break time of one cycle have been developed for special applications in American grids. In this special design, the mechanical opening time of 12 ms and maximum arcing time of about 8 ms has been achieved. This type of breaker has been used for special applications wherein several super thermal power stations located in close vicinity to each other have been inter-connected. These one-cycle breakers form the inter-connecting link which, in the event of a fault in an outgoing feeder of a particular station, isolates the station from the neighbouring ones thereby limiting the fault level for the feeder breaker.

It has long been recognised that short line fault duty is a major problem with air blast breakers, especially where the fault current is high. Earlier tripping resistors were used to damp the short line fault oscillations. Later on, a better understanding of arc and model studies made it possible to develop air blast breakers without the aid of damping resistors. However, till today, this duty continues to be the most critical and usually dictates the number of interrupters to be used per pole for a particular fault rating.

Figure 1.12a shows the construction of a typical air blast interrupter. Each interrupter consists of a porcelain insulator, mounted on the air inlet manifold, with the exhaust chamber fixed at the opposite end of the porcelain. The exhaust chamber consists of a casting with a curved hood for weather protection and slots on the underside for directional exhaust of the compressed air to atmosphere.

The moving contact assembly consists of a chromium-copper contact coupled to two pistons by means of an insulated tie rod that moves inside the contact tube. The main current in the moving contact is transferred to the contact tube by means of transfer contact fingers. An arcing tip is provided at the end of the moving contact. The moving contact is maintained in a normally closed position with the fixed contact by springs. The details of construction of the interrupter head vary with its interruption rating. Figure 1.12a shows the closed position of interrupter. Figure 1.12b shows contacts in partially open position with arcing between fixed and moving contact. Figure 1.12c shows contacts in full open position with arc extinguished.

The interrupting capability of air blast breaker is usually increased by increasing the normal pressure range. Normally the pressure level is around 30 to 35 bars. In order to maintain the insulation level and reliability of operation, it is also necessary for the condition of the air to be very dry.

Currently, however, SF6 circuit breakers have practically eliminated the use of this technology.

1.2.5 SF6 Gas Circuit Breakers

A circuit breaker in which the contacts open and close in sulphur hexafluoride (SF6) gas is known as an SF6 circuit breaker (Fig. 1.13). These circuit breakers are available for complete range of medium voltage and high voltage application up to 800 kV and above. This medium is most suitable for metal-clad and hybrid HV sub-stations. In fact, in the past this medium was primarily developed for metal-clad switchgear at HV range. However, with the development of single pressure puffer design for medium fault currents, the conventional open terminal designs for EHV/UHV transmission application have become economically very attractive. At present, it is possible to design a single-interrupter 245 kV, two-interrupter 420 kV and four-interrupter 800 kV circuit breaker.

Since SF6 gas has been identified as a greenhouse gas, safety regulations are being introduced in many countries in order to prevent its release into the atmosphere. Hence the HV circuit breaker would be designed to ensure that there is minimum leakage during the service period and that the utilities let out the least amount of SF6 gas into the atmosphere during maintenance. Conventionally, SF6 circuit breakers with a puffer type design need a high mechanical energy, which is almost five times that of a minimum oil breaker of equivalent rating. For this reason, most of the designs use a high-energy hydraulic or pneumatic system.

An innovation in the puffer type design of an SF6 circuit breaker is the 'thermal assist' principle. In this type of design, the arc energy is utilised to develop pressure in the arcing chamber

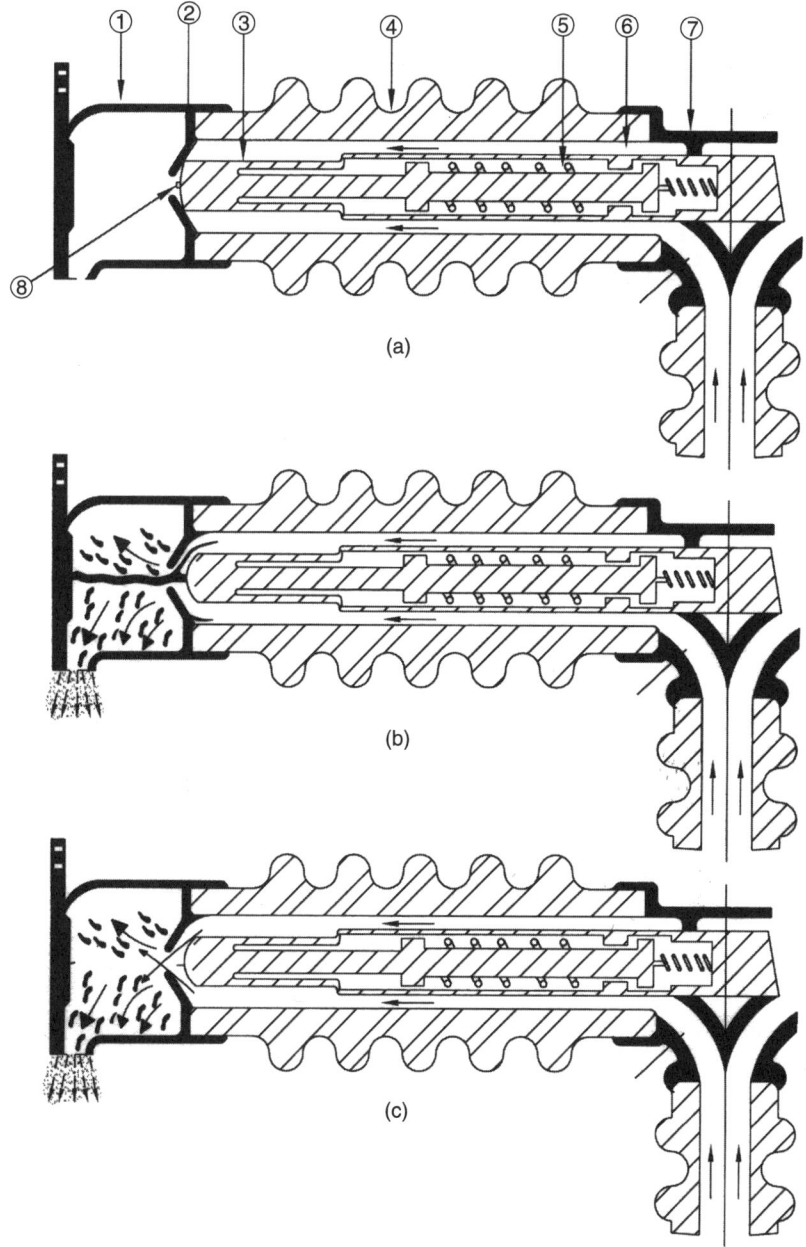

1. Exhaust Chamber
2. Fixed Contact
3. Moving Contact
4. Porcelain Weather Costing
5. Tie Rod
6. Contact Tube
7. Air Inlet Manifold
8. Arcing Tip

Fig. 1.12: (a) Contacts closed: Air blast commencing Arc struck, (b) Contacts partially open: arc extinguished, (c) Contacts fully opened: Arc extinguished

Fig. 1.13: SF6 circuit breaker

for arc quenching. Since this innovation has reduced the energy requirement of the operating mechanism it has become possible to use a spring mechanism for application in SF6 circuit breakers up to 245 kV. Although the share of SF6 circuit breakers in medium voltage application is gradually shrinking, this technology is still emerging as the only choice in the HV range. This technology is discussed in detail in Chapter 4.

1.2.6 Vacuum Circuit Breakers (VCBs)

In a vacuum circuit breaker, the arc interruption takes place in vacuum. This technology has been found to be most suitable for medium voltage application though the experimental interrupters for 72.5 kV and 145 kV have been developed, they were not found to be commercially viable.

In principle, a vacuum interrupter has a steel arc chamber in the centre and symmetrically arranged ceramic insulators. Refer Fig. 1.14 showing the main parts of a typical vacuum interrupter. Modern constructions of this interrupter have a metal shield surrounding the arcing contacts. The diameters of the contacts and their stems are matched with those of the arc chamber and insulators. The moving contacts are made movable by the use of metallic bellows. The arc chamber is welded to the housing flanges, which, in turn, are brazed to metallised ceramic insulators, thus giving a hermetically sealed interrupter. The vacuum pressure is generally 10^{-6} bar.

The material of the contacts plays a very important role in the overall performance of the vacuum circuit breaker (VCB), and, in fact, the entire development of vacuum technology is centred around the use of proper material and geometry of the contact. The most common

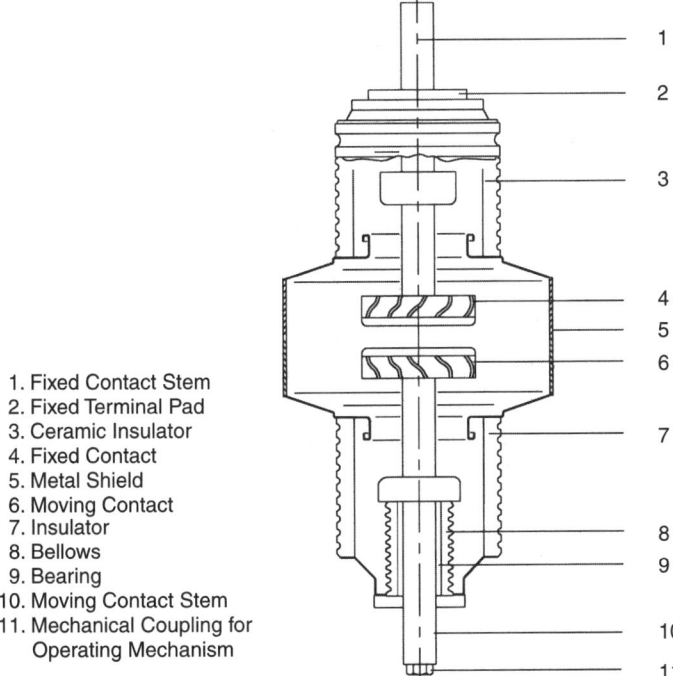

1. Fixed Contact Stem
2. Fixed Terminal Pad
3. Ceramic Insulator
4. Fixed Contact
5. Metal Shield
6. Moving Contact
7. Insulator
8. Bellows
9. Bearing
10. Moving Contact Stem
11. Mechanical Coupling for Operating Mechanism

Fig. 1.14: Vacuum interrupter

materials used today are CuBi, CuCr, or CuAg. Of these, CuCr offers the most ideal solution for all ratings of interrupters from 8 kA to 63 kA. With this material, the current chopping levels have been kept down to as low as 2 to 3 amps.

Even though modern vacuum interrupter (VI) technology was developed in the early 1960s, it is still considered to be an evolving technology, as continuous improvements and innovations are taking place in this area. For example, the size and capability of modern VI bears no relationship to the one built in the 1960s. Figure 1.15 shows an example of 12 kV, 12.5 kA VI wherein it can be seen that the diameter of the VI has been reduced from 7 inches in 1967 to 2 inches today. This size reduction is the result of major developments in vacuum technology, vacuum processing, contact material development and evolution of vacuum interrupter design. The contact geometry started from plain butt contacts in the 1960s and gradually progressed to spiral, cup and axial magnetic fields (see Fig. 1.16). The butt contact is the simplest, but is only suitable for interrupting the low current diffuse vacuum arc. It thus finds application in load break switches and contactors which are required to interrupt currents less than 4.5 kA. The transverse magnetic field (TMF) contacts have spiral or skew slitted cup type construction. These contacts use the transverse magnetic field generated by the circuit current flowing in the spiral arms or skew slits of contacts to drive the high current, columnar, vacuum arc rapidly over the contacts surfaces. This results in two effects: (a) the contact surface has uniform erosion and is left in a relatively smooth condition after high current arcing, and (b) the column cannot sustain itself when the current falls to current

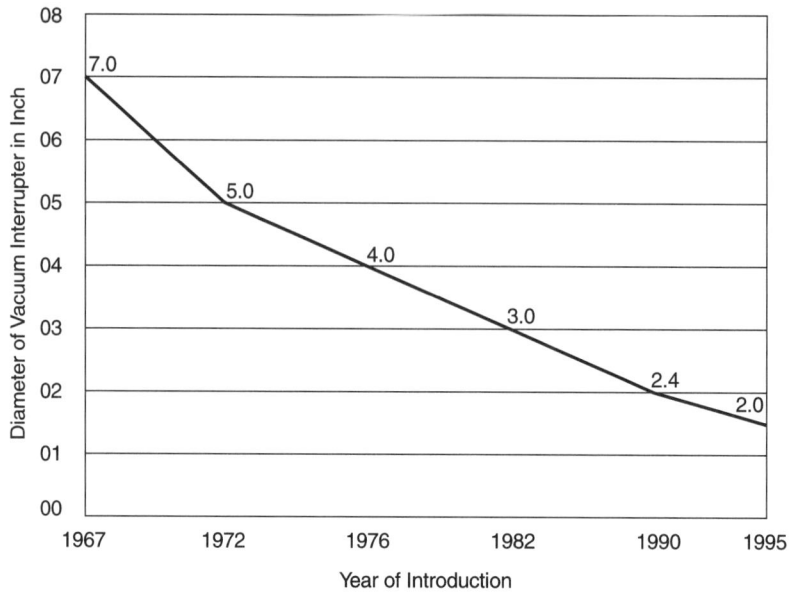

Fig. 1.15: Reduction in size of vacuum interrupter for 12 kV, 12.5 kA rating

zero in an ac circuit. It thus returns to the diffuse mode, which is easily interrupted at current zero.

The application of the VI to protect power distribution circuits has grown during this period (see Fig. 1.17). In fact, the vacuum circuit breaker (VCB) is today recognised as the most reliable, and it also needs the lowest maintenance among all the technologies available to control and protect distribution circuits. The present consensus all over the world is that the VCB will dominate medium voltage technology in the twenty-first century.

It is becoming increasingly difficult for SF6 to compete with VCB technology in the following three major areas:

(a) **Long life:** it is now possible to produce cost-effective VI designs with electrical lives that exceed the required mechanical life of the circuit breakers, and that will even be able to satisfy a recent requirement of extended short-circuit operating life.

(b) **Environmentally benign:** VIs are constructed from environmentally benign materials and do not pose the potential health risk that exposure to arced SF6 gas does. Also VIs do not need special hazardous waste handling that SF6 interrupters require when routine maintenance is performed or when the SF6 interrupter is disposed of at the end of its life.

(c) **Overall superior performance:** Besides above, the construction of VI's is very user friendly for easy mounting in this circuit breakers. Replacement of VI's whenever required is also very convenient. This is a direct result of the extensive R&D work undertaken up by universities and the manufacturers of VI's.

The details of this technology are covered in Chapter 3.

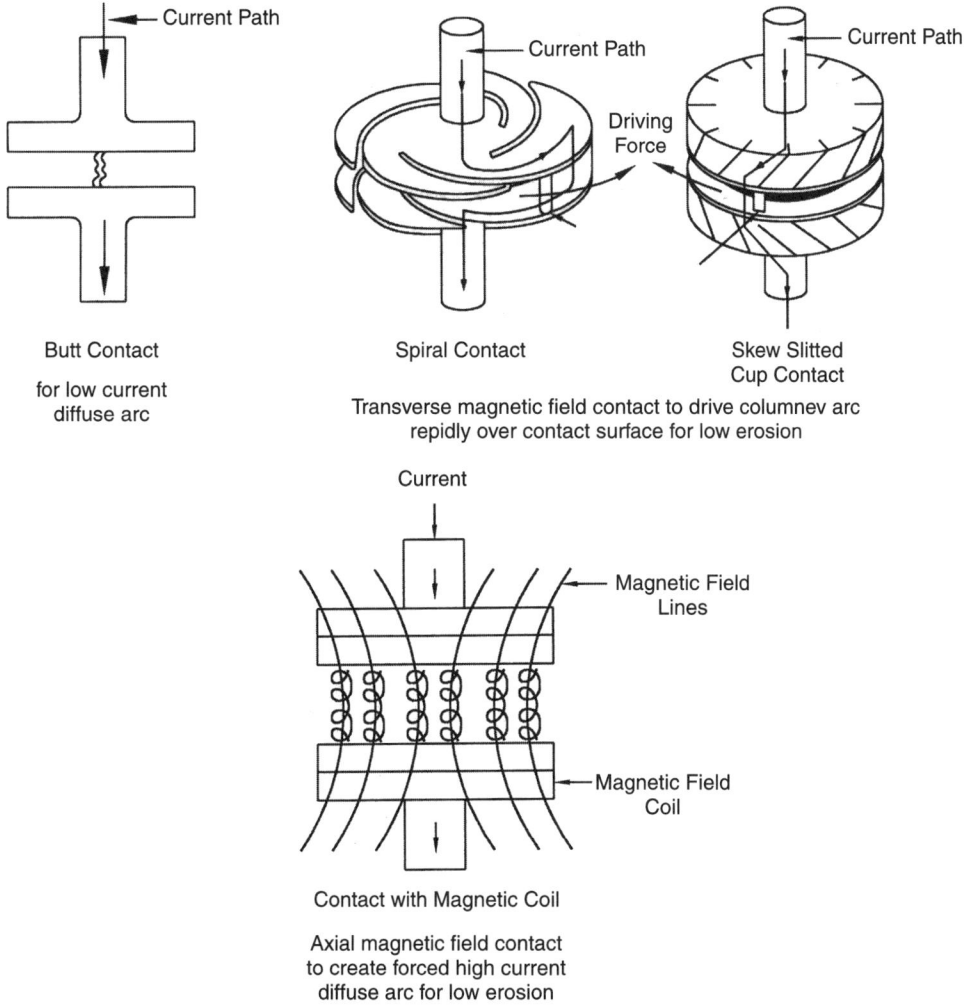

Fig. 1.16: Contact geometry and their effect

1.2.7 Gas Insulated Switchgear (GIS)

Gas insulated switchgears are metal enclosed types of switchgear with sulphur hexafluoride (SF6) gas acting as an insulation between the live parts and the earthed metal enclosure. The high voltage equipment is contained in a gas-tight container filled with the insulant gas, SF6. This type of equipment is available in the voltage range from 12 kV to 800 kV class. Medium voltage GISs up to 52 kV class are generally filled with SF6 gas at pressure below 2.5 bar. This range of GIS is covered in standard IEC 60298/IEC 62271-100, applicable for metal enclosed switchgear. GIS for voltage class 72.5 kV and above are covered in standard IEC 60517/IEC 62271-203. Gas insulated switchgear is a major component of gas insulated sub-stations. The principal gas insulated mod-

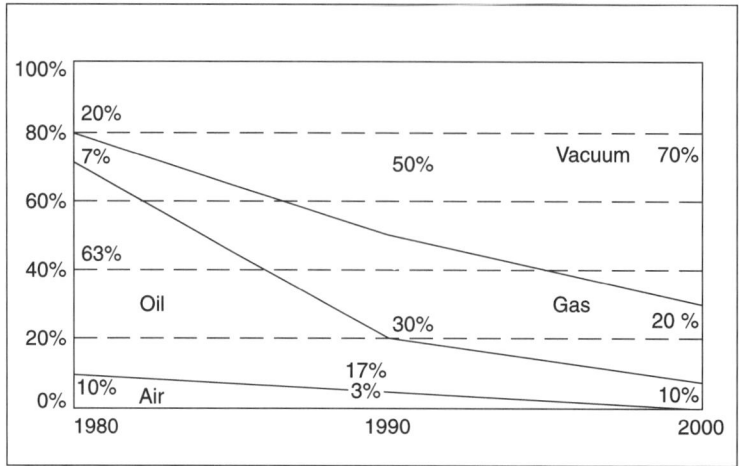

Fig. 1.17: Percentage market share trend for various technologies in medium voltage switchgear

ules for a sub-station are bus-bar, disconnector or isolator, circuit breaker, current transformer, and earth switch.

The auxiliary gas insulated module or accessories required for a sub-station are terminations, voltage transformer, and surge and lighting arrester. The assembly of such equipment at a sub-station is defined as a gas insulated metal enclosed sub-station (GIMES) by the International Electro-technical Commission. This technology has now been well established for almost thirty years.

Medium voltage GIS equipment features vacuum as the interrupting medium and SF6 gas as the main insulation. Designs using the SF6 medium for both insulation and interruption are also available in medium voltage. Two operating pressures are specified for such equipment, one for insulation and the other for interruption. The high voltage GIS for voltages of 72.5 kV and above are essentially two-pressure systems for non-availability of vacuum interrupters in this high voltage range. In gas insulated sub-stations, modular components are assembled together to form the desired arrangement for a section or a bay.

Gas insulated metal enclosed switchgears, which are developed in various constructional arrangements, are explained below:

- The **isolated-phase GIS** module consists of an assembly of individual circuit elements, like a pole of a circuit breaker, a single pole disconnector, one-phase assembly of a current transformer, etc. Three such circuits, arranged side by side, form a complete three-phase GIS bay. The circuits require larger bay width as compared to the other GIS configurations.

- Some of the designs encapsulate all the three-phase elements of individual modules, like three poles of a circuit breaker, three poles (isolations) in a disconnector or three-phase current transformer assemblies, in an individual enclosure, thereby forming a **three-phase module** for the element. The total number of the enclosures is thus, reduced to one-third of the Isolated Phase GIS.

- In another configuration called **hybrid systems**, a suitable combination of isolated-phase and three-phase common elements is used, like three-phase bus bar and single-phase elements, to achieve an optimal techno-commercial solution. While the three-phase common bus bar system simplifies the connections from the bus bar, the isolated-phase equipment prevents phase-to-phase faults in active modules, like the circuit breaker. The savings in space vary with the design and configuration of the section. Hybrid GIS technology has gained popularity, specifically in the medium and high voltage range, wherein the technology has helped to reduce the sub-station size.
- **Compact GIS** systems are essentially three-phase common systems, with more than one functional element in one enclosure. A single enclosure housing a three-phase circuit breaker, current transformer, and earth switches, supports the bus bar and the other feeder elements. The depth of the section is considerably reduced in this configuration as compared to the three-phase modules.
- The **Highly Integrated Systems** (HIS), introduced in 2000, are single unit metal encapsulated and gas insulated sub-stations, and are gaining user appreciation as this equipment provides a total sub-station solution for outdoor/yard sub-stations. The foundation work is limited to just one equipment, which helps to substantially save installation time. GIS is discussed in detail in Chapter 5.

1.3 CIRCUIT BREAKER OPERATING MECHANISMS

The primary function of a circuit breaker mechanism is to provide the means for opening and closing the contacts. Initially this seems to be a rather simple and straightforward requirement. However, when one considers the fact that most circuit breakers, once placed into service, will remain in the closed position for long periods of time, and yet on the few occasions when they are called upon to open or close, they must do so reliably, without any delay or sluggishness, then one realises that the demands on the mechanisms are not as simple as was first thought.

Opening and closing velocities, as well as stroke, or travel distance are the most important characteristics of a circuit breaker. They are dictated primarily by the requirement imposed by contacts. Opening and closing velocities are important to the contacts in order to avoid both contact erosion and contact welding. Circuit breaker stroke is primarily related to the ability of the circuit breaker to withstand the required operating dielectric stresses, which, in turn, is directly related to the total contact gap and the rate at which this gap increases. The operating characteristics need to be specified for each mechanism along with the circuit breaker as per IEC62271-100.

The two classes of circuit breakers, based on mechanical operations as per IEC62271-100, are as follows:

- Class M1 for 2000 numbers operation as per Table 1.3; and
- Class M2 for special service requirements of 10,000 numbers operation.

The typical circuit breaker operating characteristic curves, showing various characteristic quantities for the purpose of understanding the terms and their measurements for SF6, air blast or oil

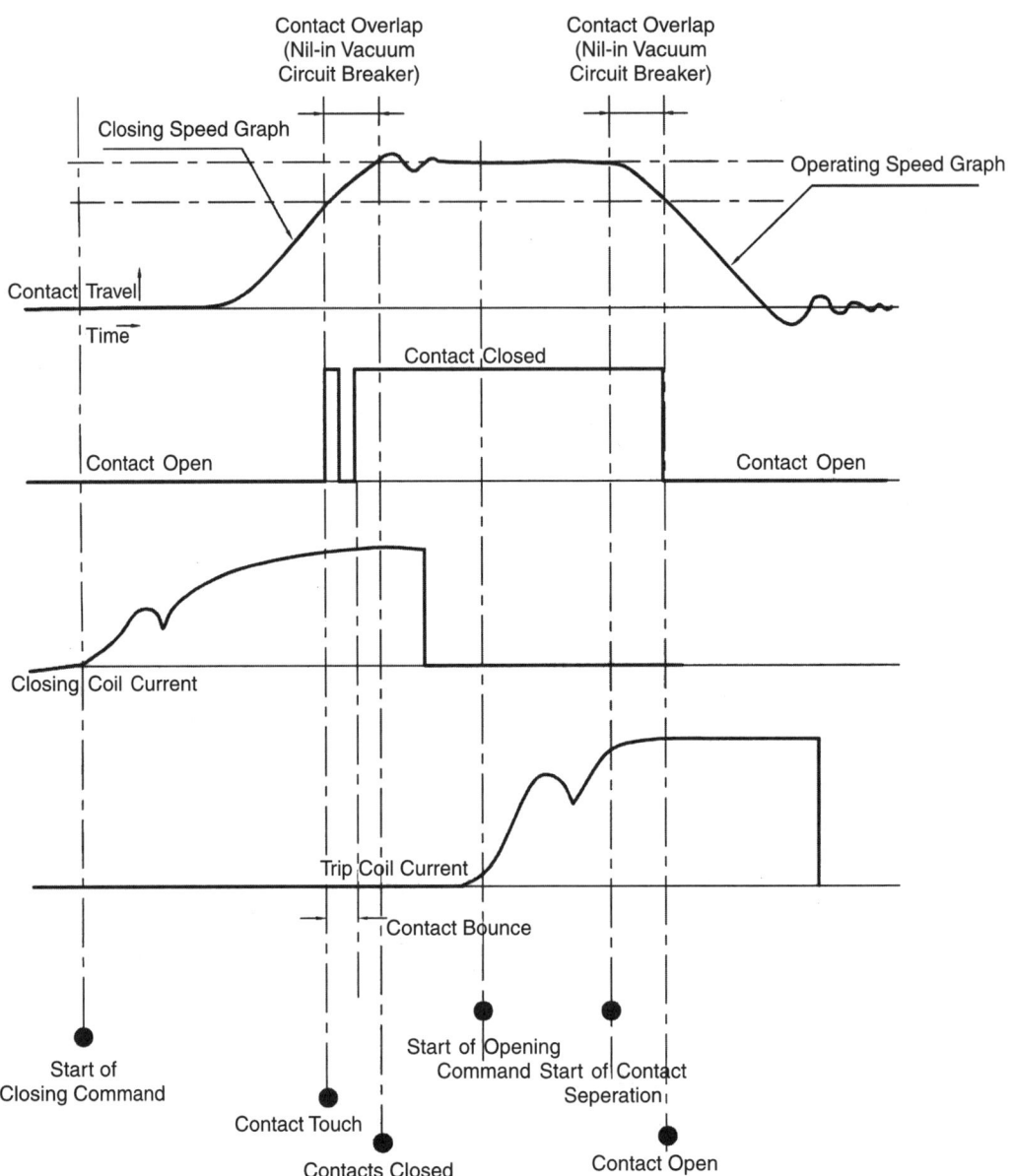

Fig. 1.18: Operating characteristic curves

type circuit breakers are shown in Fig. 1.18. Operating characteristic curves for vacuum circuit breakers get modified slightly as vacuum circuit breakers have butt type contacts with no overlap part in the moving and fixed contacts.

Operating Sequence	Control Voltage and Operating Pressure	Number of Operating Sequences	
		Circuit Breaker for Auto-reclosing	Circuit Breaker not for Auto-reclosing
C—t_a-O—t_a	Minimum	500	500
	Rated	500	500
	Maximum	500	500
C-t-CO—t_a-C-t_a	Rated	250	—
CO—t_a	Rated	—	500

Table 1.3 Mechanical Duty Requirement of Circuit Breaker Operating Mechanism

where O = opening, C = closing, CO = closing operation followed by an opening operation.

t_a = time between two operations which is necessary to restore the initial conditions and/or to prevent undue heating of parts of the circuit breaker.

t = 0.3 sec for circuit breaker intended for fast auto-reclosing duty, if not otherwise specified.

Three types of operating sequences mentioned in Table 1.3 represent the actual operation faced by circuit breaker mechanism in service. For example C—t_a—O—t_a represent opening and closing operations with long time gap in between. The next sequence represent duty of auto recloser circuit breaker and the last sequence of Table 1.3 represent closing of circuit breaker on fault and immediate opening thereafter.

1.3.1 Circuit Breaker Opening Requirements

The two basic requirements for the total opening operation of a circuit breaker are the opening speed and the total travel distance of the contacts. The opening speed requirements are dictated by the need to ensure that the separation of the contacts is done as rapidly as possible for two reasons, firstly to limit contact erosion and secondly to control the total fault duration which is dictated by the system co-ordination requirements. The total travel distance is not necessarily the distance needed to interrupt the current but rather the gap needed to withstand the normal dielectric stresses and lighting impulse waves that may appear across the contacts of a breaker which is connected to a system while in the open position.

The need for carrying the continuous current and for withstanding a period of arcing makes it necessary to use two sets of contacts in parallel, one, the primary contact, which is always made of a high conductive material such as copper and the other, the arcing contact, made of arc resistance materials such as tungsten or molybdenum, which have a much lower conductivity than those used for primary contacts. When the circuit breaker opens to interrupt the current, the primary contacts open before the arcing contacts.

However, due to the differences in the resistance and the inductance of the electrical paths of primary and arcing contacts, a finite time is required to attain total current commutation, that is, from the primary or main contact branch to the arcing contact.

One can appreciate the significance of the commutation time when one considers that in the worst case, commutation may not take place until the next current zero is reached and that during this time, the arc is eroding the copper of the main contacts. Arc erosion of the contacts not only limits the life of the contacts but can also lead to dielectric failures by creating an ionised conducting path between the contacts, thereby limiting the interrupting capability of the circuit breaker. It is also important to realise that commutation must be completed before the arcing contacts separate, otherwise, the arc is likely to remain at the main contacts.

While the opening operation continues and as the contact gap increases, a critical contact position is reached. The new position represents the minimum contact opening where interruption may be accomplished at the next current zero. The remainder of the travel is needed only for dielectric and deceleration purposes.

1.3.2 Circuit Breaker Closing Requirements

During the closing operation, the mechanism is supposed to meet the following requirements:

- Moving of contacts from open to close position at the required speed in order to reduce pre-arcing;
- Supply of energy to overcome repulsive force due closing on fault;
- Supply of energy to compress fluid in the arcing chamber, in MOCB, SF6 and air blast circuit breakers; and
- Supply of energy to charge the opening spring, if required.

The typical opening and closing energy requirements of an oil circuit breaker are given below.

Type of Equipment	Closing Energy* (kgM)	Opening Energy (kgM)
11 kV 250 MVA 400 A CB	8.25	2.17
33 kV 1000 MVA 1200 A CB	31.0	3.8
66 kV 2500 MVA 1200 A CB	98.0	28.8
132 kV 5000 MVA 1200 A CB	520	243

* This includes the energy required to charge the opening springs.

Table 1.4 Typical Opening and Closing Energy Requirements of Oil Circuit Breaker

The closing energy requirement of typical vacuum circuit breaker of 33 kV, 1000 MVA is around 20 kg-M.

1.3.3 Types of Circuit Breaker Mechanisms

Different types of mechanisms are available for different types of circuit breakers, but what is common to all is that they store potential energy in some elastic medium, which is charged from a low power source over a longer period of time.

From the energy storage point of view, the mechanisms that are used in today's circuit breakers fall in the spring, pneumatic or hydraulic categories and from the mechanical operation point of view, they are either of the cam or of the four-bar linkage type. Cams are generally used in conjunction with spring stored energy mechanisms and these cam spring-driven mechanisms are mostly used to operate medium voltage vacuum interrupters.

Cam drives are flexible in the sense that they can be tailored to provide a wide variety of motions. They are also small and compact. However, the cam is subjected to very high stresses at the point of contact, and furthermore, the cam follower must be properly constrained so that it follows the cam's contour reliably, either by a spring, which raises the stress level on the cam, or by a grooved slot, wherein a backlash may cause problems at high speeds.

Choice of operating mechanisms: There are generally three types of mechanisms to choose from, viz. spring operating, pneumatic and hydraulic mechanisms. Originally, the spring mechanisms were usually confined to MOCBs (which require about 3500/4500 NM per pole of a 420 kV circuit breaker) and SF6 circuit breakers for 36 kV to 145 kV voltages. For HV circuit breakers of 420 kV and above, especially for faster ones with total break time of two cycles, the choice is generally limited to pneumatic or hydraulic mechanisms. However, in the recent past, spring mechanisms have also been used with HV range of SF6 gas circuit breakers up to 550 kV. Spring operated mechanisms are preferred for almost all voltage class of circuit breakers due to their simple design and cost advantage over other mechanisms.

1.3.3.1 Spring Charging Mechanisms

Spring charging mechanisms are commonly used in medium voltage outdoor and indoor type circuit breakers. However, in line with recent trends, it is not uncommon to find these mechanisms on outdoor SF6 circuit breakers up to 245 kV class. As implied by its name, the energy of this mechanism is stored in closing springs. The stored energy is available for closing the circuit breaker on command following the release of the closing latch.

In its simplest form, the spring mechanism consists of a charging motor and charging ratchet or gear mechanism, a closing cam, closing springs, tripping springs, and a toggle linkage. The charging motor and ratchet assembly/gear mechanism facilitate automatic re-charging of the closing springs immediately following the closing operation as seen in Fig. 1.19.

The charged springs are held in position by the closing latch, which prevents the closing cam from rotating. In order to release the spring energy, either an electrically operated solenoid closing coil, or a manual closing lever is operated. Following the activation of the closing solenoid, a closing latch is released, which facilitates discharging of the charged closing springs thereby causing the rotation of the closing cam, which further transfers its motion to the operating rod directly or through levers and shafts. As the cam rotates, it straightens the toggle linkage, which, in turn, rotates the main shaft thus driving the contacts that are connected to the shaft by means of insulating rods.

The straightening of the toggle links loads the trip latch as they go over the centre. The trip latch then holds the circuit breaker in the closed position. In addition to closing the contacts, the closing springs supply enough energy to charge the opening spring. The opening of the contacts can be initiated electrically or manually, however, the manual operation is generally provided only for maintenance purposes. When the tripping command is given, the trip latch is released

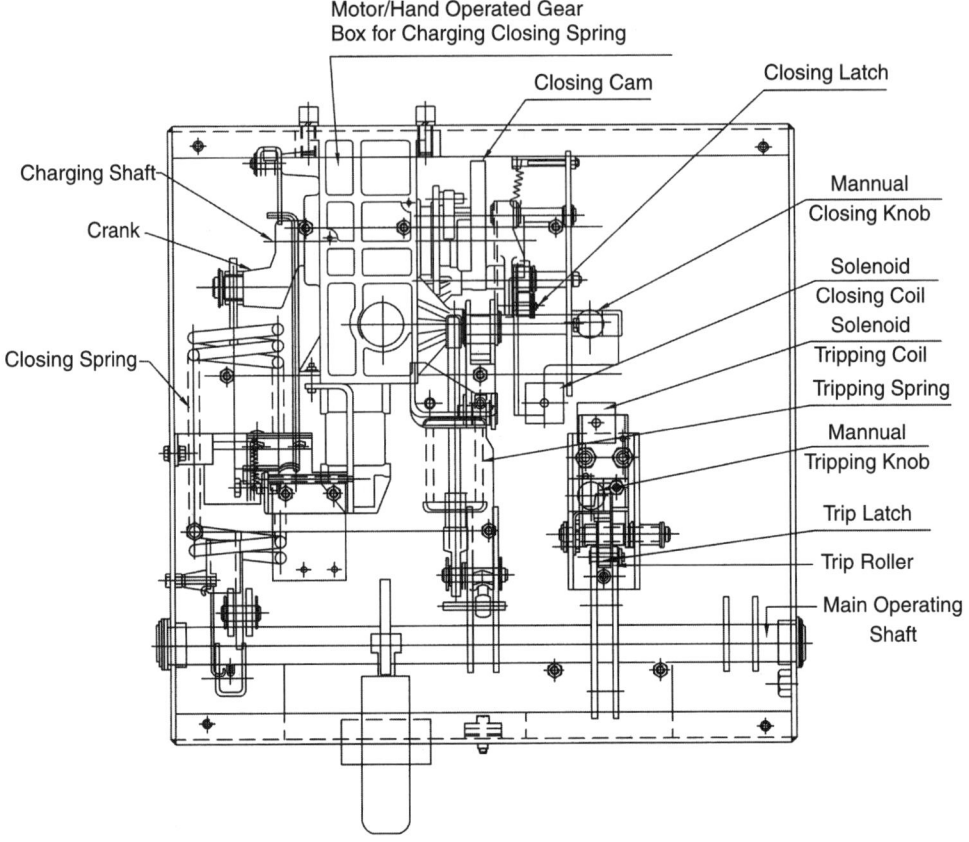

Fig. 1.19: Spring charging mechanism

thus freeing the trip roller carrier. The force produced by over the centre toggle linkage or tripping spring bias, rotates the trip roller carrier forward. The tripping springs, which are connected to the main operating shaft, provide the requisite energy to open the contacts of the circuit breaker. In vacuum circuit breakers, part of the tripping force is also provided by the contact pressure springs.

1.3.3.2 Pneumatic Mechanisms

Pneumatic mechanisms are a logical choice for air blast circuit breakers because pressurised air is already used for insulation and interruption. However, pneumatic mechanisms are not limited to air blast circuit breakers and have also been used to operate oil and SF6 circuit breakers.

These mechanisms, which are used with air blast circuit breakers, usually open and close pneumatically and in some cases, there is only a pneumatic rather than a solid link connection between the mechanism and the contacts. Other pneumatic mechanisms use an air piston to drive the closing linkage and to charge a set of opening springs. Figure 1.20 shows a typical pneumatic

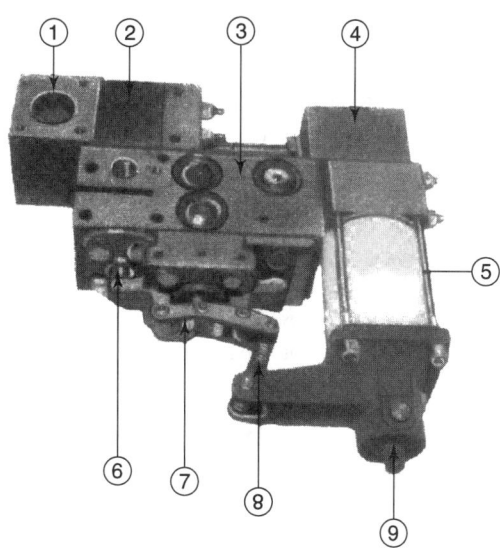

1. Air inlet valve
2. Main valve block
3. Valve body
4. Pilot valve clamping block
5. Control solenoid
6. Main valve stem
7. Toggle mechanism
8. Toggle resetting spring
9. Plunger extension for hand operation of pilot valve

Fig. 1.20: Pneumatic mechanism

mechanism used in a 220 kV class Air blast circuit breaker. Closing sequence of this mechanism has been explained here under:

Air is supplied through a filter in the air-inlet block 1 to the inlet manifold and main valve block 2, and through a connecting pipe to the pilot valve block 4. Under normal conditions with all valves closed, there is no pressure on the main body of the unit.

Initiated at the control panel, a closing operation energizes the solenoid 5 and the pilot valve is opened. Air under pressure enters the body 3 and forces down a servo-piston on to the bell crank which through the toggle mechanism 7 raises the main valve stem 6, there by opening the main valve. This allows air to pass to the circuit breaker closing cylinder.

Once the movement has been initiated, it must be completed, and by means of a one-way ball valve, the main valve is kept open until the completion of the circuit-breaker mechanism stroke, independently of the electrical control.

At the completion of the mechanism stroke, the mechanism piston uncovers a small port, which admits air to the toggle-tripping piston, driving it down and thus collapsing the toggle links. The main valve then recloses, and the air in the circuit-breaker closing cylinder is exhausted through an exhaust port.

The air above the servo-piston escapes to atmosphere through a bleed hole, and simultaneous with the resetting of the servo-piston, the toggle-tripping piston is pushed upwards by spring pressure, permitting the spring 8 to straighten the toggle in readiness for the next operation.

— If the solenoid is energized for a period longer than that required to close the breaker, the collapsing of the toggle will still close the main valve and terminate the closing operation. The toggle will remain collapsed until the solenoid is de-energised, when the unit will reset ready for another operation. Thus, the unit is pneumatically interlocked to ensure completion of an operation once initiated, and the toggle prevents "hunting".

The circuit-breaker closing operation may be initiated manually by the plunger extension 9.

An opening operation is similar, except that the air is supplied to an intermediate valve and not direct to the operating cylinder.

In order to close the circuit breaker, high pressure air is supplied to the underside of the piston by opening a three-way valve. The piston moves upwards transmitting the closing force through a toggle arrangement that is used to provide the trip free capability to the linkage, which is connected to contacts by means of an insulating push rod. In addition to closing the contacts, the mechanism charges a set of opening springs. Once the contacts are closed, a trip latch is engaged to hold the breaker in a closed position.

Opening is achieved by energising a trip solenoid, which releases the trip latch thus allowing the discharge of the opening spring, which, in turn, forces the contacts into the open position.

Another variation of a pneumatic mechanism is one wherein the pneumatic force is used for both the closing and opening operations. The direction is controlled by activation of the independent opening or closing port of the three-way valves.

1.3.3.3 Hydraulic Mechanisms

Hydraulic mechanisms constitute only a variation of the pneumatic mechanism. The energy, in most cases, is stored in a nitrogen gas accumulator, and the incompressible hydraulic fluid becomes a fluid operating link that is interposed between the accumulator and a linkage system. This fluid operating link is no different from that used in conjunction with pneumatic mechanisms.

The storage method of the energy can be either the nitrogen gas accumulator or a disk spring assembly, which acts as an energy accumulator. The disk spring assembly has significant advantages in that it is smaller, there is no chance of a gas leakage from the accumulator and the effects of the ambient temperature upon the stored energy are eliminated. The details of a typical hydraulic operating mechanism are given below and shown in Fig. 1.21a and Fig. 1.21b.

The electro-hydraulic operating mechanism consists of a drive cylinder assembly, oil tank and accumulator. Each pole of the circuit breaker is driven by an individual electro-hydraulic operating mechanism. The drive cylinder consists of a cylinder, differential piston and valve block, which contains a set of valves.

A common control unit accommodates a hydraulic pump-set, hydraulic monitoring unit, SF6 monitoring unit and terminal connectors.

The hydraulic system of all the three poles is inter-connected. A closing solenoid and two opening solenoids are mounted on the valve block. The oil tank is bolted to the valve block. It has an oil level indicator and a heater to prevent water condensation in the valve chamber and the oil tank. From the oil tank, a pipe runs to the low-pressure side of the oil pump in the control unit, which is mounted on the centre pole. The lever connected to the driver actuates the auxiliary switch and ON/OFF indicator units.

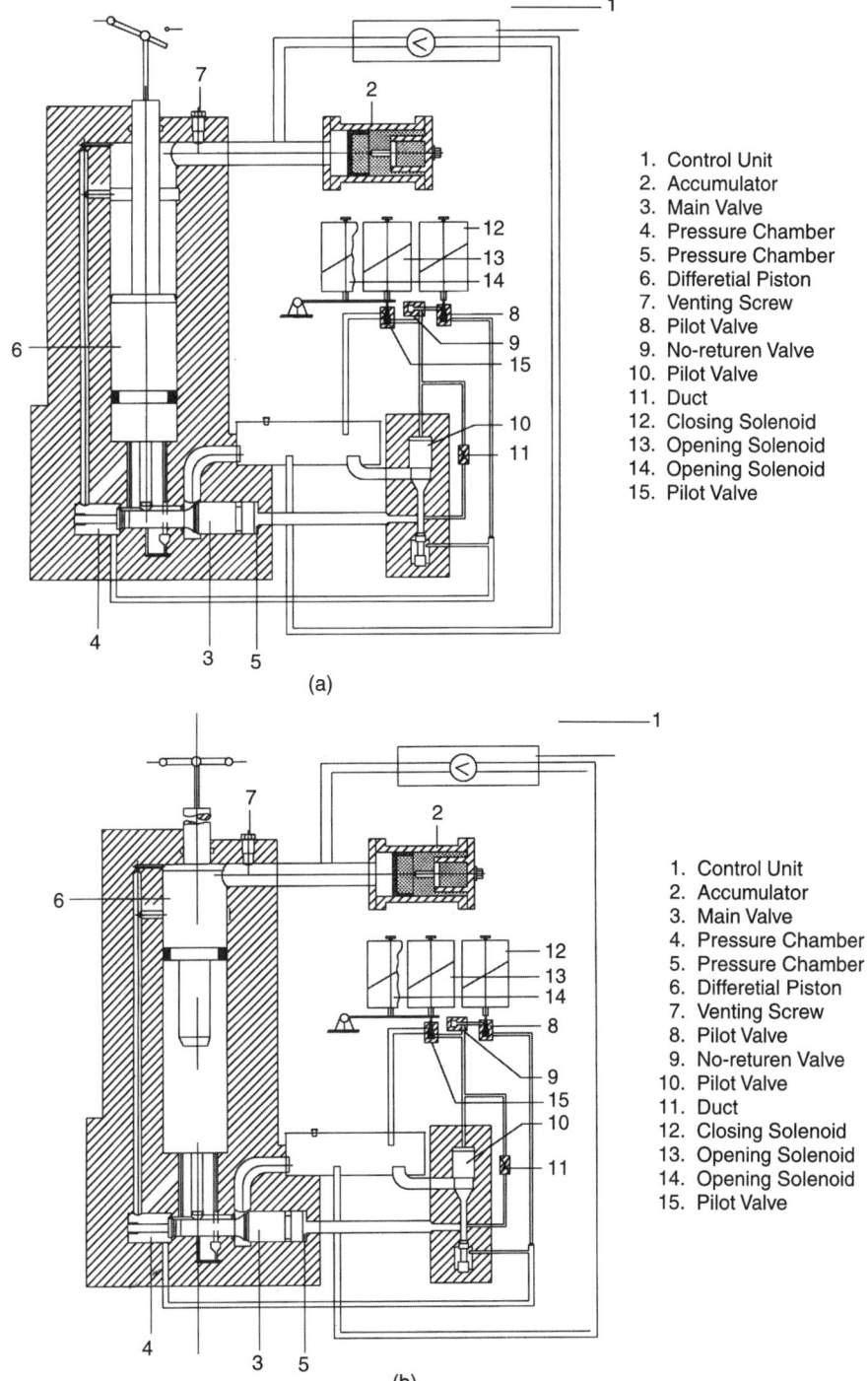

Fig. 1.21: (a) Electrohydraulic operating mechanism closing operation (b) Electro-hydraulic operating mechanism opening operation

The closing operation is effected by pressurising the longer area surface of the piston as shown in Fig. 1.21a. When the closing solenoid 12 is energised, the pilot valve 8 opens and admits oil to pilot valve 10 via non return (check) valve 9.

The pilot valve 10 opens and locks in hydraulically via the duct 11 and changes over the main valve 3. Now, it blocks the pipe connecting the larger area surface side of the differential piston with the oil tank and simultaneously opens the pressure chamber 4 towards the larger area surface side. The operating mechanism then closes the breaker.

The opening operation is effected by depressurising the larger area surface of the differential piston. Opening operation is schematically represented in Fig. 1.21b. When the opening solenoid 13 is energised, the pilot valve 15 opens and allows the self locking pressure to collapse. Now the pilot valve 10 closes off the high pressure side and the pressure in the pressure chamber 5 in the main valve opens the pipe connecting the operating cylinder and the oil tank. Thus the pressure on the larger surface of the differential piston is released and the operating mechanism opens the circuit breaker.

1.3.3.4 Solenoid and Magnetically Actuated Mechanism

Magnetically actuated circuit breakers derive their movements of contact for the closing/opening operations from a combination of permanent magnet and solenoid operated plungers (Fig. 1.22). This concept eliminates the complex linkage arrangement and makes the mechanism simple to use and more reliable. This concept has already been used for commercial switchgear design for medium voltage application. The mechanical drive is mainly a linear prime mover used to provide the desired movement to the interrupter's moving contact. The two states of drive (open and closed positions) are mechanically latched in a circuit breaker to impart mechanical stability and to economise on the latching/holding energy (e.g. low rating contactors and relays are generally held in a closed position by an operating coil that is energised continuously).

The emergence of vacuum interrupters having a low initial energy requirement (Fig. 1.22) for their smaller masses and moderate energy at the end of the stroke to recuperate the contact force requirement of the interrupter, is a major breakthrough in the switchgear industry. This has altered energy requirements to match the permanent magnet/solenoid-based actuators. The magnet-based drives deliver increased energy as the magnetic pole-gap reduces. The magnetic actuator, in addition to its linear force-distance curve, exhibits excellent latching characteristics for the above property. This feature eliminates the requirement of a mandatory mechanical latch in conventional spring or solenoid mechanisms.

It is obvious that magnetically actuated drives increase the mechanical life of the operating gear while reducing the need for maintenance.

Since they closely match the energy requirement of vacuum interrupters, magnetically actuated drives are specifically suited to vacuum circuit breakers. Their use with conventional oil, air or SF6 mediums may, however, not be feasible.

1.3.4 Motor-driven Mechanisms

In anticipation of future trends, efforts are under way to introduce Stepper motor-driven mechanisms with the least number of components that can be directly mounted on the linkage housing of a pole column. These mechanisms are already in advanced stages of development.

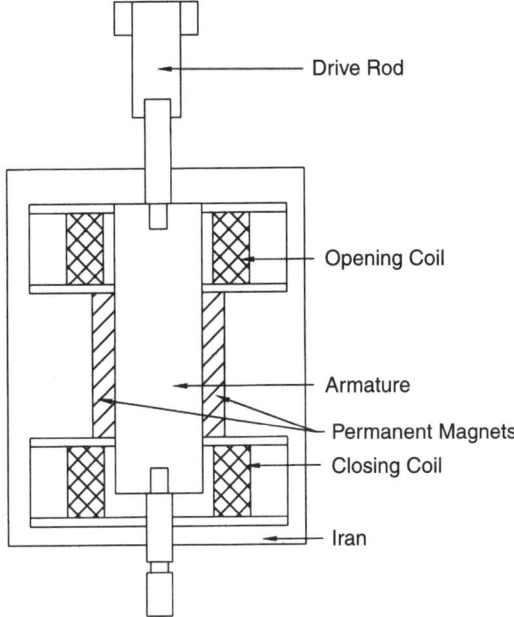

Fig. 1.22: Permanent magnet/solenoid based mechanism

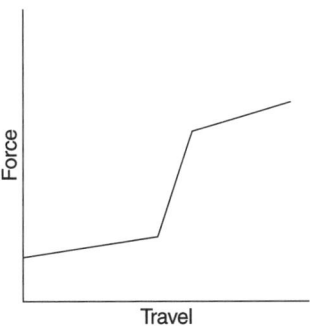

Fig. 1.23: Typical force-travel requirement for mechanism of vacuum circuit breaker

Reference

1. IEE Monograph Series 17, "Power Circuit Breaker Theory and Design", Edited by C.H. Flurscheim.

Chapter 2

LOW VOLTAGE SWITCHGEAR

D.K. Dikshit,
Murtaza Hussain

As per IEC 60947, switchgears with rated voltages up to 1000 V ac and 1500 V dc are termed as low voltage (LV) switchgear. This chapter covers the application, role and features of LV switchgear for load controlling and protection based on the location in the LV distribution network. The term 'switchgear' is a generic term encompassing a wide range of products like circuit breakers, switches, switch fuse units, off-load isolators, HRC fuses, contactors, earth leakage circuit breakers (ELCBs), miniature circuit breakers (MCBs), and moulded case circuit breakers (MCCBs), among others.

Figure 2.1 shows the commonly followed network combination in LV distribution boards. It is a combination of power control centres and motor control centres or load distribution boards including lighting distribution boards. The incomer/sub-incomer/distribution network generally depends on the capacity of the source and the distribution of load centres. The characteristics and feature of load controlling and protecting devices vary on the basis of the locations. The system being most commonly followed currently is detailed below.

2.1 CONVENTIONAL INCOMER

The devices used in the incomer should be capable of:

1. Switching and carrying normal currents (generally above 1200 A);
2. Withstanding abnormal currents for a short duration in order to allow downstream devices to operate;
3. Interrupting the maximum value of the fault current generated in the system;
4. Ensuring the safety to the operating personnel;

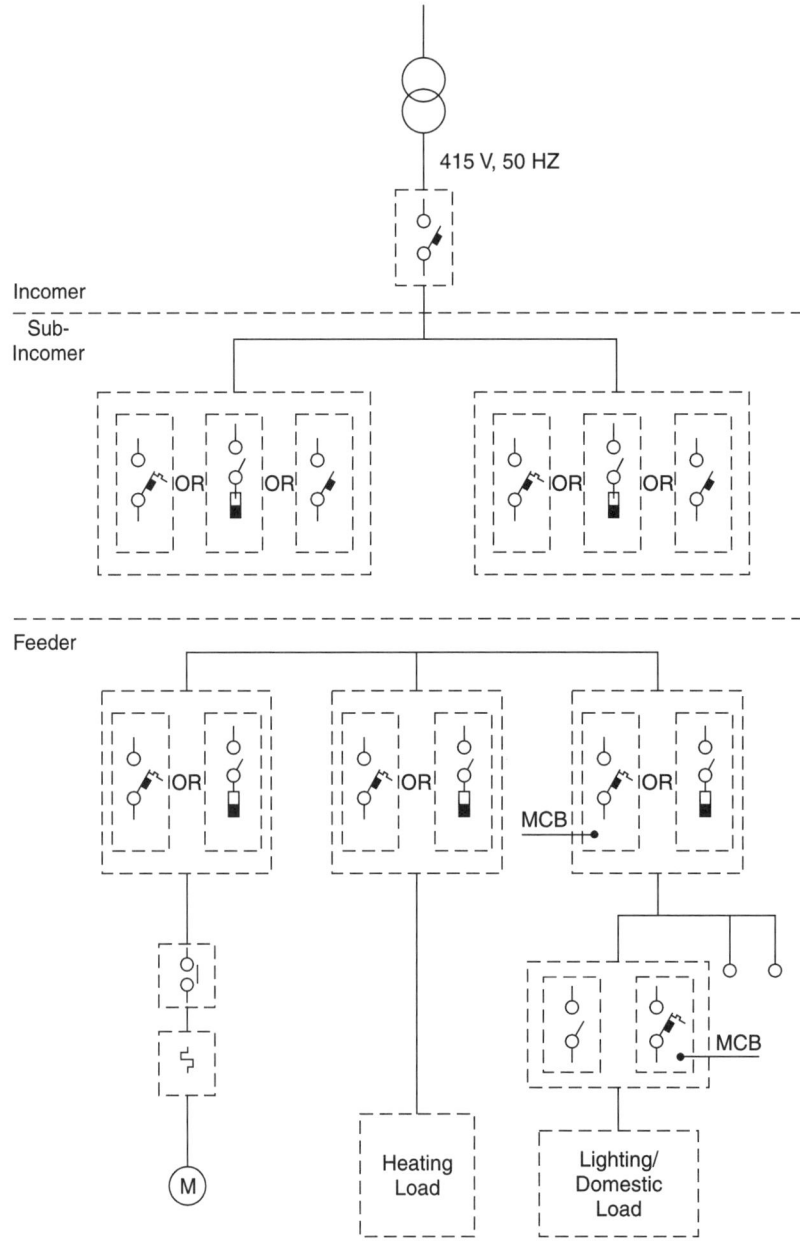

Fig. 2.1: Conventional low voltage distribution network

5. Inter-locking with downstream equipment; and
6. Facilitating easy maintenance.

In the past, oil circuit breakers (OCBs), re-wireable fused isolators and air circuit breakers (ACBs) were the commonly used devices. However, ACB has been acknowledged as an ideal device for incomer in terms of the safety, reliability and maintenance needs of the system. This is mainly due to its various characteristics like quick-make, quick-break stored energy type reliable mechanism, safety interlocks/indications, ease of maintenance and its ability to withstand fault currents for a specified duration (1–3 secs) thereby allowing the feeder device to isolate the faulty branch of the network and ensuring reliable supply to healthy areas.

The following features make ACBs popular amomg users:

- Simplicity;
- Efficient performance;
- High normal current rating up to 6000 A; and
- High fault withstanding capacity upto 63 kA.

Limitations

The following are the limitations of ACBs:

- Long tripping time;
- Bulky in size;
- High initial cost; and
- Requirement of additional floor space due to their bulky size.

2.2 CONVENTIONAL SUB-INCOMER

The devices installed as parts of a sub-incomer should have the following characteristics:

- Ability to achieve economy without sacrificing protection and safety;
- Capability to withstand abnormal currents; and
- Need for relatively less number of inter-locking indicating accessories since it covers a limited area of network.

ACBs and switch fuse units (SFUs) are, to a large extent, being used as sub-incomers along with modern devices like moulded case circuit breakers (MCCBs).

2.3 CONVENTIONAL FEEDER PROTECTION

Feeder protection covers all load centres like motor control centres, lighting switchboards and industrial load centres. The choice of feeder protection device based on the different conventional feeder load centres are discussed below.

2.3.1 Motor Control

The motor feeder needs to be protected against the following eventualities in addition to normal switching control:

- Short-circuit;
- Over-currents up to locked rotor condition; and
- Single-phasing.

The various requirements of motor control as addressed by different devices are given in Table 2.1.

Sl. No	Characteristics	Devices
1	Switching normal currents	Contactors
2	Single-phasing sensing	Bi-metallic Relay
3	Over-current (up to locked rotor condition) sensing	Bi-metallic Relay
4	Switching over-currents including single-phasing current as above	Contactor
5	Short-circuit current	MCCB/SFU
6	Logic	Timers/Auxiliary Contactors and other Accessories

Table 2.1 Requirements of Motor Control

2.3.2 Other Industrial Load Control

Loads like oven, pre-treatment and electroplating baths fall under this category and the feeders need to be protected against faulty over-currents. Presently MCCBs and SFUs are being commonly used for this purpose.

2.3.3 Lighting/Domestic Load Control

The requirements of domestic load control are similar to those listed in other industrial load control with the addition of earth leakage current protection in order to reduce any damages to life and property that could be caused by harmful leakages of electric current and fire.

In a low voltage power distribution system, electrical appliances are protected against damages from over-loads or short-circuits by fuses or circuit breakers. However, the human operator is not adequately protected when a fault occurs within the appliance itself. Hence the need for fast acting ELCBs operating on low leakage currents arises.

The device, which detects leakage current as low as 100 mA and is capable of disconnecting equipment in less than 100 msec is called an earth leakage circuit breaker (ELCB). The following two types of ELCBs are used depending upon the parameters to be detected:

1. Voltage-operated ELCB; and
2. Current-operated ELCB.

The selection and application of ELCBs depends upon the process requirement, continuity of supply desired in the system and the rating of the section to be protected. LV distribution networks are fairly simple and are dominated by conventional devices like re-wireable fuses, SFUs, OCBs and ACBs for incomer with modern devices like MCCBs, MCBs and ELCBs for sub-incomer and feeder control.

2.4 EMERGING TRENDS

Being bulky in size and weight, and consequently needing a lot of care and various spares for maintenance, conventional devices were unable to sustain themselves. They have thus been replaced by modern load controlling and protection devices, which take into account changing consumer needs and incorporate advancements in material, product and process technologies to offer the following advantages:

1. Economy through lightweight and compact devices/systems;
2. Reliability through integration of devices to achieve multiple functions;
3. Enhanced productivity through reduction in inter-connections;
4. Enhanced safety;
5. Flexibility to fine-tune the system at the commissioning stage;
6. Reduced lifecycle cost (including installation, operation and maintenance costs); and
7. Environment-friendliness.

These products, which have built-in features to meet changing consumer needs, have revolutionised trends in the design of low voltage distribution networks. These trends are discussed below (see Fig. 2.2).

2.4.1 Present Trends in Incomer

Rapid industrialisation, urbanisation and expansion in the core sectors have necessitated the development of higher capacity sources (mainly transformers for LV distribution). Thus transformers with capacities of 2.5 MVA and above are becoming very common. This has necessitated the upgradation of air circuit breakers, enabling them to offer higher normal currents up to 6000 A and fault current ratings up to 150 kA. Air circuit breakers providing normal protection against overload and short-circuit are now employing microprocessor-based relays that afford greater flexibility by offering the following advantages:

1. Protection curve selection to suit specific application needs;
2. Measurements and monitoring of circuit parameters; and
3. Interface with computers for the optimisation of energy, wherever necessary.

Thus air circuit breakers up to 6000 A rating with microprocessor-based releases are likely to be an ideal solution for main incomer.

For special application areas like (a) mining, petrochemicals, etc., where the exposed arc is not desirable, and (b) systems with higher rated operational voltages (typically 1000 V), however, LV

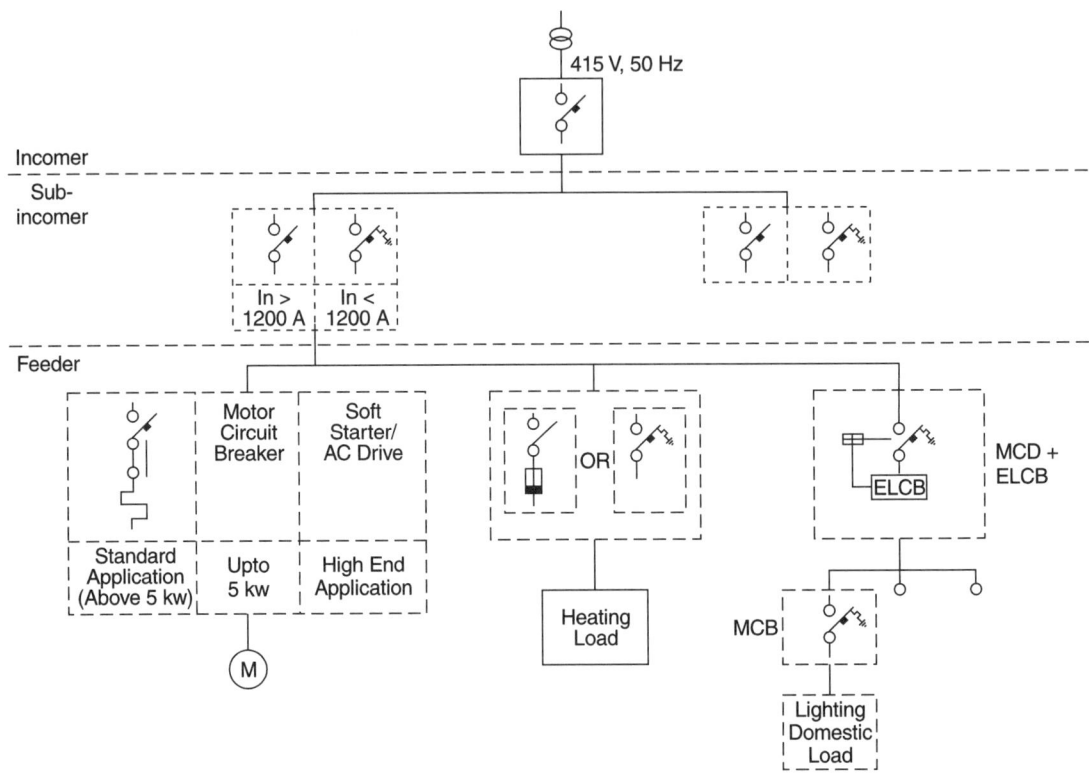

Fig. 2.2: Emerging trend of low voltage distribution network

vacuum circuit breakers, would be more acceptable. Although the maintenance-free switching life of an LV vacuum circuit breaker is superior to that of an ACB, its higher initial cost limits its usage as an incomer on a large scale.

2.4.2 Present Trends in Sub-Incomer

As discussed above, ACBs and SFUs have been the most commonly used devices in this segment. While ACBs are technically superior for incomer application, they automatically fulfil all the requirements of sub-incomers also. However, their relatively higher cost and size make MCCBs and SFUs alternate choices, especially in less than 1200 A applications. MCCBs are also rapidly emerging as a strong alternative to SFUs as the following features make them inherently superior to the letter:

1. Precision protection against over-load, short-circuit and under-voltage;
2. Remote tripping/control;
3. No downtime;
4. No inventories like fuse links, etc.; and
5. Higher number of indicating/inter-locking contacts.

2.4.3 Present Trends in Feeder Protection

As per current trends, short-circuit protection devices, contactors and bi-metallic relays are being used for motor control application. MCCBs are also replacing SFUs as short-circuit protection devices. Contactors are being used as the main devices for frequent switching of motor feeders. However, new generation contactors are increasingly becoming popular as they offer the following features, which provide flexibility at the system design and commissioning stages:

1. Compact and low coil VA;
2. Ease of maintenance;
3. Site-installability, add-on accessories like auxiliary switches, timers and on-position latch;
4. Dc control; and
5. Interface with PLC.

The following new trends are thus emerging in LV distribution networks:

- ACBs continue to dominate as sub-incomers above a rating of 1200 A; and
- Application areas up to 1200 A rating will be dominated by MCCBs, which will replace SFUs in most of the applications, except non-critical ones, where the initial investment cost remains a primary factor in decision making.

Chapter 3

MEDIUM VOLTAGE SWITCHGEAR

D.K. Dikshit

3.1 HISTORICAL BACKGROUND

With the advent of electricity and its commercial production, the need to control it also became evident. Historically, the initial form of a circuit breaker used by scientists was a mercury switch in which two rods of electrodes were dipped in mercury and moved in and out to make or break the circuit current. Interestingly, however, this apparatus was not named as a 'circuit breaker' at the time of usage.

A quick-break knife switch was perhaps the earliest recognised form of circuit breaker. Such switches, however, had no formal arc control other than that provided by the switch operator who used an insulated hatchet to chop the arc. Later during the early twentieth century, more effective switches where developed by dipping the quick-break knife switches into a tank containing insulating transformer oil. Thus was born the plain break oil circuit breaker.

During the early stage, as the power increased, it was recognised that a paramount requirement for circuit breakers was the ability to interrupt not only load currents but also the current which occurred during a short-circuit when the currents could reach a magnitude which was many times that of the full load. Initial developments carried out in the arc interruption capability of switchgears were based on in-house experiments using production facilities of industries or actual field testing. Standard test houses for verifying the claims made by manufacturers were not available. This gave rise to many difficulties and uncertainties about the performance of switchgears being bought by the utilities.

More scientific developments in circuit breaking started in the year 1925 when special short-circuit test plants came into existence followed by the development of cathode ray oscillographs for the proper recording and measurement of test quantities.

Bulk oil circuit breakers (Fig. 3.1) were developed with many improved features such as multi-break interrupter and axial flow blast interrupter to enhance the interrupting capacity. Bulk oil circuit breakers were widely used by industry in the medium voltage range from 1920 to 1970 until the advent of other improved technologies.

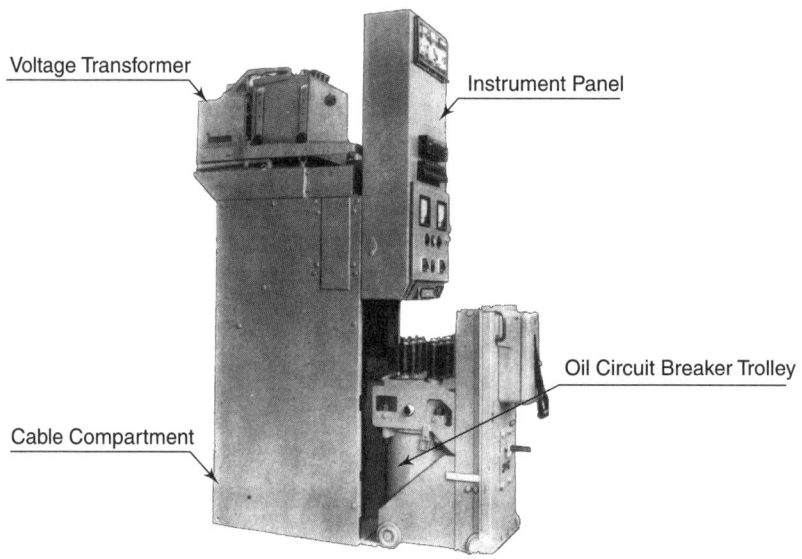

Fig. 3.1: Medium voltage bulk oil switchgear

The development of both minimum oil circuit breakers (Fig. 3.2) and air circuit breakers started almost simultaneously around the period 1930–35. New circuit breakers using very small volumes of oil were developed primarily in Europe, in order to tackle the problems of scarcity of space and high cost of oil. These breakers used alternative insulating material to achieve insulation between live and earth parts, and a small volume of oil for arc interruption, which is why the name minimum oil circuit breaker was used for these breakers.

One of the earliest successful approaches to high power air break circuit breakers was that developed by Slepian at Westinghouse in 1929. Contact separation and arc extinction in these breakers took place in air at atmospheric pressure. As the two current-carrying contacts are separated, an arc is drawn between them. The extinguishing force is generated in these breakers by elongating, cooling, splitting and/or constraining the arc (Fig. 3.3). Accordingly, they are called plain break, de-ion or air magnetic circuit breakers.

While the plain type circuit breaker is used in low voltage applications only, the other two types are used in medium voltage applications. However, the advent of compact and external arc free designs of modern switchgears has virtually abolished the use of these air break switchgears

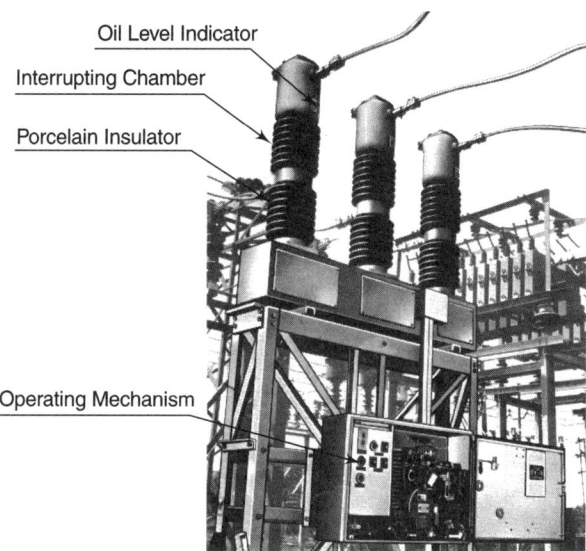

Fig. 3.2: Medium voltage minimum oil switchgear

in medium voltage application in the industry. Air blast circuit breaker designs, developed during the period 1935–45, were more popular in the EHV range and did not find a place in medium voltage application due to the complexity of their design, high cost and problems of maintenance in compressed air systems.

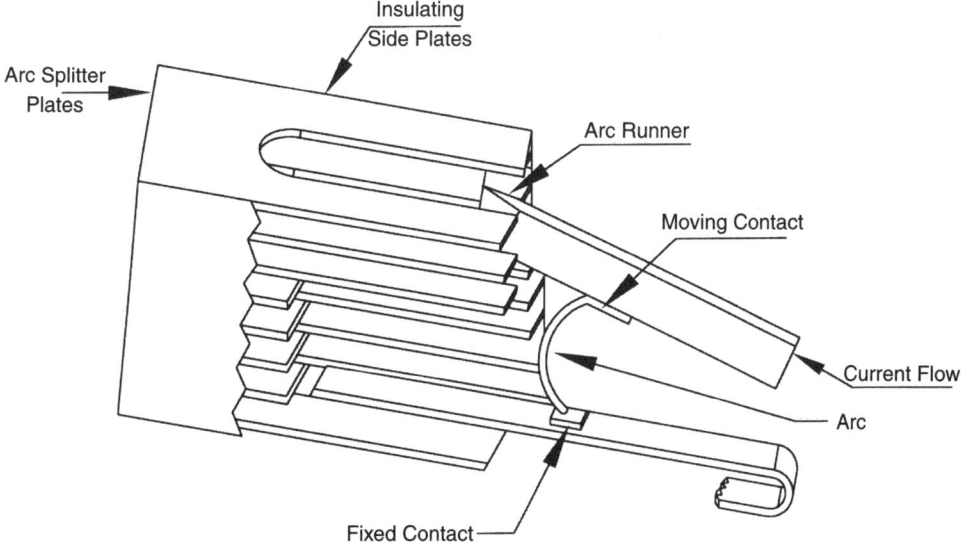

Fig. 3.3: Air break circuit breakers

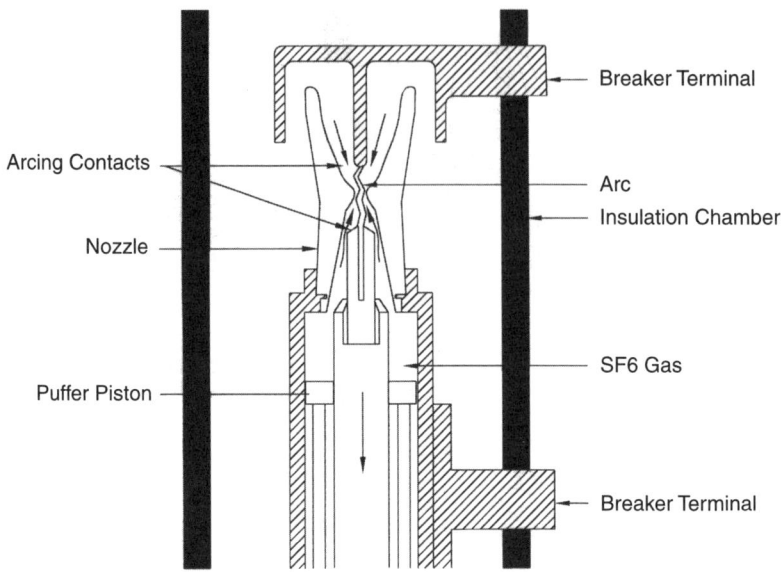

Fig. 3.4: SF6 puffer type circuit breaker

Development work on circuit breakers using sulphur hexafluoride (SF6) gas started in the 1950s and a few initial designs were made in early 1960s. The excellent dielectric properties of SF6 gas helped in the rapid development of circuit breaker designs for medium voltage and extra high voltage (Fig. 3.4). This technology assumed significance immediately after its introduction in the high voltage range of switchgear beyond 72.5 kV and above.

However, in medium voltage application, vacuum switchgear and other contemporary technologies continued to be used along with SF6 technology for quite some time. Finally vacuum switchgear, because of its simple and rugged design, low energy mechanism requirement, large number of switching operations and environment-friendly arc interruption feature, gained importance over the others. Vacuum switchgears (Fig. 3.5) virtually eliminated all other technologies in medium voltage application except SF6 switchgears, which also found only limited usage.

Although development work on the use of vacuum as a switching medium started in the 1920s, the commercial vacuum interrupter for medium voltage application entered the market only in the 1960s. The major areas of work were related to research on contact materials and their geometry, and problems of maintaining vacuum levels over the life span of the interrupter. Gradually as the manufacturing techniques of vacuum interrupter improved and the volume of production increased, the prices of vacuum interrupter started coming down. Its affordable price coupled with ease of handling and improved arc interruption behaviour made it the leading technology in medium voltage switchgear, and the usage of oil and air circuit breakers was virtually eliminated. The two superior technologies, i.e. switching in vacuum and SF6, continued to be used simultaneously in medium voltage application in the 1970s. However, the excellent

Fig. 3.5: Medium voltage vacuum switchgear

switching property of vacuum and its limitation in the design of high voltage interrupters restricted its use to medium voltage range. The SF6 gas-based circuit breakers due to its excellent switching properties at high voltage levels, SF6 gas-based technology became the best choice for the voltage class of 72.5 kV and above, but difficulties in handling the gas and its harmful arc products made SF6 unpopular for use in the medium voltage range.

3.2 PRESENT STATUS

The present global market trend shows a clear dominance of vacuum switchgear in medium voltage application (Chapter 1, Fig. 1.17).

Developments in the field of vacuum and SF6 circuit breakers have facilitated reduction in the mass of material used, thereby optimising the size of the switchgear. At the same time, the interrupter capacity has also been enhanced so as to handle larger fault levels, which occur due to the connectivity of parallel networks. Reductions in the moving masses of interrupters have led to the use of simpler, low energy mechanisms, thereby increasing the life span of the mechanism. Today, however, the main emphasis is on reliability and maintenance-free designs. These requirements have been made specific in the recently issued circuit breaker standard IEC 62271-100. Circuit breakers of class M1 must qualify for 2,000 mechanical operations, while circuit breakers for class M2 must meet the mechanical operation requirement of 10,000 numbers. Similar criteria have also been upgraded in the standards for medium voltage range upto 52 kV for short-circuit current make break duties, thus defining class E1 and class E2 circuit breakers.

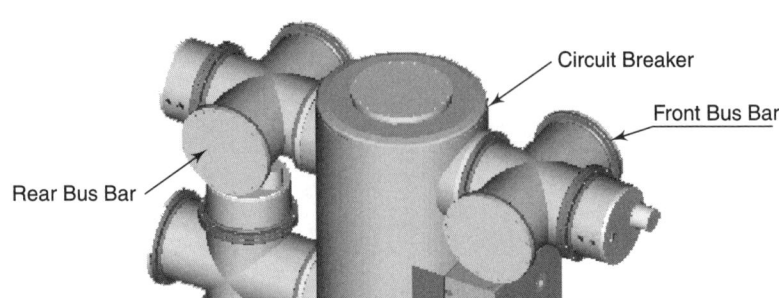

Fig. 3.6: Medium voltage gas insulated switchgear

Gas insulated switchgear constructions (Fig. 3.6) are increasingly being adopted by utilities wherein reliability is important. This concept is initially finding popularity mostly in metropolitan cities where space is a constraint. But it is very likely that most future installations would have this configuration.

3.3 FUTURE TRENDS

Work on solid state switching devices is in experimental stage for medium voltage application and is yet to find usage in commercial application. Sealed for life breakers are becoming a reality today. Synchronous switching and condition monitoring are two aspects being explored to improve switching performance and facilitate preventive maintenance. Synchronous switching and condition monitoring have acquired a great deal of relevance not only because of their potential for enhancing reliability and for helping to improve the overall power quality of the electric systems, but also for economic reasons. These concepts can be instrumental in minimising the use of auxiliary components such as pre-insertion resistors, in reducing equipment wear and unnecessary maintenance, and thus in reducing the total cost of ownership throughout the life span of the equipment.

Synchronous switching is not a new idea as the feasibility of synchronised switching is being studied for the last 30 years. Condition monitoring is another concept resulting from recent developments of electronic sensors and data acquisition equipment, which have made this idea not only technically feasible but also economically attractive. Condition monitoring is an essential component for synchronous switching because a lot depends on how accurately the operating characteristics of the circuit breaker can be controlled. It is well known that the operating characteristic can be affected by extreme ambient temperatures, and by other prevailing conditions such as the mechanism operating energy levels, control voltages, operating frequency of the equipment, its chronological age, and its maintenance history among others. Collection of information about these variations would help to build a data source from which suitable correction factors may be selected to compensate for those operating deviations, which are critical for an accurate synchronous operation.

Due to usage the development of information technology, coupled with the use of current and voltage sensors, dedicated numerical relays and condition monitoring devices, has led to the evolution of intelligent switchgears. Future switchgears are likely to be sealed for life, as plug-in type of devices need no maintenance. The advent of solid state switchgears can also not be ruled out. Such developments would completely redefine the subject and present international switchgear standards would have to undergo a major change to cover them.

3.4 CONSTRUCTIONAL FORMS

Medium voltage switchgears are available in a variety of constructional forms in consonance with the application requirements. These switchgears are installed in indoor as well as outdoor sub-stations. Live tank design and dead tank design also affect the constructional aspects. All the constructional forms can be covered under the category of metal enclosed design, live tank design and dead tank design. These are explained below.

3.4.1 Metal-enclosed Switchgears

Metal-enclosed switchgear assemblies have earthed metallic enclosures. These are complete, ready-to-install assemblies requiring high voltage and low voltage cable connections. These switchgears generally have three high voltage compartments, namely circuit breaker compartment, bus bar compartment, and current transformer and cable compartment, which are separated by partitions.

Switchgears which have compartments with metal partitions that are intended to be earthed are called metal-clad switchgears. Switchgears with one or more non-metallic partitions are known as compartmented switchgears. Metal-enclosed switchgear, other than metal-clad and compartmented switchgear, is termed as cubicle switchgear.

3.4.1.1 Design and Construction

Metal-enclosed switchgears are designed such that normal service, inspection and maintenance operations, including the usual checking of the phase sequence, earthing of connected cables, and voltage tests on connected cables or other apparatus can be carried out safely. All components of the same rating and construction that may need to be replaced are designed to be interchangeable.

If there are removable parts with different ratings and if parts are interchangeable within the assembly of metal-enclosed switchgear, any possible combination of removable and fixed parts is designed to withstand the rated insulation level of the switchgear. The various components contained within the switchgear are subject to the individual specifications applying to them.

Interlocks between different components of the equipment are provided to ensure safety and the desired sequence of the operation. The following provisions are mandatory for the main circuits.

(a) Metal-enclosed switchgear with removable parts:
- The withdrawal or engagement of a circuit breaker should be impossible unless it is in the open position.
- The operation of a circuit breaker should be impossible unless it is in the service, disconnected, removed, test or earthing position.
- It should be impossible to close the circuit breaker in the service position unless it is connected to the auxiliary circuit.

(b) Metal-enclosed switchgear without removable parts and provided with a disconnector:
- Interlocks should be provided to prevent the operation of disconnector under conditions other than those they are intended for.
- The operation of a disconnector should be impossible unless the associated circuit breaker is in the open position.
- The operation of the circuit breaker should be impossible unless the associated disconnector is in the closed, open or earthing position.
- The provision of additional or alternative interlocks is subject to an agreement between the manufacturer and the user. Wherever practical, preference is given to mechanical interlocks.

3.4.1.1.1 Enclosure Enclosures are earthed sheet metal constructions housing the switchgear components. When the metal-enclosed switchgear is installed, the enclosure is provided with at least the degree of protection specified in Table 3.1 (see section 3.4.1.1.8). The floor surface, even if it is not metallic, may be considered as part of the enclosure. Enclosures of a gas insulated switchgear (gas-filled compartments) are designed to withstanding the normal and transient pressures to which they are subjected in service. While these compartments are permanently pressurized in service, they are subjected to particular conditions of service, which distinguish them from compressed air receivers and similar storage vessels. These conditions are:

- Gas-filled compartments enclose the main circuit to prevent a hazardous approach to live or moving parts. In addition, they are shaped in such a way as to ensure that the rated insulation level for the equipment is achieved at or above the minimum functional pressure.
- Gas-filled compartments are normally filled with a non-corrosive inert gas (such as SF6 or nitrogen), in a thoroughly dried state. Since this condition is fundamental to the operation of the switchgear and since the compartments will not be subjected to internal corrosion, there is no need to make allowances for these factors in determining the design of the compartments.
- The service pressure is relatively low (less than 2.5 bar as per IEC 298).

For outdoor installation, the influence of climatic conditions is taken into account.

The design of a gas-filled compartment is based on the design temperature and on the design pressure as defined in this standard.

- The design temperature of the gas-filled compartment is generally the upper limit of the ambient air temperature. The design pressure of the enclosure is at least the upper limit of the pressure reached within the enclosure at the design temperature.
- The permissible gas leakage rate for the gas-filled compartments is ensured as per the standards.

3.4.1.1.2 Covers and Doors Covers and doors, which are parts of the enclosure, are required to facilitate operation and maintenance as and when needed. These are made of sheet metal construction. When they are closed, they provide the degree of protection specified for the enclosure.

The following two categories of covers or doors are recognised with regard to access to high-voltage compartments:

(a) Those, which need not be opened for the normal purposes of operation or maintenance (*fixed covers*). It should not be possible for them to be opened, dismantled or removed without the use of tools; and

(b) Those, which need to be opened for the normal purposes of operation (*removable covers, doors*). These should not require tools for their opening or removal. They should be provided with locking facilities (for example, provision for padlocks), unless the safety of persons is assured by a suitable interlocking device.

In the case of metal-clad or compartmented switchgear, covers or doors should be opened only when the part of the main circuit contained in the compartment being made accessible is dead.

3.4.1.1.3 Inspection Windows Inspection windows are provided to view the position of withdrawable parts and disconnector, etc. during routine inspection and operation. Inspection windows are designed to provide at least the degree of protection specified for the enclosure.

They should be covered by a transparent sheet of mechanical strength comparable to that of the enclosure. Precautions should be taken to prevent the formation of dangerous electrostatic charges, either by clearance or by electrostatic shielding.

3.4.1.1.4 Ventilating Openings Ventilating openings are provided in the switchgear to allow venting of gas or vapour that is escaping under pressure, without harming the operator. Sometimes these openings are also provided for thermal management. Ventilating openings and vent outlets are arranged or shielded in such a way that the same degree of protection as that specified for the enclosure is obtained. Such openings may make use of wire mesh provided it is of suitable mechanical strength.

Wherever pressure relief devices are provided, they are arranged so as to minimise the danger to an operator during the time that he is performing his normal operating duties, if gases or vapour are escaping under pressure.

In certain designs, pressure relief may be achieved by allowing the arc to burn through the enclosure at designated points. Where such means are employed, the resultant hole is deemed to be a pressure relief device.

3.4.1.1.5 Partitions and Shutters Partitions are provided in switchgear assemblies to segregate them into three compartments to help carry out the desired functions of operation and maintenance safely and smoothly. They also prevent the fault from spreading to another compartment.

Partitions and shutters should provide at least the degree of protection specified in Table 3.1 (see Section 3.4.1.1.8).

Openings in the enclosure of switchgear and in the partitions of metal-clad or compartmented switchgear through which contacts or removable parts engage fixed contacts are provided with automatic shutters that are properly operated in normal service operations in order to ensure the protection of persons in any of the positions of the withdrawable part.

Partitions of metal-clad switchgear should be metallic and earthed. Partitions of compartmented and cubicle switchgear may be non-metallic. Partitions between two gas-filled compartments or between a gas-filled compartment and another compartment may be of insulating material. The shutters of the three types of metal-enclosed switchgear may be either metallic or non-metallic, and are designed to meet the conditions of switchgear standards.

3.4.1.1.6 Disconnectors and Earthing Switches Disconnectors and earthing switches are provided to meet the isolation and earthing requirements of the circuit configuration. The device for ensuring the isolating distance between the high-voltage conductors is called a disconnector. The device for ensuring the earthing of circuit is called the earthing switch. The requirement that it should be possible to know the operating position of the disconnector or earthing switch, is met if one of the following conditions is fulfilled:

- The isolating distance is visible;
- The position of the withdrawable part, in relation to the fixed part, is clearly visible and the positions corresponding to full connection and full isolation are clearly identified; and
- A reliable indicating device indicates the position of the disconnector or earthing switch.

3.4.1.1.7 Earthing In order to ensure safety while maintenance is being carried out, all parts of the main circuit to which access is required or provided should be capable of being earthed prior to becoming accessible. This does not apply to withdrawable and removable parts, which become accessible after being separated from the switchgear.

This requirement is met by providing an earthing conductor extending across the entire length of the metal-enclosed switchgear. The current density in the earthing conductor, if it is made of copper, is kept within 200 A/mm^2 under the specified earth fault conditions; however, its cross-section area is always kept more than 30 mm^2. It is terminated by an adequate terminal intended for connection to the earth system of the installation.

In general, the continuity of the earth system is ensured by taking into account the thermal and mechanical stresses, which may be caused by the current that it may have to carry. The maximum value of earth fault currents depends upon the type of system neutral earthing employed and is indicated by the user.

The enclosure of each functional unit is connected to this earthing conductor. All the metallic parts intended to be earthed and not belonging to the main or auxiliary circuit, should also be connected to the earthing conductor directly or through metallic structural parts.

For inter-connection within the functional unit, fastening by bolting or welding is acceptable for providing electrical continuity between the frames, covers, doors, partitions or other structural parts. Doors of the high-voltage compartments are connected to the frame by adequate means. The metallic parts of a withdrawable part, which are normally earthed, should also remain earth-connected in the test and disconnected positions under the prescribed conditions for the isolating distance, and also in any intermediate position whilst the auxiliary circuits are not totally disconnected.

3.4.1.1.8 Degree of Protection The degree of protection needed to prevent persons from coming into contact with live parts of main circuits and with any moving parts (other than smooth rotating shafts and moving linkages) should be indicated by means of the designation specified in Table 3.1.

Degree of protection	Protection against approach to live parts and contact with moving parts to be verified with the following test object
1P2X	By fingers or similar objects of diameter greater than 12 mm
IP3X	By tools, wires, etc., of diameter or thickness greater than 2.5 mm
IP4X	By wires of diameter or strips of thickness greater than 1.0 mm

Table 3.1 Specifications for Degree of Protection

For metal-clad and for compartmented switchgear, the degree of protection should be specified separately for the enclosure and for partitions.

For cubicle switchgear, it is necessary to specify only the degree of protection for the enclosure. For main circuits of gas-filled compartments, no degree of protection needs to be specified.

3.4.1.2 Metal-enclosed Switchgears (Constructional Forms)

The commonly used constructional forms for indoor and outdoor applications are explained below.

3.4.1.2.1 Metal-enclosed Switchgears (Indoor) Metal-enclosed switchgears are generally designed in a horizontal draw-out pattern though some manufacturers also follow a vertical isolation pattern. These designs are made suitable for extension of switchboard on both sides. The design incorporates the single/double bus bar system as per the requirement.

A typical switchgear panel consists of a fixed portion (and a withdrawable portion) having three high voltage chambers, namely a breaker chamber, a bus bar chamber, and a current transformer (CT) and cable chamber (Fig. 3.8). The instrument panel is a separate low voltage chamber. The withdrawable portion comprises a wheel-mounted truck fitted with an operating mechanism, interrupters and isolating contacts. Generally there is a manual charging provision for springs to tackle the possibility of failure of auxiliary power to the spring charging motor.

The main breaker chamber of earthed sheet steel construction is designed so as to accept the withdrawable portion at the floor level. However, some designs have the withdrawable portion at a height. These designs have a trolley for placing the withdrawable portion in the breaker chamber. The primary isolating contact system comprises a self-aligning type female contact on the moving portion and male contact on the fixed portion of the switchgear panel. This chamber also includes features like a secondary isolating contacts socket, guides for the withdrawable portion, an earthing contact which mates with an earthing strip on the moving portion, and safety shutters.

The withdrawable portion with circuit breaker (CB) generally has three positions, i.e. the 'service', 'test' and 'isolated' positions. The position of the withdrawable portion can be seen through the glass window on the door.

The bus bar chamber of earthed metal construction houses the bus bars, which consist of multiple parallel bars of aluminium or copper as per the requirement of current. Bus bars are generally supported on epoxy support insulators of adequate strength and voltage class. The bus bars are either bare or of the insulated type. The bus bar chamber is generally provided with a partition between two adjacent bus bar chambers. The current transformer (CT) and cable

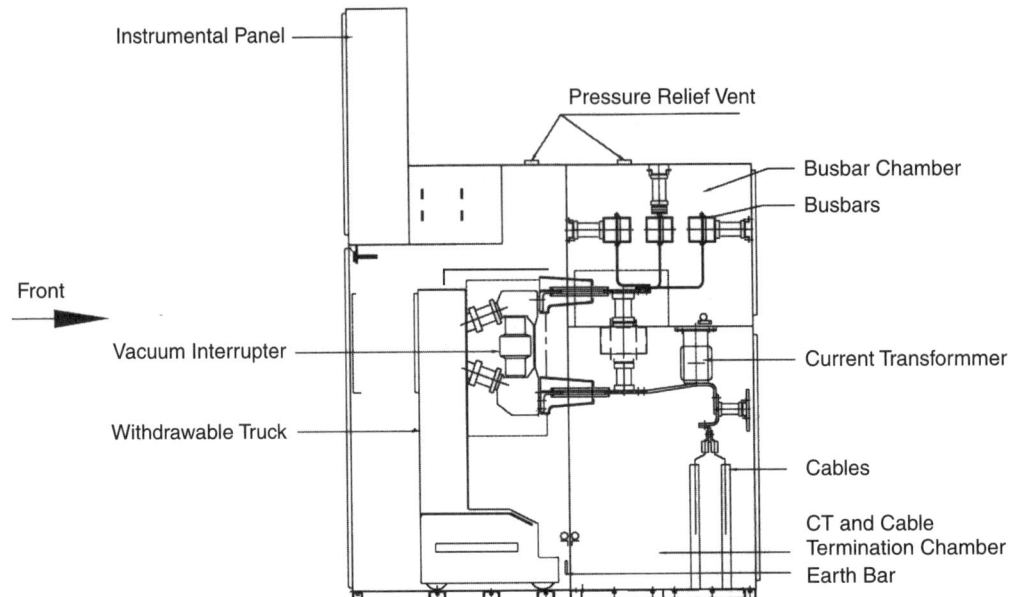

Fig. 3.7: Typical switchgear panel

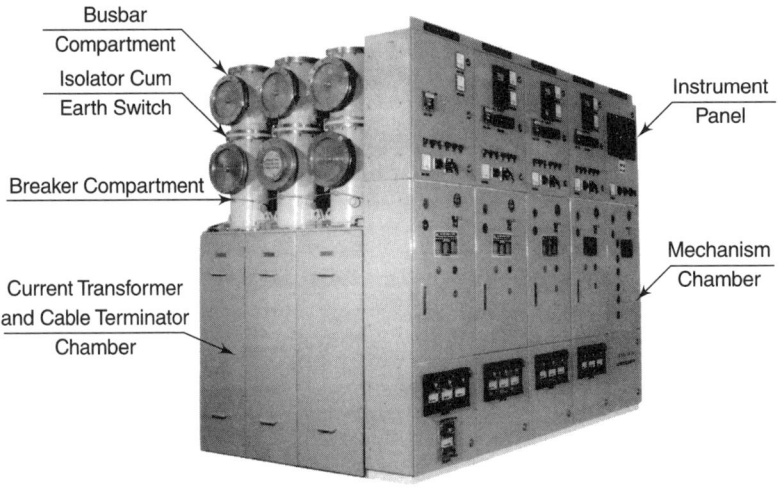

Fig. 3.8: Medium voltage gas insulated switchgear

termination chamber of earthed metal construction are designed to be able to mount the CT and have a provision for cable termination. The withdrawable part consists of a truck frame with four wheels on which three interrupters and the operating mechanism are mounted.

The interlocks are designed to meet the requirements mentioned in Section 3.4.1.1.1 above.

The instrument and relay chambers are earthed metal constructions with a hinged instrument panel door suitable for flush mounting of instruments in the front in order to provide better access for operation and maintenance.

MV GIS Construction These are fixed type metal enclosed designs and do not have withdrawable parts. Generally these designs have two major compartments, namely the busbar and breaker compartments. The busbar compartment generally has a three-position switch, marked with the 'service' 'isolated' and 'earth' positions. The operation of the switch is interlocked with the circuit breaker. The breaker compartment houses three interrupters. It has a provision for mounting ring type CT and terminating cables. The common practice is to provide the plug-in type of cable termination in GIS. Two types of constructions are followed by designers, namely 'isolated phase construction' and 'three-phase construction'. The interrupter used is mostly of the vacuum type with very few designs offering SF6 interrupters. Figure 3.8 shows the typical GIS bay. Details of this product are covered in Chapter 5.

3.4.1.2.2 Metal-enclosed Switchgears (Outdoor) The design of these switchgears is similar to that of indoor metal-enclosed switchgears in all internal features except the external housing. The enclosure is normally a welded sheet steel fabrication suitable for outdoor application having slanting roof and rain shields, etc. (See Fig. 3.9). Although this type of design is not commonly used, some utilities find it convenient to use it for city distribution networks with underground cable systems. Such installations are set up at easily accessible locations in the city.

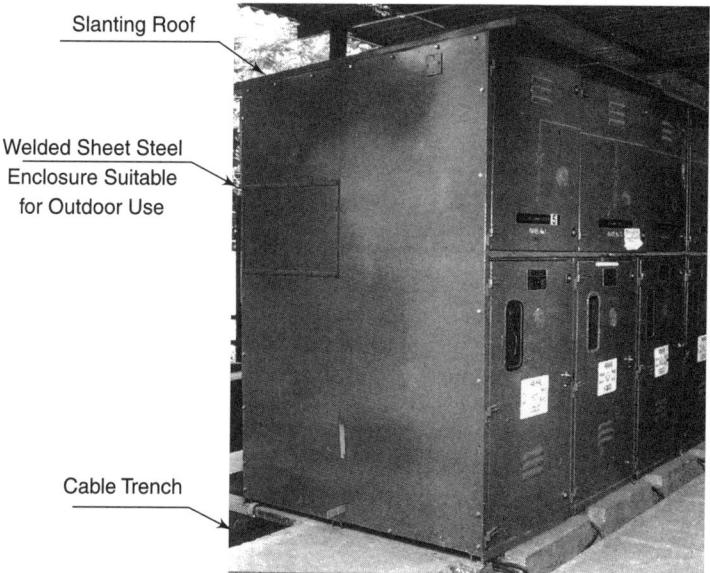

Fig. 3.9: Outdoor metal-enclosed switchgear

3.4.2 Live Tank Switchgears (Outdoor Porcelain-clad Construction)

Porcelain-clad construction is a live tank design for power distribution applications suitable for outdoor sub-stations. This design has interrupters in porcelain housing having current collection flanges at both the ends. The fixed contact of the interrupter is directly secured to the top flange and the moving contact has a current transfer arrangement so as to allow movement of the contact and also to transfer current to the other flange. This porcelain chamber is mounted on the hollow insulator to withstand system voltage requirements. The hollow insulator carries an insulating operating link to facilitate opening and closing of the interrupter through an operating mechanism. The complete three-pole assembly is mounted on a structure, which also provides support to the mechanism (see Fig. 3.10). Add-on structures are suitably designed to support the current transformer, voltage transformer, and the control and relay cabinet as per the sub-station requirement.

A similar live tank design for single-phase application is also used for railway trackside power supply systems at 25 kV.

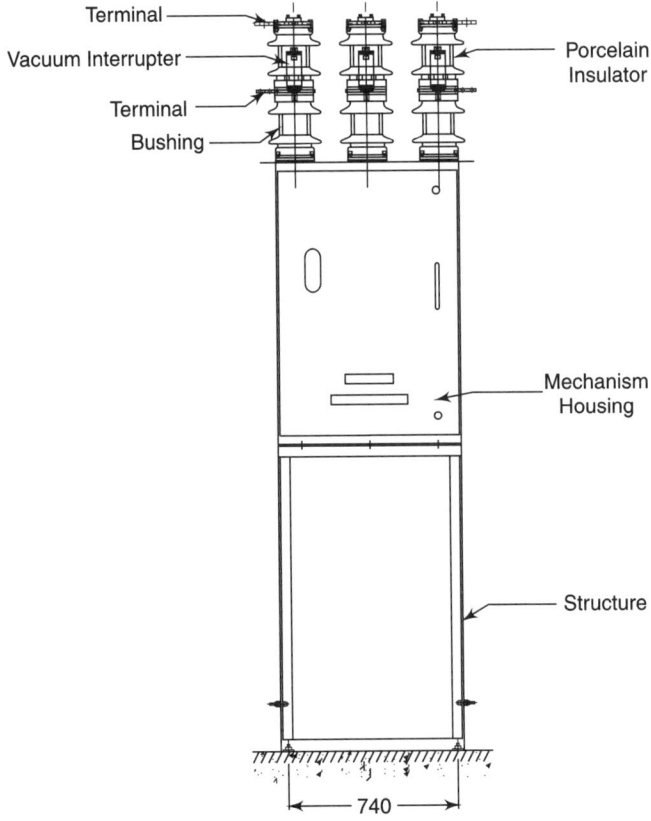

Fig. 3.10: Live tank switchgears (outdoor porcelain clad-contruction)

3.4.2.1 Dead Tank Switchgears

The two commonly used types of dead tank designs are the kiosk type and the pole-mounted auto-reclosure. These are briefly explained below.

The outdoor kiosk type of design (see Fig. 3.11) is a very simple configuration. It has a welded sheet steel-earthed enclosure suitable for outdoor application. Six roof bushings are provided for connecting to the incoming and outgoing power circuits. The kiosk houses the circuit breaker, current transformer, potential transformer and sometimes even the isolators. This becomes a complete unit ready for commissioning. At times, the control and relay cabinets are also housed in this outdoor design. These designs mostly have a fixed type of construction but designs with withdrawable circuit breakers are also available.

Auto-reclosure circuit breakers (see Fig. 3.12) are mostly used in rural applications where transient faults occur more frequently. These configurations are suitable for mounting on single-pole, two-pole and independent support structures. They have earthed metal enclosures housing the circuit breaker and current transformer. The control supply is mostly drawn from the high voltage line by deploying a control transformer, which can be housed inside the auto-reclosure, the enclosure or sometimes mounted outside on the extended structure. Details of this product are covered in Chapter 6.

3.5 DESIGN PARAMETERS

3.5.1 Normal Current Rating (Thermal Aspects of Design)

The normal current rating of the switchgear is decided on the basis of the value of current it can carry without exceeding the temperature rise limits specified in the switchgear standards (see Table 3.2). The thermal losses (I^2Rt) in the switchgear should be kept to a minimum value so as to maintain the temperature rise within limits. The temperature rise depends on the rate of heat generation and its dissipation. The most important characteristic of the material used as a conductor in switchgear is to have a low resistance. Excessive temperature rise due to high resistance conductors in small enclosures will reduce the life of associated insulation materials. The next important characteristic is that the material must resist corrosion. Any corrosion in the joint area increases the resistance and higher resistance leads to higher temperature, which, in turn, leads to more corrosion.

Care should also be taken while choosing the material for the conducting path. Two materials, namely copper and aluminium, are the most suitable materials for making current-carrying conductors. Copper has lower resistivity and also resists oxidation better than aluminium, particularly at temperatures below about 80°C. Aluminium has a much smaller specific gravity and though larger cross-sections of conductors are needed to give the same resistance as the equivalent of that obtained by using copper, the weight and, therefore, the cost of material often

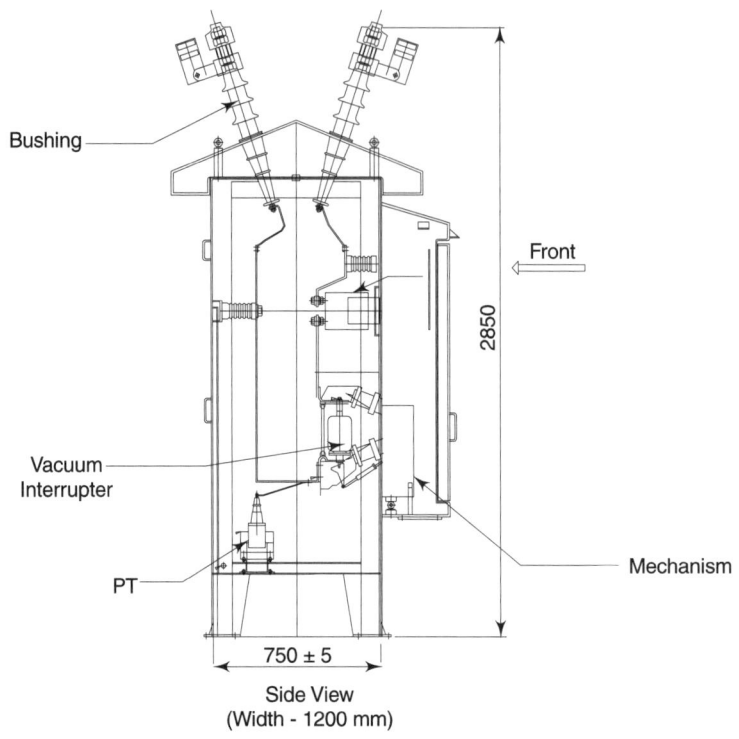

Fig. 3.11: Dead tank switchgear (Outdoor kiosk type)

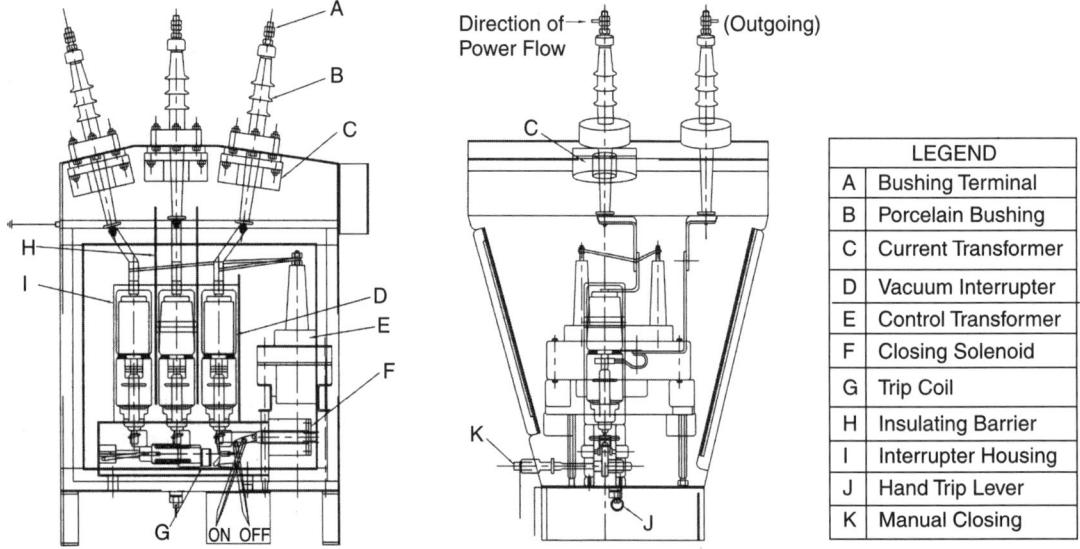

Fig. 3.12: Deed tank switchgears (Pole mounted auto reclosed)

LEGEND	
A	Bushing Terminal
B	Porcelain Bushing
C	Current Transformer
D	Vacuum Interrupter
E	Control Transformer
F	Closing Solenoid
G	Trip Coil
H	Insulating Barrier
I	Interrupter Housing
J	Hand Trip Lever
K	Manual Closing

Nature of the Part of the Material and of the Dielectric	Maximum Value	
	Temperature at 40°C Amb.	Temperature Rise at Ambient Temperature of 40°C
	C	K
1 Contacts		
Bare copper or bare copper alloy	75	35
-in air	105	65
-in SF6 (sulphur hexafluoride)	80	40
-in oil		
Silver-coated or nickel-coated	105	65
-in air	105	65
-in SF6	90	50
-in oil		
Tin-coated		
-in air	90	50
-in SF6	90	50
-in oil	90	50
2 Connection, bolted or equivalent		
Bare copper, bare copper alloy or bare aluminium alloy	90	50
-in air	115	75
-in SF6	100	60
-in oil		
Silver-coated or nickel-coated	115	75
-in air	115	75
-in SF6	115	75
-in oil		
Tin-coated	105	65
-in air	105	65
-in SF6	100	60
-in oil		
3 Terminals for the connection to external conductors by screws or bolts		
-bare	90	50
-silver, nickel or tin-coated	105	65
4 Materials used as insulation and metal parts in contact with insulation of the following classes	90	50
-Y	105	65
-A	120	80
-E	130	90
-B	155	115
-F	100	60
-Enamel: oil base	120	80
synthetic	180	140
-H	Limited by surrounding parts	Limited by surrounding parts

(Contd.)

(Contd.)

Nature of the Part of the Material And of the Dielectric	Maximum Value	
	Temperature at 40°C Amb.	Temperature Rise at Ambient Temperature of 40°C
	C	K
-C Other insulation material		
5 Any part of metal or of insulating material in contact With oil. Except contacts	100	60
6 Accessible parts		
-expected to be touched in normal operation	70	30
-which need not to be touched in normal operation	80	40

Table 3.2 Temperature Rise Limits

prove to be an advantage over copper. Table 3.3 shows the differences between the relative properties of copper and aluminium.

Conductivity for Equal Areas Copper	Copper	Aluminium
Electrical	1.0	0.61
Thermal	1.0	0.56
Tensile strength (hard-drawn)	1.0	0.40
Hardness (hard-drawn)	1.0	0.44
Modulus of elasticity	1.0	0.55
Coefficient of thermal expansion	1.0	1.39
Melting point	1.0	0.61

Table 3.3 Differences in Relative Properties of Copper and Aluminium

For achieving equal conductivity, aluminium is lighter in weight than copper. In many switchboard applications, where space considerations are more important than weight, copper is preferred over aluminium.

Silver plating finds extensive use in switchgear equipment as a thin coating in the range of 0.02–0.05 mm. Coating of greater thickness is used when a sliding motion takes place between the plated surfaces. The use of silver coating permits the conductor to be used for higher currents with the same temperature rise as the base material. It also prevents oxide formation, which ensures longer life of the conductor. It is possible to electroplate aluminium with either copper or silver by using a special process. This is required to prevent electrochemical corrosion if aluminium conductors are used in association with copper.

3.5.1.1 Resistance of the Main Circuit

The current-carrying paths of the switchgear have many joints to meet the design parameter. Each joint will have joining resistance. Care should be taken to minimise the total resistance of the circuit, i.e. the total watt loss should be kept minimum to ensure low temperature rise. The measured value will set the limiting parameter during a routine test.

The resistance of the main circuit depends on factors such as the material of the current-carrying conductor, its cross-section, type, quality and number of joints in the main circuit. The values of conductivity and cross-section of the material chosen for the current-carrying conductor should be such that while the normal current is being required, the temperature rise is maintained within limits. The higher the cross-section of the material, the smaller will be the value of the resistance. While selecting the material for current path in the switchgear, care should be taken to ensure that the conductivity and cross-section of the material is sufficient to carry the rated normal current without over-heating. The higher the number of joints, the higher will be the resistance. Also while making the bolted joints of conductors, sufficient contact pressures should be maintained to ensure the minimum possible resistance.

3.5.1.2 Joints

There are mainly three types of joints between two contacts, viz. permanent, semi-permanent and temporary joints, depending upon the possible need to break a joint between two conductors. The joint can be bolted or brazed depending upon the material. The most common method of making joints is by bolting or clamping the conductor. Clamping requires more space and is used for outdoor applications, it precludes the need for making holes in the conductor. The surfaces of the conductors must be cleaned before making the joints. In the case of flat conductors, the overlap should be equal or greater than the width of the bars or ten times the bar thickness, whichever is greater. In a joint of any design, the current is transferred through a number of point contacts formed by irregularities of the surface. The effectiveness of the joint depends upon the pressure applied and its distribution. Before the joints are made, it must be ensured that the surface is clean and oxide-free as far as possible. The bolt diameter needs to be such that the necessary pressure to crush the oxide layer does not lead to over-stressing of the bolt itself.

Since aluminium oxide is a good insulator, it is important to follow good jointing practices when aluminium conductors are being bolted together. When bolt pressure is applied, the resistance is initially reduced as the oxide filler is crushed and the area of contact around each bolt increases. Then a limit is reached beyond which further tightening causes very little reduction in resistance, so a properly designed joint does not need excessive tightening.

3.5.1.3 Contact Design

The 'temporary' joints used in switchgear for the contacts of switching devices, and for withdrawable circuit breakers in metal-enclosed units fall into two groups: butt contacts and sliding or wiping contacts. Each group has many variations and the different characteristics of these types have a bearing on the design of the mechanical lives of the joints. Oil circuit breakers, SF6 circuit breakers and air blast circuit breakers need wiping contacts but vacuum circuit breakers must have butt contacts.

3.5.1.4 Transfer Contacts

Another area of contact design which is worth mentioning is that of providing for current transfer to a moving contact. This is an essential requirement in any circuit breaker and many varieties have been developed. The simplest form is a bolted connection through a flexible conductor, usually laminated from a sufficient number of thin strips for the current rating, in order to ensure adequate flexibility. If these are arranged so that the movement is controlled and does not impose any frictional forces, then a contact system with low mechanical and low electrical resistance can be devised.

One of the recent additions to the methods used for performing this function is the Multilam type of sliding transfer contact. It is made of a silver-plated copper alloy, and is punched out and shaped in such a way that there exist a large number of parallel, individual contact bridges between the two surfaces. Because of the large number of transfer points, this contact does not require much contact pressure on each point, so it provides a very compact contact system with a high current rating and low friction.

3.5.2 Short-circuit Current

3.5.2.1 Thermal Aspects

When short-circuit current flows through the circuit breaker, different current-carrying parts are subjected to mechanical and thermal stresses. If the conductor area is insufficient to carry the fault current, the temperature of the circuit may increase to a dangerously high level as a result of which insulators in the vicinity of the conductor are severely thermally stressed. The contacts experience temperature stresses proportional to I^2RT. The resistance, R depends on the contact pressure and surface condition.

The thermal stresses depend on the r.m.s. value of the fault current and the time period for which this current is carried by the switchgear, which, in turn, depends on the total breaking time of the interrupting device, which operates to isolate the fault. During short-circuit, it is assumed that all the heat is absorbed by the conductor, there is no sufficient time for convection and radiation. The temperature rise is calculated by using the formula:

$$T = m \,(I/A)^2 \,(1 + \varepsilon)$$

where T = temperature rise/second in degree C

I = current, r.m.s. symmetrical (ampere)

A = area of cross-section of the conductor (mm^2)

ε = temperature coefficient of conductor resistivity at 20°C

During short-circuit, at a temperature of about 160°C aluminium becomes soft and loses its mechanical strength. This sets a limit on the permissible temperature rise during a short-circuit.

3.5.2.2 Short-circuit Forces

3.5.2.2.1 Single-phase Short-circuit Stresses The electromagnetic force developed between two straight parallel conductors of a circular cross-section and length of 1, each carrying a current of 1 ampere, may be calculated by using the following formulae:

$$F_{max} = 4.5 \times I^2 \times L/S \times 10^{-8} \qquad (1)$$

where F_{max} = The force of attraction or repulsion in lb
  L = length of conductors in inches
  S = distance between conductors in inches
  I = short-circuit current in amperes

In the case of conductors carrying direct current, I should be the maximum value of the short-circuit current. In the case of alternating currents, special consideration has to be given to the choice of a suitable value for I. In most cases, it is probably sufficient to assume that the maximum forces which a busbar structure is likely to be subjected to would be those calculated by taking a value of I equal to about 1.75 times the initial r.m.s. value of the symmetrical short-circuit current wave.

However, in certain circumstances it is possible that forces greater than these may develop, such as, for instance, in the case of very rigid bars or due to resonance in the case of bars liable to mechanical vibration. Experiments have also shown that the reactions produced in a non-resonating structure by an alternating current at the instant of application or removal of the forces may exceed the reactions experienced while the current is flowing. Thus it is advisable to err on the side of safety and to allow for all contingencies, for which one should take into account the maximum force which could be developed by the initial peak value of the asymmetrical short-circuit current. This force may be taken as having a value which is twice of that calculated from formula (1).

Formula (1) is usually found to be adequate for all practical purposes. However, one must take into account two slight modifications in certain cases, since formula (1) strictly applies only to conductors of circular cross-section and of infinite length.

Correction for End-effect For the purpose of bus bar design, the assumption that the conductors are of an infinite length does not, in general, lead to any great error. However it must be borne in mind that there is a great difference between the flux density near the ends and that near the middle of a long straight conductor carrying a current. With relatively short conductors, its effect may be considerable. Thus if the above formula is used when dealing with comparatively short bars, the forces calculated may be too high.

It has been seen that correct results are given in all cases by a formula in which

$\sqrt{\{(L/S)^2 + 1\} - 1}$ is substituted for L/S.

The formula then becomes

$$F_{max} = 4.5 \times I^2 \times \left[\sqrt{\{(L/S)^2 + 1\}} - 1\right] \times 10^{-8} \qquad (2)$$

In most other cases, sufficiently accurate results can be obtained by employing the following formula:

$$F_{max} = 4.5 \times I^2 \times (L/S - 1) \times 10^{-8} \qquad (3)$$

In general, for all practical purposes, formula (1) may be used without serious error as long as the ratio of L/S is greater than, say, 20. When the ratio lies between, say, 20 and 4, formula (3) should be used, and when the ratio is less than 4, formula (2) should be employed in all cases.

Shape Factor of Bars As already mentioned, the previous formulae apply strictly only to conductors of circular cross-section and it is thus necessary to apply a correction when conductors of rectangular cross-section are used.

Except in cases when conductors are very small, or are spaced a considerable distance apart, a correction factor, K has to be introduced into the formula, which then becomes

$$F_{max} = 4.5 \times I^2 \times L/S \times K \times 10^{-8}$$

The effect of the conductor shape decreases rapidly with increased spacing; while it is maximum for strip conductors of small thickness, it is practically negligible for bars of square cross-section. For a circular conductor, K symbolises unity.

3.5.2.2.2 Balanced Three-phase Short-circuits In the case of a balanced three-phase short-circuit, the currents in the conductor retain their normal 120° phase displacement, and as the relative direction of current flow in any two conductors with respect to that in the third conductor is constantly changing, so the magnitude and direction of the short-circuit forces also changes.

The maximum force of attraction between any two bars of a short-circuited three-phase bus bar system (centre line of three bus bars in one plane) when the effect of the third is neglected, is

$$F_A = 1/8 \, (4.5 \times I^2 \times L/S) \, K \times 10^{-8} \tag{4}$$

Similarly the simultaneous maximum force of repulsion between two bars when the effect of the third is neglected, would be

$$F_R = 3/4 \, (4.5 \times I^2 \times I/S) \, K \times 10^{-8} \tag{5}$$

In the above equations, S is the distances between the bars under consideration.

When the component forces due to the currents in all the three bars are taken into consideration, it can be shown that the maximum resultant force on any bar is

$$F_{max} = 3.375 \times I^2 \times L/S \times K \times 10^{-8} \tag{6}$$

where S is the distance between adjacent bars.

It will be seen that the above force is less than that produced by a single-phase short-circuit, and since single-phase short-circuits are usually both more numerous and severe than those involving all three phases, the design of the bus bar structure should be based upon the stresses set up by them.

3.5.2.2.3 Vibrational Stresses The mechanical stresses to which bus bars are subjected under short-circuit conditions may, in certain circumstances, be materially increased by vibration, particularly under conditions of resonance. Although such stresses may not seriously affect the conductors themselves, they are very important from the point of view of the strength of the supporting structures and insulators. The load on insulators may be increased to as much as five or even more times its normal short-circuit value, if favourable conditions for resonance prevail.

In the case of enclosed compound insulated busbars, the damping effect of the compound is usually sufficient to prevent the establishment of serious vibrations, but in the case of air-insulated bars, the damping is generally not sufficient to prevent the building up of vibrations, should favourable conditions exist.

No supporting insulator is absolutely rigid and since all types of insulators deflect in some degree when a transverse load is applied to them, they have a natural period of vibration, which varies with their size and design. In general, the strength of an insulator does not increase as rapidly as its stiffness, and the ideal insulator support for busbar purposes is one, which has a maximum strength and a low natural period of vibration. Just as the conductor supports alone have a natural period of vibration, so do the conductors. Thus the busbar structure as a whole will have natural frequencies of vibration, which are a combination of the natural frequencies of the bars and their supports. There are usually two dominant frequencies of importance, the fundamental and the first harmonic, and these may differ considerably from either of the original component frequencies.

The natural frequency of busbar structures may vary considerably according to the design over a wide range of anything from, say, 3 to 300 vibrations per second, and there is always a danger of partial resonance if the frequency of the applied electromagnetic forces is within about 30 per cent of the natural frequencies of the structure. Owing to the great variation in the detail design of busbars, it is not possible to give any universally applicable data to help readily ascertain the possible increase in stresses due to vibration. Every case has to be considered separately.

In practically all cases, it will be found that bars and supporting structures are strong enough to withstand increased stresses due to possible vibrations, if they are designed on the assumption that the greatest force to which they may be subjected is that produced by the maximum value of the asymmetrical short-circuit current. In this connection, it should be pointed out that the dc component of the asymmetrical short-circuit current disappears rapidly, usually within five or six cycles, with the duration of the high initial asymmetrical short-circuit current being very brief. There is, therefore, no time for the vibration of the conductors to be built up before the current has declined to a lower and more steady value. If the current wave is initially asymmetrical, the electromagnetic forces set up by it also change rapidly in frequency as well as in magnitude, and it is unlikely that serious vibrational stresses will occur coincident with the high initial value of the short-circuit current.

When short-circuit current flows through the circuit breaker, different current-carrying parts are subjected to mechanical and thermal stresses. The dynamic force occurs at the peak of the first major loop on the short-circuit. Perpendicular conductors tend to straighten out due to electromagnetic forces. The insulators are selected by considering mechanical bending load occurring at that instant of peak short-circuit current. During a short-circuit in the system, short-circuit current flows through the bus bars. The insulator supporting the bus bars experiences the bending force. The insulator should have enough cantilever strength to withstand load dynamic force occurring during a short-circuit.

When current flows in two parallel conductors, there is mechanical force between the conductors of attraction, when currents are in the same direction, and of repulsion, when currents are in opposite directions. While carrying normal current, these forces are negligible but under short-circuit conditions, they may be appreciable and it is necessary to ensure that the bus bar supports can withstand these forces even though they may be of momentary duration. It is important for the support insulator to be arranged so that insulators are not subjected to lateral forces. The busbar configuration should be designed in such a fashion that minimum forces are generated during a short-circuit.

The electrodynamic forces of attraction and repulsion between adjacent phase conductors are proportional to phase spacing. During short time current, the insulator support experiences an impact cantilever force due to electrodynamic forces.

The busbars are generally selected on the basis of considerations of mechanical strength of supporting arrangement. When spacing in busbars is small, mechanical forces become significant.

When circuit breakers are subjected to a short time current test, the contacts may have a tendency to open at the peak current. The adequate contact pressure should avoid an undesirable opening during the test.

3.5.3 Insulation Aspects of Design

An electrical insulation system is the heart of any electrical equipment. The life of the equipment can be increased if the life of the insulation system is increased.

Switchgear parts of different electric potentials are separated by various insulations to prevent flashovers and to ensure the safety of the personnel working on the switchgear, and reliability of its operations. The insulation serves three main purposes. It provides insulation:

1. Between the current-carrying live parts and earth;
2. In the contact gap during the 'breaker open' condition; and
3. Between the current-carrying live parts of different phases.

While selecting any insulating material for switchgear, the user should have a wide knowledge of the properties and behaviour of the materials in different environmental situations for their appropriate use. Large varieties of composite insulating materials are used in the seven classes of insulation systems including Y, A, E, B, F, H and C.

During their manufacture and service life, the insulating materials are subjected to several types of stresses as such as electrical, thermal, and mechanical stresses. Although the primary role of the insulation is to withstand electrical stresses, the capability of the materials to fulfil this requirement depends upon their ability to withstand other stresses as well.

A wide range of materials is used for insulation. These materials can be solid, liquid or gas. They are used to fill spaces, support live parts and to extinguish the arc, depending on their physical properties and form. The insulations widely used in switchgear design are detailed below.

3.5.3.1 Gaseous Materials

Air is the most commonly used gaseous material, which is composed of 80 per cent nitrogen and 20 per cent oxygen. Its properties are therefore close to nitrogen. Air clearances between various phases and between various phases to earth as mentioned in standards, are listed in Table 3.4. Switchgear enclosures are filled with SF6 in GIS, the dielectric strength which is approximately three times that of air. It functions both as an insulation medium as well as an arc-quenching medium in SF6 breakers. Of all the gases, air is the only insulating material which can be used effectively at atmospheric pressure.

Rated Voltage Upto (kV rms)	Minimum Clearance Between Phases and Phase to Earth in Air (mm)	
3.3	50	50
6.6	90	90
11	120	120
33	320	320

Table 3.4   Air Clearances

3.5.3.2   Fluids

A range of fluids has also been used for insulation in switchgear. Hydrogen carbon oil, often referred to as 'transformer oil', has been used in bulk oil CB. It serves the dual purpose of acting both as an insulator as well as an arc-extinguishing medium. Its dielectric strength is three times that of SF6 at atmospheric pressure. However, it is coming under increasing scrutiny from the safety point of view because of its flammability.

3.5.3.3   Vacuum

Another medium used for insulation in switchgear is vacuum. But since it is very expensive, its use in distribution switchgear is restricted to the circuit interruption devices only.

3.5.3.4   Solid Materials

Solid insulating material include those of natural origin such as mica, asbestos, and slate or those derived from natural material such as porcelain and shellac varnish. Solids are mainly used in switchgear to insulate conductors where they pass through the walls of metal enclosures apart from supporting the conductors. The resin is usually epoxy. One major advantage of the resin casting technique is the virtually limitless range of its shapes.

3.6   VACUUM SWITCHGEAR

In medium voltage (3–36 kV) power system applications, recent trends in both international and Indian markets show that conventional types like oil, air and SF6 circuit breakers are being mostly replaced by vacuum technologies. This technology will completely dominate the industry in coming years due to the fact that a vacuum circuit breaker represents an ideal solution to the widely varying demands of switching functions in medium voltage applications.

The following section discusses in depth the usage of vacuum technology to meet the switching requirements of MV applications, in distribution, in the industrial and power segments of the market.

3.6.1 Vacuum as an Interrupting Medium

The performance of a circuit breaker depends upon the properties of the dielectric medium and the design of the arc-quenching system. The superior performance of the vacuum circuit breaker is purely due to the inherent characteristics of vacuum as an interrupting medium. This is due to certain important features, which are detailed below.

For a given contact gap, the dielectric strength of high vacuum is approximately eight times that of air or four times that of SF6 (one bar). This higher dielectric strength makes it possible to quench an arc with a very small contact gap. Vacuum has the fastest recovery strength after full arc interruption to its full dielectric value at current zero. This makes it ideally suitable for capacitor switching.

The arc energy dissipated in vacuum for a given interrupting current is approximately one-tenth of that in oil or one-fourth of that in SF6. This is due to the short clearance times and low arc length in vacuum. The low arc energy keeps the contact erosion to a minimum, thereby giving a maintenance-free interruption system. For a given breaking current, the drive energy required in vacuum switchgear is the least as compared to that for other types of breakers (Fig. 3.13).

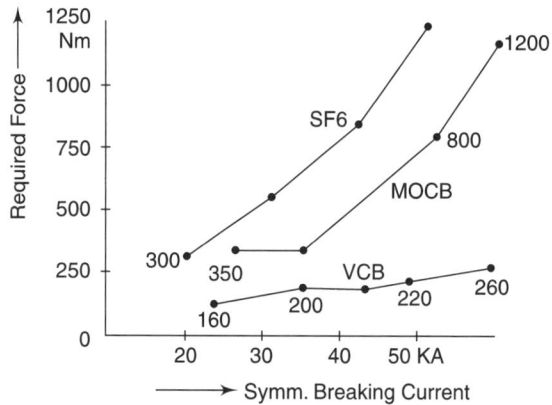

Fig. 3.13: Comparison of energy requirement

This low drive energy is due to the low contact mass, short contact travel and the fact that no compression of medium is required in vacuum switchgear. This low drive energy facilitates a simple mechanism with minimum linkages requiring little maintenance, and gives a high mechanical life of upto 60,000 operations in properly designed mechanisms.

When the contacts open in vacuum, the current to be interrupted initiates a metal vapour arc discharge and continuous flowing through this plasma until the next zero. The arc is extinguished near the current zero and the conductive metal vapour re-condenses on the contact surfaces in a matter of micro-seconds. Only one-hundredth condenses in the arc chamber wall provided for this purpose. As a result, the dielectric strength of the breaker recovers very rapidly and contact erosion is almost negligible.

There are two types of arc shapes. Up to about 10 kA, the arc remains diffused. It takes the form of vapour discharge and covers the entire contact surface. Above 10 kA, the diffused arc is constricted considerably by its own magnetic field and it contracts. The phenomenon gives rise to over-heating of contacts at its centre. In order to prevent this, the design of the contacts should be such that the arc does not remain stationary but keeps travelling by its own magnetic field. Various manufacturers have used different designs to achieve this. Specially designed contact configurations make the constricted stationary arc travel along the surface of the contacts, thereby causing minimum and uniform contact erosion.

3.6.2 Demands Made on Circuit Breakers

The demands made on MV power systems, irrespective of the place of use of the circuit breakers, vary widely in nature. All these functions cannot obviously be performed with the same efficiency. Thus, finally any circuit breaker design can only be a compromise. The main function of circuit breakers is to interrupt short-circuit currents and protect their loads against the effect of such faults. This function must be carried out with a high degree of safety and reliability.

3.6.2.1 Switching Functions

In an MV power system, apart from the normal ON/OFF operations, the switching operations can be divided into the following:
- Short-circuit Interruptions;
- Switching of Capacitive Currents;
- Switching of Inductive Currents; and
- Special Applications.

3.6.2.1.1 Short-circuit Interruptions All circuit breakers are designed to interrupt rated short-circuit currents. The number of short-circuit interruptions seen in the life of a circuit breaker depends on its location in the system, quality of the system design, environmental conditions, etc. In countries like India, where most of the distribution is by means of overhead lines (except in the case of large urban cities where underground cable networks are used), the distribution switchgear is subjected to a large number of short-circuit interruptions. This is due to bird hits, snapping of lines, thunderstorms, etc. Vacuum switchgear, which can interrupt rated short-circuit current up to 25 kA 100 times without requiring maintenance, is the obvious choice for this application as compared to other circuit breakers which can do so for only 15–20 operations.

The other important criterion to be considered in distribution through overhead lines is the auto-reclosing feature required due to the transient nature of the faults. The VCBs are ideally suited for auto-reclosing duties because they act fast and can be reclosed with a minimum time interval between two closings as cooling of the dielectric is not involved.

Most of the sub-stations in rural areas are outdoor and situated in far-off places. This necessitates the setting up of outdoor maintenance-free circuit breakers. Porcelain-clad vacuum circuit breakers meet this demand reliably as against the conventional indoor BOCBs used in outdoor kiosks. Moreover, the availability of vacuum technology for auto-reclosers and sectionalisers has given birth to highly economical/unattended sub-stations for rural application.

3.6.2.1.2 Switching of Capacitive Currents This includes switching of capacitors, unloaded cables and overhead lines as well as single and parallel capacitor banks. Most of the electricity utilities use capacitors in their systems to improve the power factor and to tackle voltage drops. Vacuum circuit breakers disconnect these loads safely without re-ignition and thus without the associated over-voltages.

When switching in capacitor bank, and especially when paralleling them, very high making currents and high rates of rise of current occur. Conventional circuit breakers with liquid-quenching medium and tulip contacts may suffer from contact pin retardation. Thus additional measures are usually needed for these breakers to reduce such effects. Very low energy loss in the contact gap owing to low arcing during the short pre-arcing time, and flat contact surfaces of vacuum breakers obviate the necessity for additional measures. VCB's are suitable for switching in single capacitor bank as well as paralleling of Multi-Capacitor banks at much higher currents and rates of rise than MOCBs and SF6 circuit breakers, without the need for damping reactors.

3.6.2.1.3 Switching of Inductive Currents The following values are likely to be encountered in various cases of switching inductive currents.

Application	*Range of Inductive Current*
Distribution Industrial Segments	Very low inductive currents like transformer on no-load up to 20 A
	Low inductive current like compensating reactors up to 2000 A
Industries/Power Stations	Motors in operation up to 1000 A
	Motors during starting up to 5000 A

Table 3.5 Typical Values of Inductive Current Switched by Medium Voltage Switchgear

Vacuum circuit breakers have proved to be adequately suitable for these applications.

Switching of Very Low Inductive Currents Older vacuum circuit breaker designs had current chopping levels that were as high as 20 A. When these breakers were used to switch transformers, special surge protection devices were required. Other types of circuit breakers also have high chopping currents of, say, 20 A. Modern Vacuum Circuit breakers with very low chopping currents of 2 to 4 A are capable of switching unloaded transformers (most difficult inductive loads) with very low over-voltage, thereby requiring no surge limiters.

Switching of Low Inductive Currents Two main applications fall in this category. These are switching of compensating reactors and switching of motors under stalled rotor starting conditions. In these special cases, high over-voltages may occur due to multiple re-ignitions. Vacuum interrupter manufacturers have systematically investigated the switching of motors during starting and made the following observations:

1. High voltage motors may be safely operated under start-stop conditions by vacuum circuit breakers where surge suppressors are provided to limit associated switching over-voltages.

2. The front time of the switching impulses is determined by the system configurations mainly by the surge impedance of busbars and cables, and current transformer inductances. These tests were conducted to meet the most severe conditions and it was observed that the occurrence of a front time of less than one micro-second is most unlikely in service. This ensures the safe operation of high voltage motors which are tested with a standard lightning surge wave of 1.2/50 micro-seconds.

3. Switching off of motors having starting currents of more than 600 A during starting generates low switching over-voltages and the motors do not require protection by surge suppressors even in 11 kV cable systems.

It has also been observed that when other types of breakers are used to handle such switching operations, high over-voltages may occur due to multiple re-ignition. Moreover, it has now been found that when motors are switched 'ON', high over-voltage surges can occur which are independent of the arc-quenching medium including in SF6 circuit breakers. Thus, vacuum circuit breakers, which are provided with surge limiters, offer protection to the motor even when it is switched 'ON.'

3.6.2.1.4 Special Applications Special applications like arc furnaces involve a large number of operations per day. Track-side sub-stations require the switchgear to operate in varying set of conditions from switching the charging current of catenary systems to transformer magnetising currents to a whole range of load currents and fault currents varying from 2 kA to 12 kA. These applications also require the switchgear to have the capability to withstand a range of voltages of various waveforms from sinusoidal to steep fronted surges throughout its useful life. Vacuum circuit breakers have been found to be superior to all conventional and SF6 circuit breakers for such applications.

(a) *Arc Furnace Applications* Electric arc furnaces with ratings up to 100 MVA generally employ special circuit breakers, for which purpose air blast circuit breakers have been used till now. In this application, the falling scrap in the furnace causes short-circuits between the electrodes during the melting process, and the currents to be switched lie between zero and 1–8 times the rated current of the furnace transformer. The frequency of switching can be as high as 100 operations per day with rated currents up to 2000 A. Normal oil, air and SF6 circuit breakers are found to be unsuitable for this application. Special air circuit breakers are very expensive due to stringent requirements for relatively small quantities. The standard vacuum circuit breaker offers an economical and reliable solution for this application.

(b) *Use in Traction System* The system used for electric traction generally has voltages between 15 and 25 kV and frequency of 16-2/3, 50 and 60 Hz. The main function of the single-phase traction circuit breakers is quick interruption of short-circuits on the overhead catenary system, which occur frequently and are usually transient in nature. Since VCBs have short contact travel and shorter arcing times, the total break time is quite less and thus meets the special requirements of short breaking times easily.

In case of 16-2/3 Hz traction systems, the current zero occurs in every 30 milli-seconds resulting in an arcing time of around 33 milli-seconds. Although the arc energy in the contact gap for a single-phase breaker is much greater than that for three-phase circuit breakers, it is still much lower in a vacuum circuit breaker than in conventional circuit breakers owing to the low arc voltage. Figure 3.14 shows vacuum circuit breakers for use in traction systems.

Fig. 3.14: 25 kV outdoor track-side vacuum switchgear

The number of short-circuits occurring on an overhead catenary system is much higher than those occurring on transmission lines. Thus the higher permissible cumulative current of vacuum circuit breakers, i.e. up to 100 operations at rated short-circuit current or 30,000 operations at rated normal current, makes them especially suitable for this application.

(c) Use on Ships Circuit breakers meant for use on ships have to fulfil the following special requirements:

- They must remain operational even in an inclined position; and
- Circuit breakers used on ships are subjected to vibrations in actual use.

Circuit breaker manufacturers are able to comply with these requirements more easily in the case of vacuum circuit breakers.

References

1. IEC: 62271-100, "High Voltage Alternating Current Circuit Breakers".
2. IEC: 62271-200, "Metal Enclosed Smitchgear and Control Gear for Rated Voltage upto and including 38kV".
3. IEC: 60694, "Common Specifications".

Chapter 4

HIGH VOLTAGE CIRCUIT BREAKER

Yellaiah Babu B.

High voltage circuit breaker technology has changed radically in the last 15 years. Most utility systems utilise a combination of bulk oil, minimum oil, vacuum, air blast, SF6 two-pressure, and SF6 single-pressure circuit breakers. The SF6 single-pressure system circuit breaker has become the current 'state of the art' technology at transmission voltages. Further improvements of this single-pressure version are being introduced with thermal assisted arc quenching, in order to facilitate reduction in operating energy requirements leading to cost reduction and reliability improvement. Integration of various sub-station equipments like breaker, disconnectors, instrument transformers, and earth-switch into single equipment (Integrated Switchgear), or reducing the spacing between them, is also being considered—resulting in compact sub-station. These types of compact sub-stations will be more suitable in case of modernisation or refurbishment of existing sub-stations. Introduction of simpler and reliable drive mechanisms for EHV breakers has also become an important aspect, as most of the breaker failures are with reference to the operating-mechanism of the breaker.

As SF6 gas has an impact 23 times stronger than CO_2 on the Greenhouse effect, in order to minimise the emission of SF6 gas, the N_2–SF6 and CF_4–SF6 gas mixtures use may be more predominant in future as an alternative to using pure SF6 gas. Also, equipment like SF6 gas circuit breakers shall be designed so that the gas leakages are minimum during their service period, utilities shall let out the least amount of SF6 gas into atmosphere during maintenance.

First, a major overhaul of the SF6 gas breakers installed in 1980s, is the major task for some of the utilities, as it was due after 2000. While most utilities have personnel, experienced in the maintenance of older breakers like ABCBs and MOCBs, newer SF6 gas breaker technology is very different and requires special procedures. Since SF6 gas has been identified as a Greenhouse gas, safety regulations are being introduced in many countries in order to minimise its release into the atmosphere. Hence, suitable protective gear and handling equipments should be used during maintenance work, to minimise the SF6 gas release into atmosphere.

It is important that for the maintenance of SF6 gas equipment like circuit breakers, the technical personnel involved should have a complete understanding of the system. They should not only know how to perform maintenance tasks, but also why and when these tasks should be performed. Condition-based maintenance (CBM) and Reliability centred maintenance (RCM) are being proposed to replace periodic maintenance of EHV circuit breakers, in order to minimise the maintenance costs.

4.1 CLASSIFICATION OF CIRCUIT BREAKERS

Circuit breakers can be arbitrarily classified using criteria such as, intended voltage application, location of installation (i.e. outdoor, indoor), their external design characteristics, or perhaps the most important, method and medium used for the current interruption. This is shown in Fig. 4.1.

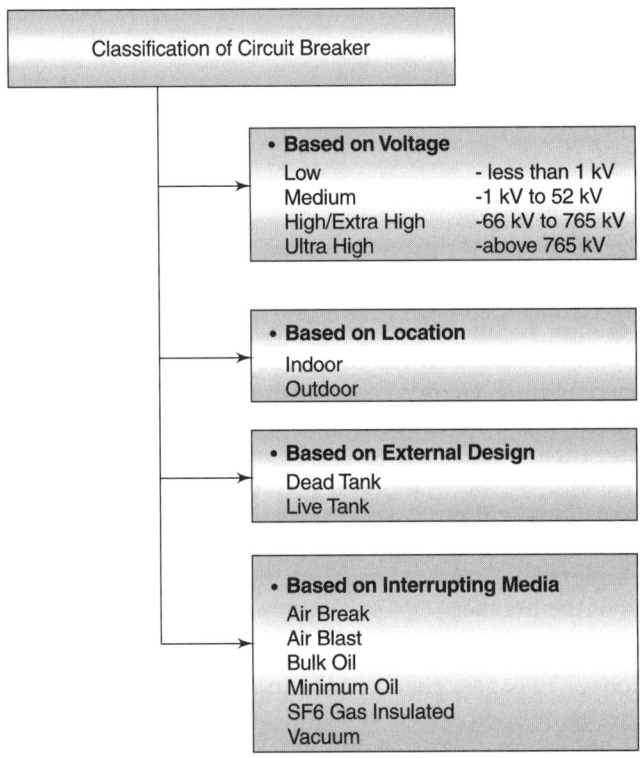

Fig. 4.1: Classification of circuit breakers

4.1.1 Classification by Voltage

The classification of circuit breakers by its intended application voltage, is normally as given in Fig. 4.1.

High Voltage Circuit Breaker **4.3**

4.1.2 Classification by Location

Switchgears, based on where they are located are classified as, indoor and outdoor types.

Medium and low voltage switchgears, and high voltage Gas Insulated Switchgears (GIS) are mostly categorised as Indoor switchgears, whereas the switchgears which have air as an external insulating medium, i.e. Air Insulated Switchgear (AIS), are categorised as Outdoor Switchgears. These are shown in Fig. 4.2.

33 kV Indoor Type (GIS)   Outdoor Type (AIS)

Dead tank breaker (courtesy ABB)   Live bank breaker (Courtesy BHEL)

Fig. 4.2: Switchgear based on location

4.1.3 Classification by External Design

Outdoor circuit breakers can be identified as either dead-tank or live-tank type of circuit breakers, from the point of view of their physical structural design.

In dead-tank circuit breakers, the switching device is located, with suitable insulator supports, inside a metallic vessel(s) at ground potential and filled with insulating medium. In dead-tank circuit breakers, the incoming and outgoing conductors are taken out through suitable insulator bushings, and low voltage type current transformers are located at the lower end of both insulator bushings, i.e. at the line side and the load side.

In live-tank circuit breakers, the interrupter(s) is located in an insulator bushing, at a potential above ground potential. The live-tank circuit breakers are cheaper (with no current transformer), and require less mounting space.

4.1.4 Classification by Interrupting Media

The interrupting medium has been the vital factor in the evolution of circuit breakers. It dictates the overall design parameters of the breaker. The choice of air and oil, as the interrupting media, was predominant till late 70s. But today, vacuum and Sulphur hexafluoride (SF6) are the only dominant interrupting technologies, for medium and high voltage segments of circuit breaker design respectively.

4.1.4.1 Air Blast Circuit Breaker

In the design of air blast circuit breakers, the interrupting process is initiated by establishing an arc between two parting contacts and by simultaneously opening a pneumatic valve. A blast of high pressure air is blown over the arc column, thereby subjecting it to the intense cooling, and quenching the arc between the contacts when the arc column ionisation is least at current zero. Whenever current at higher voltages needs to be interrupted, more breaking units are used, in series. Dry and clean air supply is one of the essential requirements for the operation of the ABCBs.

4.1.4.2 Oil Circuit Breaker

Oil circuit breakers can be further classified with reference to the quantity of oil used as:

1. Bulk oil circuit breakers (BOCBs), and
2. Minimum oil circuit breakers (MOCBs).

In BOCBs, the interrupting unit is placed in a tank of oil at earth potential and the incoming and outgoing conductors are connected through insulator bushings. In MOCBs on the other hand, the oil requirement can be minimised by placing the interrupting units, in insulating chambers at live potential, on an insulator column.

In an oil circuit breaker, the arc quenching process is entirely dependent on arc energy generated. The arc drawn across the contacts is contained inside the interrupting pot, and thus the hydrogen gas formed by the vaporised oil (gas) is also contained inside the chamber. As the contacts continue to move, and the moving contact rod separates itself from the orifice of the chamber, an exit similar to a nozzle allows escape of the hydrogen trapped inside the interrupting chamber. The escaping high-pressure hydrogen gas, having a high thermal conductivity, takes away the heat, thus making the contact gap cool and free from ionisation, immediately after current zero. However the oil breakers have prolonged arcing times, due to insufficient vapour pressure generated at lower interrupting currents. Periodic monitoring of oil di-electric condition and its maintenance is important for the effective operation of MOCB. However, maintenance required for oil circuit breakers is of low technology.

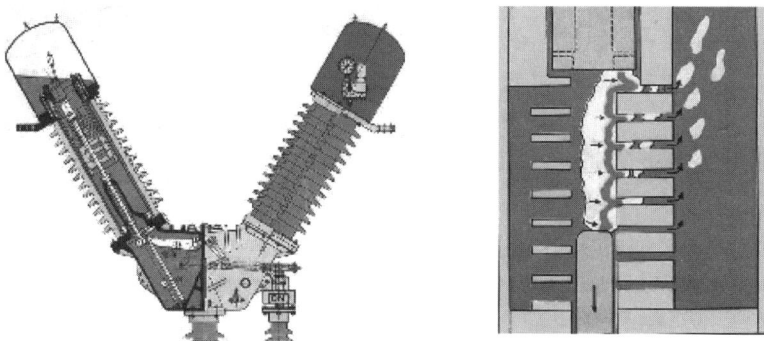

Fig. 4.3: Cross sectional view of type HLR MOCB (courtesy BHEL)

4.1.4.3 Vacuum Circuit Breakers

The vacuum circuit breaker takes the advantage of non-sustainability of electric arc in vacuum, and employs the principle of contact separation under vacuum where there is no ionization due to medium. The initial arc caused by field and thermionic emissions during contact separation, will die away soon, as there is no further ionization because of vacuum. The contact design in vacuum breakers is such that the arc becomes elongated during contact separation process. Vacuum breakers are cost effective in the medium voltage range. Vacuum interrupters are sealed units and maintenance-free for 10,000 normal load operations.

Fig. 4.4: Cross section of vacuum interrupter

4.1.4.4 SF6 Gas Circuit Breakers

SF6 gas circuit breakers, are widely used for EHV applications today, as Sulphur hexafluoride (SF6) gas is an electronegative gas having excellent di-electric and arc-quenching properties, resulting in many advantages such as compactness and less maintenance of EHV circuit breakers.

The excellent insulating properties of SF6 gas make it possible to design circuit breakers with smaller overall dimensions, shorter contact gaps, which help in the construction of outdoor breakers with fewer interrupters, and evolution of metal enclosed SF6 gas insulated switchgear (GIS).

The behaviour of the arc during current zero and immediately afterwards, is of decisive importance for the reliable interruption of the current. In SF6 gas, the diameter of an arc is relative small. As a consequence, the thermal time constant of the arc (which is a function of the square of the arc radius), approaching current zero, is almost 100 times smaller than in air. The di-electric strength of the break increases rapidly after current zero, so that the breaker is able to control even extreme rates of rise of the transient recovery voltage (TRV), in case of short line faults.

4.2 DEVELOPMENT OF SINGLE-PRESSURE PUFFER BREAKERS

The first generation of SF6 gas circuit breakers were of the double pressure type developed in the mid 60s in Europe and USA. These breakers did not become very popular as outdoor circuit breakers, from the point of view of cost and the number of interrupters used, as they were developed by simply replacing air in ABCBs with SF6 gas. But with the development of metal-clad gas insulated sub-stations, SF6 gas medium became popular, where utilities decided to go in for metal clad stations, mainly due to the limitation of space and environmental constraints.

During the 1970s, single-pressure designs were developed, ushering in remarkable simplicity in construction and a drastic reduction of number of interrupters per pole and weight (kg/GVA) of the circuit breakers. Modern live-tank SF6 puffer circuit breakers have been developed upto 245 kV in a single interrupter design, upto 420 kV with two interrupters per pole and 800 kV with four interrupters per pole, whereas the dead-tank versions are available with two interrupters upto 800 kV. Any further reduction in number of interrupters does not seem feasible as the nature of the circuit breaker is now generally dictated by the requirement of open air clearances needed for external insulation.

With the development of fast operating mechanisms, the puffer design breakers can now operate with a total break time of two cycles.

4.2.1 First Generation SF6 Circuit Breakers

The first generation of SF6 gas circuit breakers used the dual pressure principle of air blast breakers. The SF6 gas is compressed and stored in a stationary high-pressure receiver, much as in the earlier air blast circuit breakers. To interrupt the current, gas is blown into the breaking chamber, from there it is collected in a low-pressure receiver and subsequently pumped back to the high-pressure receiver.

The major disadvantage of this approach, is liquification of SF6 gas at the storage pressures at low temperatures. Hence, the high-pressure storage tanks are provided with additional heaters to keep SF6 in gaseous state.

4.2.2 Second Generation SF6 Circuit Breakers

Second-generation SF6 circuit breakers work on the single-pressure principle, i.e. the breaker is filled with SF6 gas at rated pressure and the differential quenching pressure required for extinguishing the arc, is generated during the opening movement of contact system. Figure 4.5 explains

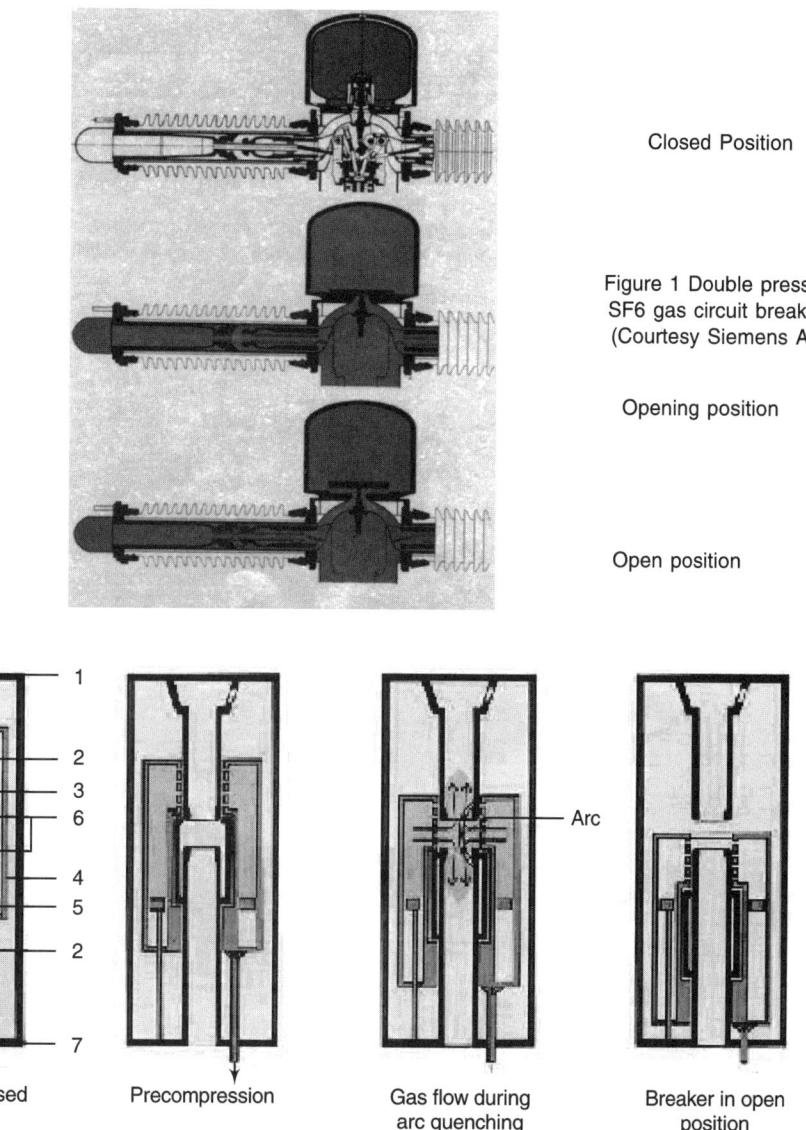

Figure 1 Double pressure SF6 gas circuit breakers (Courtesy Siemens AG)

Fig. 4.5: Single-pressure interrupter principle of operation

the process of current interruption. The compression cylinder moves along with the contact system against a fixed piston, during the breaking operation, thus generating the required quenching pressure inside the compression cylinder. As the contacts separate, an arc established between them, is extinguished at current zero. At the same instant, the quenching gas flowing out of the nozzle at high speed, de-ionises the contact gap, and thus prevents the re-ignition of the arc. In open condition the breaker contact gap with SF6 gas, forms the insulation to withstand the recovery voltage of the system.

In this approach, the operating mechanism must provide the energy required not only for moving the contacts, but also for building up the quenching pressure. For this reason, circuit breakers of this type require powerful, complex operating mechanisms

4.2.3 Third Generation SF6 Circuit Breakers

In the third generation of circuit breakers (Fig. 4.6), the arc energy is effectively used for generating the quenching pressure in the pressure chamber. The operating mechanism, thus relieved to a great extent of the energy required for gas compression, uses low energy spring operated mechanisms. Reliability of the spring operating mechanism has improved to a great extent with low energy operations.

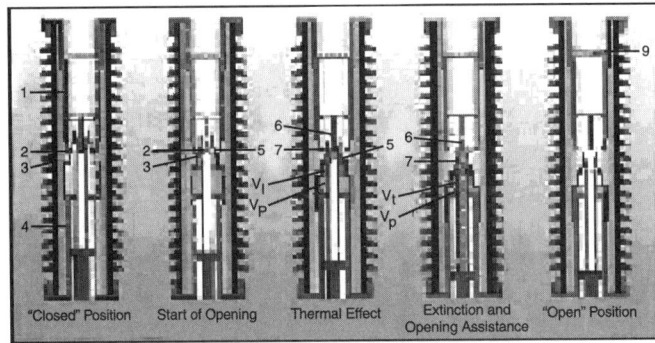

Fig. 4.6: Thermal assist interruption

At present, third generation breakers using thermal assist or self-blast interruption principle, are in service upto 245 kV. The principle may be extended upto 420 kV in future.

4.2.4 Design Features of Puffer Circuit Breaker

4.2.4.1 Interrupter Design

Two basic puffer type of interrupter designs are available. The first, with the single flow series piston arrangement, with nozzle made up of insulating material coupled with the moving contact. In this type, the moving contact tip and the insulating nozzle throat are exposed to arcing for a long time, which gives rise to higher contact burn and nozzle ablation. Of course, nozzle ablation helps in the increase of differential gas pressure inside the puffer chambers, thereby helping in the arc extinction, but this may also reduce the life of the nozzle and the contacts.

The second type, is the double flow, fixed nozzle arrangement (see Fig. 4.7). In the closed state, the moving contact, bridges the two fixed contact tubes with its spring-loaded fingers, which are arranged in a ring. The large cross-section of these tubes and the large number of contact fingers give the interrupter unit a high normal current rating.

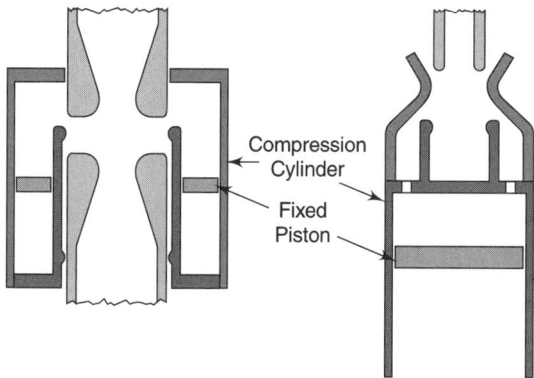

Fig. 4.7: Basic puffer designs

On opening command, the blast cylinder is moved towards the fixed piston, so that the SF6 gas in the blast chamber is compressed to the pressure required for arc quenching, with the moving contact acting as a slide valve. The gas flow is not released until the contacts separate. An insulated ring, and a ring of arc-resistant material, are at the front end of the moving contact. The insulating ring is shaped so that gas flow does not start until contact separation. As a result, small inductive currents are interrupted without high over-voltage build up. The arc, which loops between the front nozzle and the arcing ring on the moving contact, is driven into the nozzles in a matter of milli-seconds by the gas flow and by the electrodynamic forces developed. The basic pressure, compression ratio, and compressed gas volume are chosen, such that the pressure required for interrupting the short circuit current is available at the end of the compression phase (i.e. at the instant of contact separation). This enables the breaker interrupt any current in its entire range with very short arcing times.

While the arc is being interrupted, the blast cylinder encloses the arc-quenching assembly in the manner of a pressure chamber. The compressed gas takes the shortest path, i.e. it flows radially into the break, from where it is discharged axially through the quenching nozzles and contact tubes. The blast cylinder, which is made of arc-resistant insulating material, protects the porcelain enclosure of the arc chamber against any arcing effects.

Since the quenching nozzles are fixed in position in this design, their arrangement for optimum arc quenching does not vary. After current interruption, the moving contact and blast cylinder continue travelling until they reach the 'OFF' position. Any chamber elements, which may have a bridging effect or which may influence the electrical field distribution, have now cleared the gas-filled break between the quenching nozzles. The favourable shape and size of the nozzles and the excellent insulating characteristics of SF6, give the break a very high di-electric strength (see Fig. 4.8).

The quenching nozzles and arcing rings for the puffer breaker can be made of graphite, which is superior in many ways to metallic, arc-resistant materials. The outstanding features of graphite are that it is subject to almost negligible arc erosion, and that its surface quality is not affected even by heavy arcing. As a result, the breaker can handle a large number of fault interruptions without maintenance. The decomposition and reaction products of graphite and SF6 which form during arcing, have almost the same thermal and di-electric characteristics as those of pure SF6

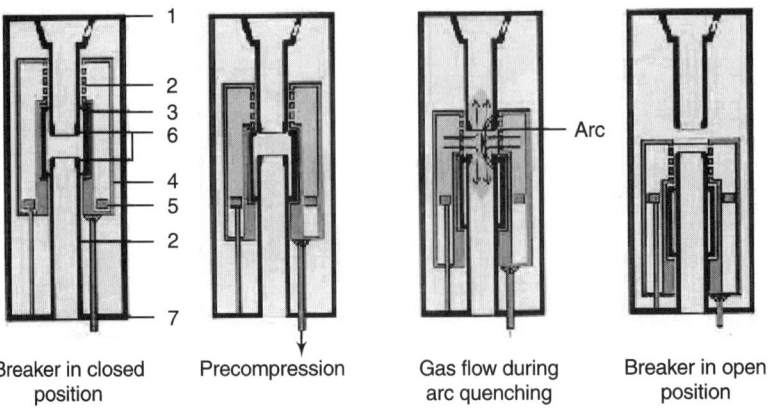

| Breaker in closed position | Precompression | Gas flow during arc quenching | Breaker in open position |

Fig. 4.8: Arc-quenching process

gas. Metallic materials have quite a different effect. The pollution of the break by metal vapour would reduce the arc-quenching capability considerably and have a detrimental effect on the switching performance.

Since graphite vapourises directly from the solid state, the surface of graphite nozzles show no significant roughness even after twenty interruptions of the full short circuit breaking current. Consequently, the high di-electric strength of the break is fully maintained.

4.2.4.2 Requirement of Operating Energy for Puffer Breakers

An analysis of energy requirement for the puffer breaker has shown that the puffer type of interrupter requires comparably high energy to build up the differential gas pressure and to maintain it throughout the interruption process. In fact, the reliability of a puffer circuit breaker primarily depends upon its ability to maintain reasonably high differential pressure throughout the arc interruption window. The gas compression process involves a considerable amount of force and energy required to drive the contact system moving mass, and to compress SF6 gas in the puffer chamber for arc interruption.

4.3 OPERATING MECHANISMS FOR CIRCUIT BREAKERS

There are generally three types of mechanisms to choose from—spring-operating, pneumatic, and hydraulic for EHV breakers. Originally, the spring mechanisms were confined to MOCBs (which require about 3500–4500 NM per pole of 420 kV circuit breaker) and SF6 circuit breakers for lower transmission voltages. For EHV/UHV circuit breakers, especially for faster ones with a total break time of two cycles, the choice is generally limited to pneumatic or hydraulic mechanisms. However, in the recent past, spring mechanisms have also been used with the EHV range of SF6 gas circuit breakers upto 550 kV.

Various type of mechanisms used for operation of circuit breakers are described below.

4.3.1 Spring Mechanism

This type of mechanism is commonly found in high or extra high voltage outdoor type circuit breakers, and practically in all medium voltage indoor type breakers. These mechanisms have been used with MOCBs and were adapted later for SF6 breakers.

As the name suggests, the energy of this mechanism is stored in the closing springs. The stored energy is available for closing the circuit breaker upon a command following the release of a closing latch. The energy distribution of the spring operated mechanism is shown in Fig. 4.9.

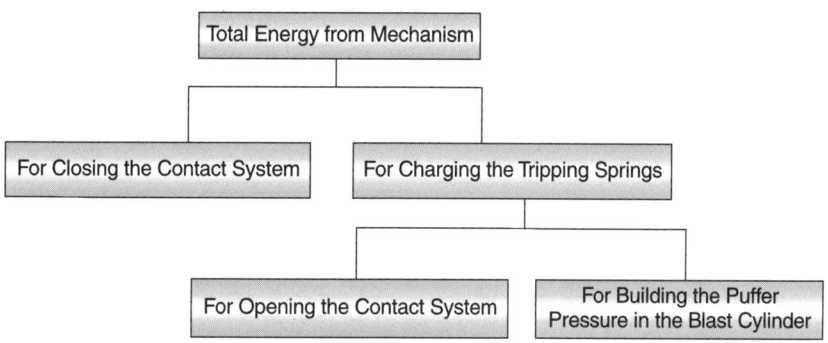

Fig. 4.9: Energy distribution in spring operated mechanism

The spring mechanism in its simplest form consists of a charging motor and a charging ratchet, a closing cam, closing springs, and a toggle linkage. The charging motor and ratchet assembly provide automatic recharging of the closing springs immediately following the closing contact sequence. In order to release the spring energy, either an electrically operated solenoid closing coil, or a manual closing lever is operated.

The opening of the contacts can be initiated either electrically or manually. The manual operation is generally provided only for maintenance purposes. When the tripping command is given, the trip latch is released, thereby freeing the trip roller carrier. The opening springs, which are connected to the main operating shaft, provide the necessary energy to open the contacts of the circuit breaker.

4.3.2 Pneumatic Mechanism

This type of mechanism is best suited for air blast circuit breakers because the pressurised air is already in use for insulating and interrupting. Pneumatic or pneumo-spring mechanisms have been adapted by a large number of manufacturers for SF6 gas circuit breakers.

In the pneumo-spring mechanism, a piston is used to drive the opening linkages and to charge a set of closing springs. In order to close the circuit breaker, high-pressure air is applied to the under-side of the piston by opening a three-way valve; the piston moves downwards while transmitting the opening force through a toggle arrangement. In addition to opening the contacts, the mechanism charges a set of closing springs. Once the contacts open, a close latch is engaged to hold the breaker in the open position. Closing is achieved by energising the closing solenoid, which in turn releases the closing latch, thus allowing the discharge of the closing springs, forcing the contacts to the closed position.

4.3.3 Hydraulic Mechanism

Hydraulic mechanisms are actually a variation of the pneumatic operation. In most cases, the energy is stored in a nitrogen gas accumulator and the incompressible hydraulic fluid becomes a fluid operating link that is interposed between the accumulator and a linkage system. Figure 4.10 explains the operation of the hydraulic mechanism.

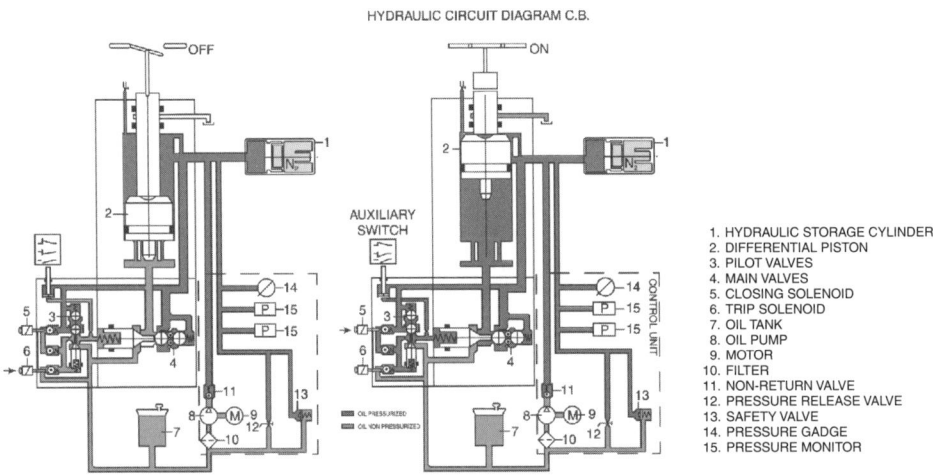

Fig. 4.10: Operation of hydraulic mechanism

Hydraulic mechanisms operate at much higher pressures than pneumatic cylinders. Leakage is an important factor to be considered because the fluid is conserved and recycled. Moving parts are protected from external corrosion by the hydraulic mechanism—as they are totally immersed in hydraulic fluid, and the hydraulic system is completely sealed.

4.3.4 SF6 Gas Dynamic Mechanism

M/s NMG of Italy developed this innovative mechanism, which is being used by them for breakers upto 420 kV. This is basically a single-pole operating mechanism as shown in Fig. 4.11.

The potential energy of the pressurised SF6 gas is used to operate the contact system of the interrupter. During closing/opening operations, the pressurised SF6 gas in the pole column is allowed into the upper or lower chamber of a differential piston cylinder to open/close the

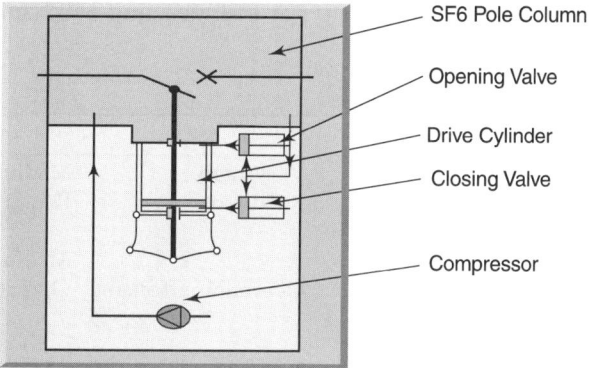

Fig. 4.11: SF6 Gas dynamic mechanism

breaker. During this operation, some of the SF6 gas is allowed to flow to a secondary chamber, which is later pumped back into pole column with the help of a small compressor.

4.3.5 Motor-driven Mechanisms

In line with future trends, stepper motor driven mechanisms are under development. This mechanism has the least number of mechanical components, and can be directly mounted on linkage housing of pole column, using capacitor banks as the energy storage device. These mechanisms will be able to drive breakers with low operating energy requirement, such as breakers with thermal assist interrupters.

4.3.6 Comparison of Operating Mechanisms

With reference to the usage of various types of mechanisms available, few manufacturers have used spring mechanisms for their entire range of outdoor circuit breakers (up to 420 kV) in the past two decades. Most manufacturers use either hydraulic or pneumatic mechanisms with second generation single-pressure SF6 gas circuit breakers, as the energy levels to be handled are high and the manufacturers do not have adequate experience pertaining to reliability of spring mechanisms with higher energies. However, with the development of thermal assist (self-blast) interrupter designs, most of the manufactures are able to meet the low energy requirements of these interrupters with spring mechanisms. Out of all the mechanisms available today, spring mechanisms are the most preferred, as they have minimum monitoring with low energy levels if the manufacture of components is done in a controlled environment.

The various characteristics of different types of Operating Mechanisms are furnished in the following table.

Sl. No.	Feature	Spring	Hydraulic	SF6 Dynamic	Pneumatic
1	Energy stored	Medium	Very high	Medium	High
2	Mechanism fluid monitoring	—	Required	Required	Required
3	Noise	Medium	Low	Medium	High
4	Mechanism fluid leakage	—	Operation affected	Operation affected	Operation may be affected
5	Corrosion of components	May be present	Not present	Not present	May be present
6	Cost	Low	High	Medium	Medium
7	Interposing mechanism	Required	Not required	Not required	Required
8	Damping device	Required	Built-in	Built-in	Required

Table 4.1 Characteristics of Different Operating Machanisms

4.4 SYSTEM REQUIREMENTS OF CIRCUIT BREAKERS

With the phenomenal growth of the power system (EHV and HV) the duties that a circuit breaker is called upon to perform, besides the ability to handle short circuits, have become complex. Recent studies in this field has thrown much light on switching duties, which often place severe demands on the breakers.

In this chapter, the most important of these duties are dealt with briefly. They are:

(a) Terminal faults

(b) Short line faults

(c) Transformer magnetising and reactor currents

(d) Energisation of long transmission lines

(e) Switching of unloaded transmission lines and capacitor banks

(f) Out of phase switching

In each of the above-mentioned switching duties, the shape of the transient recovery voltage is largely influenced by a number of network parameters. The influence of these parameters has been described below, with reference to each duty.

4.4.1 Terminal Faults

Usually the short-circuit current is inductive in nature, and the re-strike voltage appearing across the contacts at the instant of arc interruption is characterised by high frequency oscillations in the order of a few hundred cycles. The magnitude of severity imposed on the circuit breaker depends on this frequency and the peak value of voltage.

High Voltage Circuit Breaker

The recovery voltage appearing across the contacts of the breaker can be considered on two different successive time intervals: transient recovery voltage with high frequency oscillation immediately after the arc extinction, and power frequency recovery voltage after die down of the high frequency oscillations.

4.4.1.1 Transient Recovery Voltage

The transient recovery voltage is the actual voltage appearing across the circuit breaker contacts immediately after the arc interruption. The voltage will, of course, ultimately approach the open circuit voltage, but will do so in the form of high frequency voltage oscillation governed by the circuit parameters 'L' and 'C'.

This Recovery voltage is given by

$$e\,(1 - \cos(t/\sqrt{L*C}))$$

The maximum transient recovery voltage is twice **e**. The resistance present in the circuit damps this transient voltage. In practice, the transient recovery voltage is not of a single frequency but of several frequencies because of the complicated network configuration. The Transient recovery voltage after short circuit interruption is shown in Fig. 4.12a. To test the breaker, a six parameter method is applied, as shown in Fig. 4.12b.

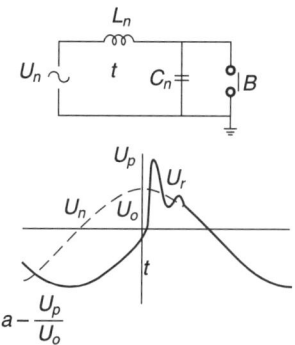

Fig. 4.12a: Transient recovery voltage after short-circuit interruption

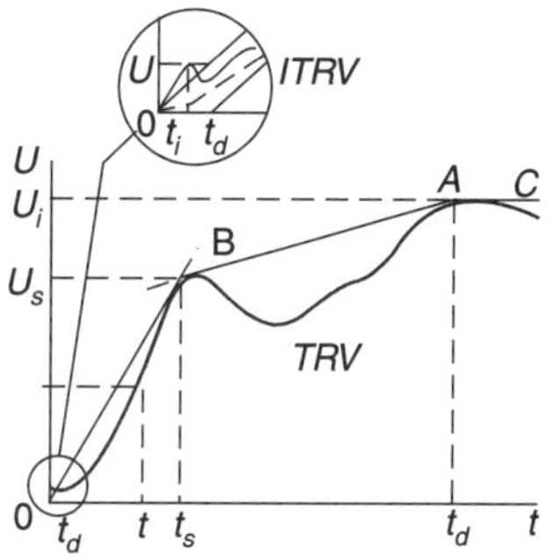

Fig. 4.12b: Transient recovery voltage represented by six parameters

4.4.1.2 Power Frequency Recovery Voltage

The power frequency recovery voltage is basically equal to the open circuit voltage of the network in the open position of the circuit breaker. Immediately after a fault is cleared in a three-phase circuit, the power frequency voltage differs in the different phases, usually being highest in the first phase. If the network neutral is not earthed, the voltage across the first pole to be cleared is 1.5 U, where U is the phase voltage. In an earthed neutral system, the voltage across the first pole to clear will be approximately 1.3 U.

By using suitable damping resistance, the magnitude of the transient recovery voltage and the rate of rise of recovery voltage (RRRV) can be limited. The di-electric recovery of the interrupting medium has great influence on the interrupting performance of the breaker with reference to the rate of rise of transient recovery voltage: faster the recovery of di-electric strength, better the performance.

Air blast circuit breakers are the most sensitive to the shape of the TRV, because the heavy ionised air de-ionises very slowly, and thus takes a comparatively long time to regain its insulation level. It is often necessary to provide such breakers with low ohm breaking resistors, in order to reduce the RRRV. On the other hand, the air blast circuit breakers are less sensitive to the initial transient recovery voltage (ITRV) because of a comparatively high arc voltage.

SF6 breakers are less sensitive to the shape of the TRV because SF6, an electronegative gas, has ions that re-combine faster than in air. However, the comparatively low arc voltage makes them more sensitive to the ITRV, and the shape of the initial voltage determines the breaking capacity of these circuit breakers.

Bulk oil and minimum oil circuit breakers, are inherently sensitive to RRRV and to ITRV. In these circuit breakers, the arc burns in heavily pressured hydrogen, which quickly regains its di-electric strength after current zero. The ability to cope with the ITRV is further enhanced by the occurrence of a comparatively high arc voltage immediately prior to current zero.

4.4.2 Short Line Faults (SLF)

Short-circuit faults occurring on a line length between 0.5 to 5 km are termed as short line faults or kilometric faults. A fault of this type imposes a highly onerous duty on the circuit breaker, thereby affecting its interrupting ability.

The basic phenomenon, is the double frequency being impressed on the circuit breaker and the difference of the source and the line side TRV. Both voltages start from instantaneous values at the position of the circuit breaker prior to the interruption. On the supply side, the voltage will oscillate at the supply frequency ultimately approaching the open circuit voltage. On the line side of the circuit breaker, the trapped charges, after interruption, initiate travelling waves on the transmission line, and these waves impose a saw tooth waveform at the circuit interruption. Since there is no driving voltage on the driving side, the voltage ultimately becomes zero because of the line losses. The network diagram, and the line side and supply side voltage wave forms are shown in Fig. 4.13.

Even though the magnitude of the voltage oscillation is considerably lower than that of the system voltage, the frequency is very high. This means that the RRRV across the circuit breaker can be much higher than the terminal short-circuit cases. Even though the short-circuit current is quite less, the high RRRV may cause the interrupter to fail. Thus, for an EHV circuit breaker, which is sensitive to RRRV, there may be a critical line length, which imposes the most severe duty on the breaker.

Very high RRRV of up to 10 kV/micro-second have been encountered during the SLF. Shorter line length gives a higher rate in rise of recovery voltage but the voltage peaks become so low that the circuit breaker clears easily. Longer length gives higher voltage peaks but the RRRV is much lower.

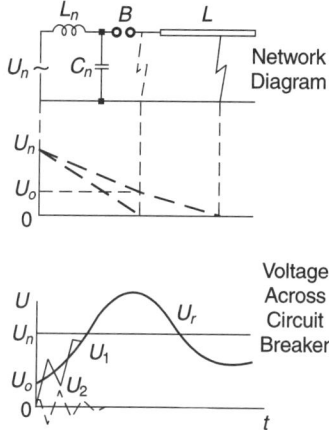

Fig. 4.13: Short line faults—basic phenomena

As in the case of the terminal fault, the use of the damper resistors reduces the RRRV. Since the SLF represents a case with very high RRRV, it is quite natural that the air blast circuit breaker will face difficulties with the interruption. Low ohm resistors are thus usually necessary.

SF6 circuit breakers are highly sensitive to the high voltage rates of rise during the first few micro-seconds of the current zero. SF6 gas breakers can easily handle the high RRRV in case of SLF duty, as the di-electric strength of the gas is very high.

4.4.3 Transformer Magnetizing and Reactor Currents

The transformer no-load current and the reactor current impose severe stresses on the circuit breaker upon their interruption because of the low magnitude of these inductive currents. In the case of externally generated pressure there is a tendency for the current to be forced to zero before its natural zero. This phenomenon is termed as current chopping. This sudden interruption of current leads to severe over-voltages on the transformer side of the breaker. These over-voltages result from the electromagnetic energy stored in the transformer, which is converted into electrostatic energy. Depending on the di-electric strength of the gap, re-ignitions or re-strikes may occur before the maximum voltage can be reached. As soon as the arc is re-struck it is immediately chopped again, but as the current has decreased during this time, a lower prospective voltage will be induced. Over-voltages, of the order of 2.5 to 3.5 times the rated voltages, are encountered in normal practice. The basic phenomenon of current chopping is shown in Fig. 4.14.

The interruption of the reactor current also has the same effect except that the severity on the breaker is more compared to magnetising current. This is so because in breaking the transformer current, only part of stored energy is released due to the hysteresis and other losses, but in the case of reactor current, the complete energy is released. In turn, the over-voltages generated in reactor switching are higher.

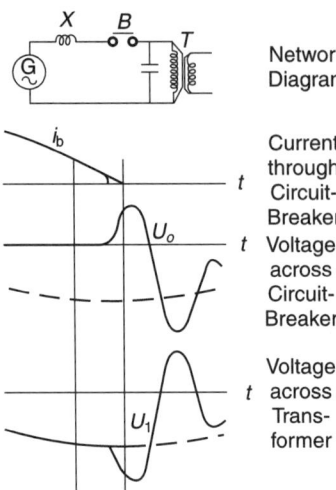

Fig. 4.14: Disconnection of unloaded transformers—basic phenomena

Severity of over-voltages is less in the case of a self-generated pressure circuit breaker (the pressure generated is proportional to the current interrupted), compared with the externally generated pressure breaker (the pressure is independent of the current).

In order to minimise the over-voltages on disconnection of transformer and reactor currents, switching resistors on the circuit breakers or surge arresters are provided. Switching resistors provide damping in the arc circuit thus reducing the tendency of the breaker to chop the current. They also provide a discharge route for the transformer energy after interruption. High voltage transformers are fitted with surge arresters for protection against switching over-voltages.

The over-voltages produced due to current chopping are high in gas blast breakers, as the gas blast is normally dimensioned for heavy short-circuit currents. Minimum oil circuit breakers on other hand, generate their own pressure, the magnitude of which is related to the magnitude of the current interrupted.

Current chopping is associated with the switching of reactors, minimum oil circuit breakers, pressurised breaker units, and puffer type SF6 circuit breakers. However, field and laboratory tests conducted on such circuit breakers, without any special means for surge reduction, have shown voltages much less than 1.5 times the rated voltages.

4.4.4 Energisation of Long Transmission Lines

When an initially uncharged, unloaded long line is charged, then over-voltages to the order of twice the normal voltage are impressed, imposing an onerous duty on the power system. Higher voltages can occur if the line has trapped charges. When a line is switched on to an energised network, a voltage wave is impressed on it. The impressed wave will be reflected at the far end of the line, and if this is open, the voltage will increase two-fold. In the case of the three-phase circuit, the situation becomes more severe if the circuit breaker poles do not close simultaneously. The wave on one phase will produce an induced wave on the other phase.

The highest over-voltages occur at the open end of the unloaded line when this is being energised or re-energised. Tests have confirmed that these voltages can exceed even 3 p.u., when no special means to reduce them are used. With the help of pre-insertion resistors, the switching voltages can be reduced to less than 2 p.u.

The best method of reducing the over-voltages generated on the energisation of the lines is to equip the circuit breaker with pre-insertion resistors (see Fig. 4.15), to ensure that the closing takes place in two stages. In the first stage, the resistor is included in series with the line to damp the waves. In the second stage, the resistor is short-circuited. The optimum value of the pre-insertion resistance is usually of the same order of magnitude as that of the surge impedance of the line and the insertion time should be at least 10 milli-seconds.

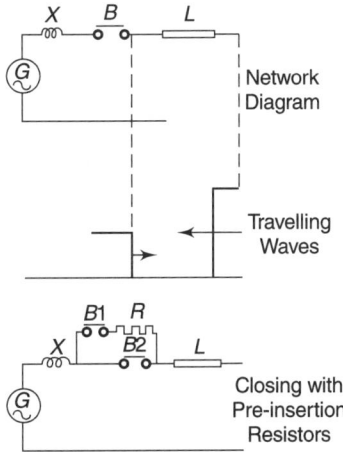

Fig. 4.15: Energisation of unloaded lines—basic phenomena

The switching over-voltages occurring with energisation and re-energisation of the lines are caused by the system phenomenon and are thus independent of the circuit breaker used. Other means like the controlled synchronous closing of the circuit breaker poles can also be used.

4.4.5 Switching of Unloaded Transmission Lines and Capacitor Banks

The disconnection of long unloaded line involves the interruption of large capacitive current. Interruption takes place at the natural zero of the current, i.e. when the voltage is at its maximum. After interruption, a direct voltage remains on the line. This disappears slowly, whereas the voltage on the supply side of the circuit breaker continues to vary with the system frequency. After a half cycle of this, the voltage across the circuit breaker may become twice the phase voltage. If the circuit breaker re-strikes at this instant, (see Fig. 4.16) the line will be discharged across the network inductance. In this manner, an oscillation will arise with a fundamental frequency. If the discharge current is interrupted at its first current zero, a direct voltage, now with opposite polarity and with twice the amplitude as before will remain on the line. After a further half cycle, the voltage across the circuit breaker may increase to three times the phase voltage with the risk of another re-strike and further escalation of the voltage on the line.

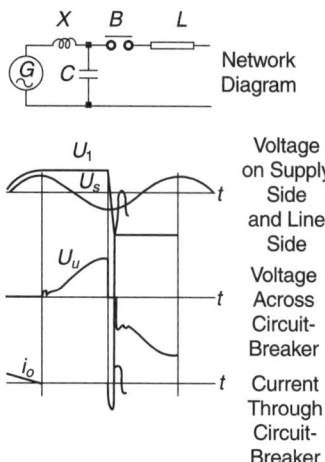

Fig. 4.16: Disconnection of unloaded lines—basic phenomena

Circuit breakers with externally generated pressure are practically re-strike free when interrupting capacitive currents. An air blast circuit breaker is a good example. On the other hand, the circuit breakers with self-generated pressure like bulk oil and minimum oil circuit breakers might re-strike, resulting in over-voltages. The reasons being that the capacitive currents are of small magnitude, generate low pressure to make the breaker re-strike free. This problem can be solved by slightly pressurising the interrupting units with nitrogen thereby rendering them re-strike free when dealing with capacitive currents. SF6 gas puffer circuit breakers have a low probability of restrike in case of line charging application.

4.4.6 Out of Phase Switching

An important requirement to be considered for the circuit breaker regarding interconnection between two different generating systems, is the ability of the circuit breaker to open the circuit when the two systems fall out of synchronism.

If the circuit breakers are rated for the total short-circuit power available at the busbar, the maximum short-circuit current in the tie between the systems under fully out of phase conditions can reach a value of 50 per cent of the rated short-circuit current of the circuit breaker.

For the circuit breaker ($B1$) in station 1, (see Fig. 4.17) this case occurs if $X2 + XL = X1$, i.e. if the short-circuit power transmitted from network 2 equals the short-circuit power of network 1. However this is not a realistic case, because the transmission line in most cases will considerably reduce the power from network 2 as also the current. Studies carried out in different countries show that the current under out of phase conditions never exceed 25 per cent of the rated short-circuit current of the circuit breaker. Hence the relevant IEC standard recommends testing at 25 per cent of the rated short-circuit current.

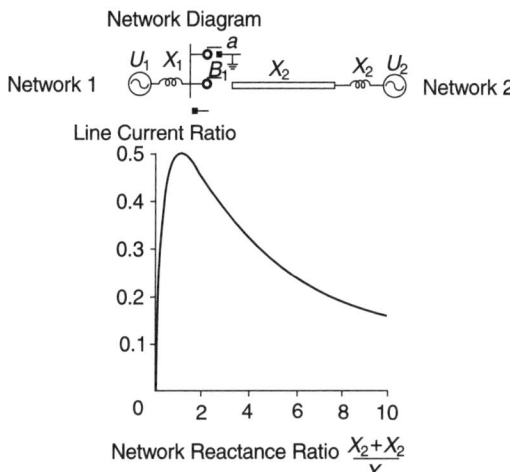

Fig. 4.17: Out of phase switching—basic phenomena

The recovery voltage across a single pole of the circuit breaker can be assumed to be twice the rated phase to earth voltage, if the system neutral is directly earthed, and 2.5 times the phase voltage if the system neutral is not earthed. These values apply to out of phase switching without any prevailing fault on the system.

Out of phase switching must be avoided as far as possible in systems and this can be achieved by proper relaying. Circuit breakers are normally tested for out of phase conditions.

Chapter 5

GAS INSULATED SUB-STATION/SWITCHGEAR (GIS)

Dr. H.S. Jain

Air insulated power transmission and distribution sub-stations suffer variations in the dielectric capability of air to withstand varying ambient conditions and deterioration of the exposed components due to oxidisation and the corrosive nature of the environment. The size of the sub-station is also substantial due to the poor dielectric strength of air. In order to enhance the life and reliability of a power transmission and distribution sub-station, it is desirable to protect the sub-station components from a corrosive and oxidising environment. Metal encapsulation of the sub-station elements provides a simple and effective solution to the problem of durability of the sub-stations. The use of a bus duct, with pressurised nitrogen gas, is a good example of devices with metal encapsulation used in power sub-stations. The size of the container is a direct function of the dielectric strength of the insulating medium. The container/enclosure sizes are thus large with a poor insulation like air or nitrogen. The use of a gaseous medium with higher dielectric strength like sulphur hexafluoride (SF6) instead of air helps in manifold reduction in the size of the sub-station component. The grounded metal encapsulation, on the other hand, makes the equipment safe, as the live components are no longer within the reach of the operator. The electric field intensity, at the enclosure surface, is reduced to zero as the enclosure is solidly grounded. Using this design philosophy, sub-station/switchyard equipment, like circuit breakers, disconnectors, earth switches, busbars and instrument transformers (both current and voltage), have been metal-encapsulated or metal-enclosed and pressurised with SF6 since 1968. The assembly of such equipment at a sub-station is defined as Gas Insulated and Metal Enclosed System (GIMES) by the International Electrotechnical Commission (IEC). The equipment is popularly known as a Gas Insulated Sub-station (GIS) system. The term GIS is also sometimes used to refer to Gas Insulated Switchgear.

5.2 Handbook of Switchgears

Medium voltage GIS equipment features vacuum as the interrupting medium and SF6 gas as the main insulation. Designs using SF6 medium for both insulation and interruption are also available. Two operating pressures are specified for such equipment (one for insulation and the other for interruption). The high and extra-high voltage GIS are essentially two-pressure systems. They are used in high and extra high voltage classes where vacuum interrupters are not available.

Metal encapsulation and SF6 gas insulation of the live high voltage sub-station components in a GIS result in reduced space requirement, to one-sixth (~15 per cent), as compared to a conventional air insulated yard sub-station. The size is reduced to about 8 per cent for a higher kV class GIS.

The size and weight of the higher kV class GIS equipment increases marginally over the 145 kV class GIS equipment, while the floor loading is limited to about 1000 kg/sq.m. This characteristic of the GIS helps to upgrade the sub-station vertically in future. The saving in space consequently results in less civil work and reduction in the installation time for an EHV sub-station with GIS equipment.

GIS are available internationally, covering the complete voltage range from 11 kV to 800 kV. The thermal current-carrying capacities and the fault-withstanding capabilities are tailored to meet all the sub-station requirements. More than 100,000 GIS bays have been in service all over the world since the introduction of such sub-station systems in the transmission and distribution field.

5.1 DESIGN

5.1.1 Single Line Diagram

A sub-station consists of many sections/bays. The main equipment in a section consists of circuit breakers, isolators or disconnectors, earth switches, current transformers, surge arrestors, etc. Figure 5.1 shows a single line diagram of a section at a sub-station identifying different compo-

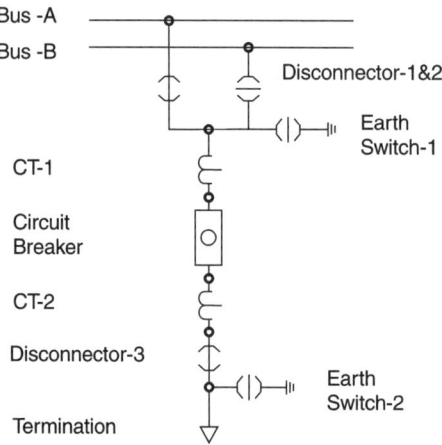

Fig. 5.1: Single line diagram for a double bus section

nents. Single busbar, double busbar and 3/2 circuit breaker are popular configurations at sub-stations.

5.2 BAY INFORMATION

In GIS, the modular components are assembled together to form a desired arrangement for a section or a bay. Figure 5.2 shows a cross-section of a double bus GIS section. Here, the constituent components are assembled side by side. The porcelains and connections (ACSR conductors), as required in a yard sub-station, are totally eliminated in this new configuration. The high voltage conductors (bus bars) are supported on simple disc insulators.

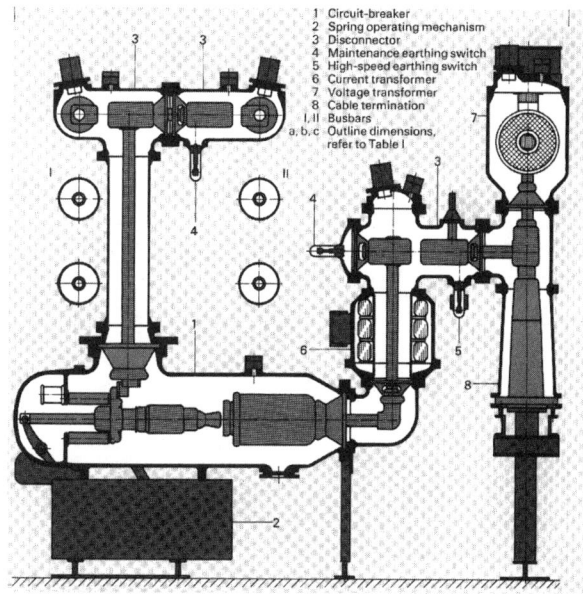

Fig. 5.2: Cross-section of a double bus GIS section

The connections to the overhead lines or underground cables are arranged through necessary air-to-gas or gas-to-cable terminations. The gas insulated instrument voltage transformer and the surge arrestor are installed directly on the gas insulated bus, using openings provided in the GIS for this purpose.

5.3 CLASSIFICATION OF GIMES

Gas Insulated Metal-enclosed Sub-station systems are classified according to the type of modules or the configuration.

The following configurations have been evolved over the years and are generally used:
- Isolated-phase (segregated phase) module
- Three-phase common modules
- Hybrid modules
- Compact modules
- Highly integrated systems

The isolated-phase GIS module consists of an assembly of individual circuit elements like a pole of a circuit breaker, a single pole disconnector, one-phase assembly of a current transformer, etc. A single-phase circuit is formed by using individual components and pressurising the elements with gas forming a leak-free gas circuit. Three such circuits, arranged side by side, form a complete three-phase GIS bay. The circuits, since assembled individually, require larger bay width (see Fig. 5.3) as compared to the other GIS configurations.

Item	Isolated Phase Type	Main Busbar Three-Phase Common Type	All Three-Phase Common Type	Compact Type
Arrangement	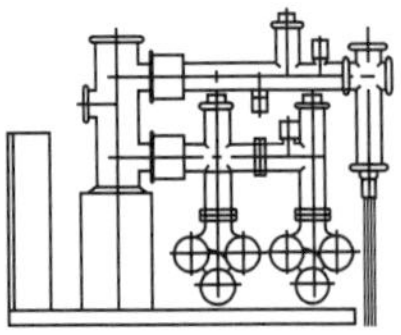 ← Length →	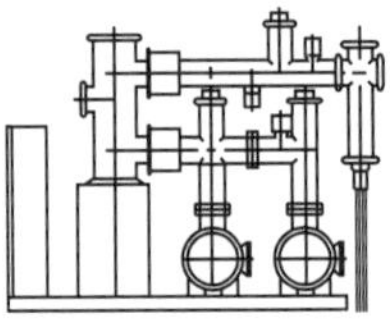	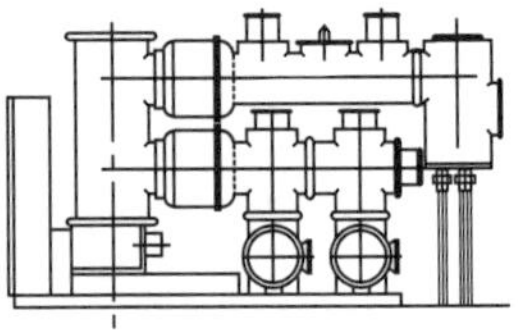	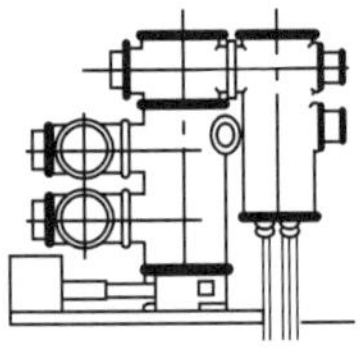
Bay Width (m)	2.5	2.5	1.4	1.4
Length (m)	4.8	4.6	6.0	4.0
Space (m³)	12.0 (100%)	11.5 (96%)	8.4 (70%)	5.6 (47%)

Fig. 5.3: Types of GIS arrangements/configurations

Designers have successfully encapsulated all the three-phase elements of individual modules, like three poles of a circuit breaker, three poles (isolations) in a disconnector or three-phase current transformer assemblies, while utilising excellent dielectric properties of SF6 gas in an individual enclosure, forming a three-phase module for the element. In the three-phase common module style, a three-phase bay is assembled using the desired numbers of three-phase elements. The total number of the enclosures is thus, reduced to one-third. The bay width is controlled (see Fig. 5.3) by the dimension of the largest three-phase module in such an assembly. Reduction up to 70 per cent in the sub-station floor area is possible by using three-phase modules.

In hybrid systems, a suitable combination of isolated-phase and three-phase common elements is used like three-phase busbar and single-phase elements, to achieve an optimal techno-commercial solution. While the three-phase common busbar system simplifies the connections from the busbar, the isolated-phase equipment prevents phase-to-phase faults in active modules like the circuit breaker. Savings in terms of space vary with the design and configuration of the section.

Hybrid GIS technology has gained popularity, specially in the medium and high voltage range, where advancements in technology have helped to reduce the sub-station size. With hybrid design, it is possible to construct techno-economical sub-stations providing additional flexibility for maintaining and expanding sub-stations in future at lower costs. In compact systems, there is flexibility of horizontal expansion (same voltage class), while replacement of the total equipment is necessary for vertical expansion (higher voltage class).

Compact GIS systems are essentially three-phase common systems, with more than one functional element in one enclosure. A single enclosure, housing a three-phase circuit breaker, current transformer, and earth switches, supports the busbar and the other feeder elements (see Fig. 5.3). The depth of the section is considerably reduced in this configuration as compared to the three-phase modules. A total reduction of up to 47 per cent in the equipment area, is possible by using this configuration.

Highly Integrated Systems (HIS), introduced in the year 2000, are single unit metal encapsulated and gas insulated sub-stations (Fig. 5.4), and are gaining user appreciation as this equipment provides a total sub-station solution for outdoor/yard sub-stations. The foundation work is limited to just one equipment, thus resulting in saving of substantial installation time.

Fig. 5.4: GIS in HIS configuration

These units are directly connected to the overhead lines. The incomer and feeder side connections (bushings) are directly mounted on the metal enclosure in this system. It is a ready-to-instal sub-station, with pre-defined circuit elements housed, sealed and pressurised in a single enclosure. A different version of one-unit GIS is also available as a Plug and Switch System (PASS). This variant can similarly be installed as a single unit, replacing the existing identical bay of a yard sub-station. A full sub-station can be built by multiplying these units.

5.4 DESIGN CONSIDERATION

Basically, tubular sections are used for high tension (HT) conductors in GIS equipment. The tubular enclosure, enveloping this HT conductor and the high pressure insulating gas, is grounded and maintained at earth potential. Disc insulators, in cone or funnel shape, are used to support the HT conductor in such a system. The metal enclosures (principally made from steel or aluminium material) and the current-carrying HT conductors (made from high conductivity copper or aluminium) are positioned concentrically. Other configurations, like rectangular cross-section enclosures, limiting the electrical stresses within safe working limits, are also used for GIS equipment of all voltage classes.

The surface profile of the HT conductors, enclosures and the insulators is carefully contoured to control the electrical stresses and to avoid charge concentration, ionisation and discharges. Two/three-dimensional (2-D and 3-D) models, using finite-difference, finite-element and charge simulation techniques, are used to model the electric field distribution of the GIS components, enclosures and high voltage conductor geometries. Commercial or in-house designed software is used for such studies. A 2-D field plot for a section of GIS is shown in Fig. 5.5. The field plot shows equi-potential lines at 5 per cent intervals, with the first line near the HT conductor, indicating a 95 per cent voltage contour. The model simulates a section of the bus bar, which is supported by an insulator and shielded at the terminal using a spherical shield. The simulation depicts electrical stresses in the insulation and on the surface of the insulator, and the maximum stresses in the simulation domain.

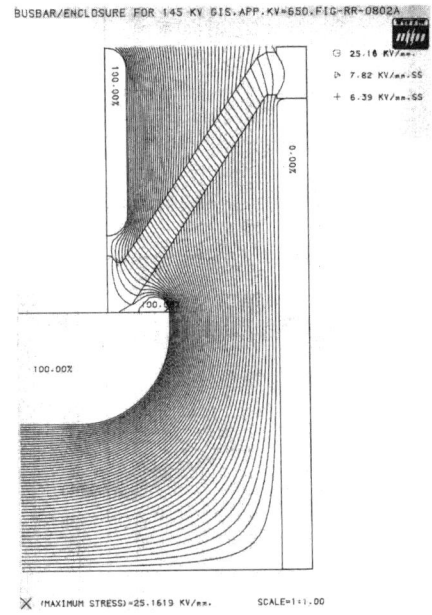

Fig. 5.5: Electrostatic field plot for a GIS busbar (with support insulator and end shield)

Apart from discharges and breakdowns caused by threshold voltage stressing, particle-initiated discharges are common to GIS. The discharges are initiated by the random movement of particles (both conducting and non-conducting) in high intensity fields. The fields are prevalent in an enclosed high voltage system similar to GIS for lower insulation gap between the HT and grounded conductors. The particle-initiated discharges are controlled in GIS by effectively trapping the discharge initiating particles in particle traps. The traps are located close to devices prone to the generation of particles. Two types of particle traps, namely 'passive' and 'active', are commonly used in GIS. Passive particle traps use gravity and adhesion to trap the particles, while the active traps use electrical fields to attract and retain the particles. The probability of particle-initiated discharges can also be reduced by:

- Using clean/particle-free assembly area (clean rooms class 10,000 or better) for the assembly of GIS as this helps to contain the number of particles in a new system prior to commissioning;
- Using designs featuring controlled/zero generation of particles during service; and
- Surface treatment of HT conductor and enclosures to get rid of loosely held particles.

Electro-polished surfaces are preferred for GIS components as this chemical process removes loosely held particles and blunts sharp protrusions generated during machining operations.

5.5 COMPONENTS/MODULES

The following are the principle gas insulated modules for a sub-station:

1. Busbar;
2. Disconnector or isolator;
3. Circuit breaker;
4. Current transformer; and
5. Earth switch.

The auxiliary gas insulated module or accessories, excluding control panel, that are required to complete a sub-station are:

1. Terminations;
2. Instrument voltage transformer; and
3. Surge and lightning arrestor.

5.5.1 Busbar

The busbar is one of the most elementary components of the GIS system. Co-axial busbars are common in isolated-phase GIS as this configuration results in an optimal stress distribution. Busbars of different lengths are used in GIS to cater to the requirement of circuit or the bay formation.

The high voltage conductor (copper/aluminium) is centrally placed in a tubular metal enclosure. The conductor is supported, at a uniform distance, by the disc or post insulator to maintain concentricity. Two sections of bus are joined by using plug-in connecting elements. Various sizes of the bus enclosures are shown in Fig. 5.6.

Fig. 5.6: Metal enclosures for GIS busbars

5.5.1.1 Connectors

The high voltage and high current electrical connections from one module to another in a gas insulated sub-station system are carried out with the help of the following two types of spring loaded plug-in contacts:

1. Spring loaded fingers or bridge contacts; and
2. Multi-lam contacts.

These two plug-in contact systems impart the maximum flexibility during assembly and dismantling. Both these contacts offer plug-in features and are suitable for tubular conductors. The connections made are reliable without the need for any additional hardware to secure their location.

The bridge contacts (see Fig. 5.7) use the conventional spring-loaded, silver-plated forged copper contact fingers as the current carriers. The forged copper fingers and the non-magnetic springs are housed on the inner periphery of a metallic casing. The contact points are pre-loaded with springs (helical or leaf) placed between the contact and the casing. The pre-loaded springs provide the requisite contact pressure under normal and dynamic fault conditions.

The multi-lam contacts are made from copper-beryllium sheet (0.15 to 0.4 mm thick). These contacts are available in the form of a ribbon on which several such contacts are pre-formed. The total thermal current is conducted in parallel paths, formed by each contact (see Fig. 5.8). This division of current helps in weakening the electro-dynamic repulsive force and in improving the short time peak-current rating of the system. The stiffness of the material is used here to replace

Fig. 5.7: Spring-loaded connector for GIS busbar

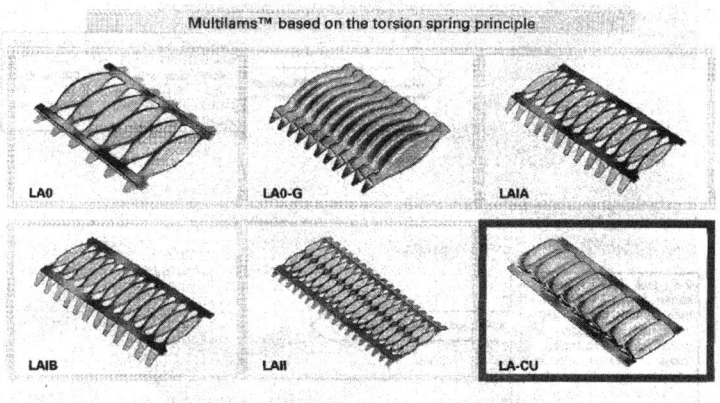

Fig. 5.8: Multi-lam contact system

springs in the conventional system. The contact is retained in a twisted form in the recess between two parent conductors (the fixed and the removable contact) in such a way that it exerts sufficient contact pressure. The physical location of the contacts on the busbar is secured mechanically by using a special groove and stainless steel wire loops. The contacts are electro-plated with silver to reduce both sliding friction and contact resistance. The multi-lam contacts are characterised by excellent electrical conductivity, high continuous current-carrying capacity, high short-circuit current-carrying capacity and low contact resistance. The multi-lam contact provides large radial tolerance and angular misalignment. The contacts are assembled easily.

A T-joint (metal casting) used for tapping a high voltage connection in GIS is shown in Fig. 5.9 The joint houses a pair of multi-lam contacts in the cavity provided for connection to the new element.

The size and diameter of the HT conductor in an isolated-phase system are governed by the ratio of the conductor and the enclosure (maintained around 1/2.82), while the thermal current

Fig. 5.9: T-joint for a GIS system

rating controls the section of this HT conductor. Tubular sections provide a large surface area for heat transfer and improve the heat dissipation rate to the gas.

5.5.1.2 Insulating Materials and Insulators

Insulating materials like sheet moulding compound (SMC), dow moulding compound (DMC), glass fibre reinforced plastics, compression and thermo-setting plastics, and refractory-based insulating materials like cordrite and alumina are commonly used in low tension (LT) and air insulated sub-station applications. Of these insulations, glass/silica-based systems are generally found unsuitable for SF6 applications due to their weak resistance to hydrofluoric acid (a by-product of moisture and decomposed SF6). Large shrinkage and instability at higher working temperatures prohibit the use of plastics in GIS. Stable polymers like PTFE (poly tetra fluoroethane) are selectively used in GIS and associated accessories. Ceramic and high-alumina ceramics are also used in GIS as solid insulation materials between the live conductor and the enclosures. However, their poor mechanical and thermal shock-withstanding capabilities and difficult processing and manufacturing cycles have limited the use of ceramics in GIS.

Insulating materials like PTFE (teflon) with very high volume resistivity retain electrical charges for long durations. This material property is sometimes undesirable and causes a deterioration in the performance of GIS (critically for direct current applications). The stagnation of charge locally modifies local potential and the electrical field. The electrical stresses in the system thus get modified unpredictably from the designed values. In an ac system, this trapped charge concentration also varies with time and adversely affects the electric field intensities. The use of materials promoting charge concentration is thus avoided in gas insulated systems.

Alumina-filled epoxy matrix is a common insulating material for GIS-related applications. The filler alumina offers good resistance to decomposed SF6 products like hydrofluoric acid (HF) as compared to silica or felspar (common fillers used with epoxy). The matrix of epoxy resin and filler, in liquid form, is readily cast in metal moulds at moderate temperatures and is gelled quickly at increased temperatures to reduce the component manufacturing cycle. The components are post-cured for obtaining good mechanical characteristics in hot air ovens.

Disc and post are the preferred shapes and types of support insulators for GIS. The disc insulators are conical in shape and are also known as funnel insulators. The disc insulators are further sub-divided into:

- Communicating; and
- Non-communicating insulators.

The two types of disc insulators are shown in Fig. 5.10 and Fig. 5.11, respectively. The communicating insulators help in establishing gas continuity between the two adjacent enclosures, while in situations where the gas communication between two sections is undesirable, a non-communicating insulator is used. Solid core epoxy insulators are commonly used to support busbars in three-phase common GIS systems. The drive insulators are fibre-reinforced epoxy and filled epoxy systems.

Fig. 5.10: Support insulator for GIS (communicating)

Fig. 5.11: Support insulator for GIS (non-communicating)

The post insulators are cylindrical or oval in shape. The surface of the insulator is smooth for minimising particle attachment. A few manufacturers also use shaded post insulators with small projections/shades.

Rib insulator, a variant of the post insulator, combines two post insulators in one. This insulator supports a tubular bus conductor and provides additional rigidity to the conductor. Since it is anchored at two ends to the enclosure, the insulator is lightly stressed during faults. Bus systems with such insulators provide better fault withstanding capabilities. Figure 5.12 shows the view of a 145 kV, three-phase bus with rib insulators. The equipment uses two rib insulators per phase for supporting the individual busbars. A three-phase disc insulator, Fig. 5.13, is used to support a three-phase busbar and to limit the pressure-rise to the affected enclosure in the conditions of an arc fault.

Fig. 5.12: Three-phase bus module with rib insulators

Fig. 5.13: Support insulator for three-phase GIS

5.6 DISCONNECTORS (ISOLATORS)

Isolators are placed in series with the circuit breaker to provide additional protection and physical isolation. In a circuit, two isolators are generally used, one on the line side and the other on the feeder side. Isolators are designed for the interruption of small currents, induced or capacitively coupled. The isolators can be motorised or driven manually. In GIS systems, motorised isolators are preferred.

A pair of fixed contacts and a moving contact form the active parts of an isolator. The fixed contacts are separated by an isolating gas gap. During the closing operation, this gap is bridged by the moving contact. The moving contact is attached to a suitable drive, which imparts the desired linear displacement to the moving contact at a pre-determined design speed. A firm contact is established between the two contacts with the help of spring-loaded fingers or the multi-lam contacts. The isolation gap is designed for the voltage class of the isolator and the safe dielectric strength of the gas. Figure 5.14 shows a cross-section of an isolated-phase GIS isolator.

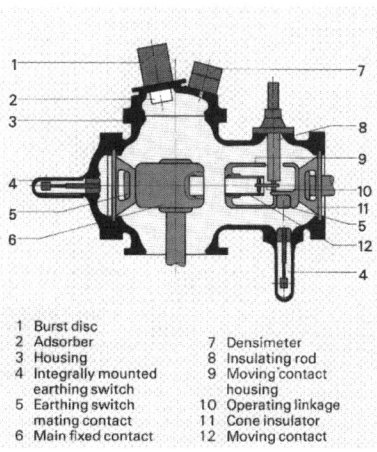

1. Burst disc
2. Adsorber
3. Housing
4. Integrally mounted earthing switch
5. Earthing switch mating contact
6. Main fixed contact
7. Densimeter
8. Insulating rod
9. Moving contact housing
10. Operating linkage
11. Cone insulator
12. Moving contact

Fig. 5.14: Cross-section of an isolated-phase GIS isolator

An insulator is used to drive the moving contact and to isolate the drive from the high voltage components of the isolator. The shape and size of the insulator are controlled by the electrical and mechanical requirements of the isolator. In three-phase ac systems, the individual phase isolators are ganged together to operate simultaneously. Leak-tight rotary seals are used in gas insulated isolators for transferring motion from external drive to the gas. Isolators in high voltage GIS operate at SF6 pressures of 0.38 MPa to 0.45 MPa. The operating speed of the isolator moving contact ranges from 0.1 to 0.3 m/sec.

The design of electrostatic shields on two fixed contacts and the earth side of the drive insulator plays an important role in ensuring the satisfactory performance of a gas insulated isolator.

5.7 CIRCUIT BREAKER

The circuit breaker is the most critical part of a gas insulated sub-station system. The circuit breaker in a gas insulated system is metal-clad and utilises SF6 gas, both for insulation and fault interruption. The SF6 gas pressure in a circuit breaker is around 0.65 MPa. The circuit breaker is directly connected to either current transformers or the isolators in gas. A barrier is maintained between the circuit breaker and the other connected equipment, operating at lower gas pressure, to maintain a pressure difference.

Puffer SF6 circuit breakers are commonly used to accomplish fault current interruption in gas insulated sub-station systems. In three-phase common modules of circuit breakers, hot gas mix-up is checked to prevent inter-phase short-circuit by electrically conducting hot gas. Spring, spring-hydraulic and pure hydraulic are the preferred drives for the circuit breakers of gas insulated sub-stations. Hydraulic drives are reliable, robust and compact as compared to their spring counterparts. Hydraulic drives can be interfaced to the circuit breaker directly without any intermediate motion seals and linkages. The spring drives are relatively cheaper and can be used only with the state-of-the-art self-blast or hybrid circuit breakers. Opening speeds in the range of 6.0–8.0 m/sec and operating energies in the range of 4500–8500 Nm are common for operating the GIS circuit breakers.

As a safety device, the circuit breaker enclosure features a rupture diaphragm or a spring-loaded plate valve. This arrangement vents high pressure gas, if it is above proof pressure, during extensive arcing or pressure build-up for some reason in the circuit breaker enclosure. The circuit breaker enclosure also serves as the main support element for the individual GIS bay. The GIS circuit breakers are oriented both in horizontal and vertical configurations, depending on the system requirements and ease of installation. A cross-section of a GIS circuit breaker is given in Fig. 5.15.

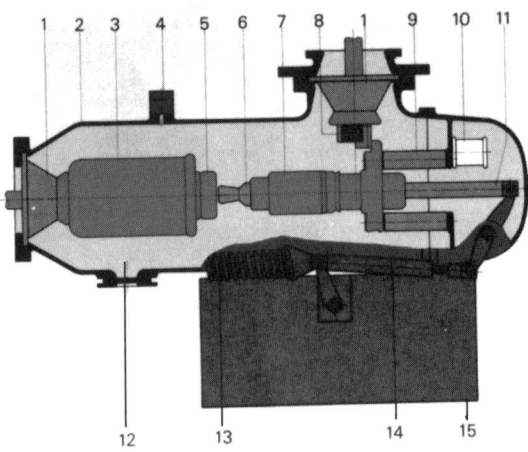

Fig. 5.15: Cross-section of a GIS circuit breaker

5.8 CURRENT TRANSFORMER

The conventional sub-stations use either live-tank or dead-tank type current transformers with oil/SF6 insulation. A porcelain insulator is used to insulate the low potential section of the current transformer from the high voltage zone. Ribbon or cut silicon steel cores are used for the magnetic circuit of the current transformer for obtaining the desired ratio and accuracy. Hairpin shaped primary conductor is the standard geometry for a dead-tank type current transformer.

The current transformers in gas insulated systems are essentially in-line current transformers. Gas insulated current transformers, with classical coaxial geometry, consist of the following parts: the tubular primary conductor; an electrostatic shield; ribbon-wound toroidal core and the gas-tight enclosure. The primary of a current transformer is a tubular metal conductor linking two gas insulated modules, placed on either side of the current transformer. Disc insulators, at either end of the current transformer enclosure, support this high voltage conductor. One end of the conductor end is solidly fastened, while the other end is provided with a sliding joint, which compensates for the thermal expansion of the conductor and simplifies the assembly of the current transformer module. A ribbon-wound silicon steel core (formed in toroidal shape) is used for the magnetic circuit of the current transformer. A coaxial electrostatic shield, at ground potential, is placed between the high voltage primary and the toroidal magnetic core of the current transformer for ensuring zero potential at the secondary of the current transformer. The electrostatic shield also helps in generating a perfect coaxial geometry and uniform electrical field in the gas gap. The secondary connections are accessed through a leak-tight terminal block or a circular connector, bolted to the current transformer enclosure by using appropriate seals and gaskets. The enclosure of a current transformer is designed to accommodate the desired number of current transformers and connectors. The length of the current transformer module thus changes with the number and types of current transformers specified. The magnetic core and the secondary winding assembly of the current transformer are supported in gas by an enclosure or a grounded support enveloping the core and the winding. The accuracy of gas insulated current transformers for their classical coaxial geometry is the highest as it meets all the standard (IEC-185) requirements.

5.9 EARTH SWITCH

Fast earth switch and maintenance earth switch are the two types of earth switches used for gas insulated sub-station systems. The maintenance earth switch is a slow device used to ground the high voltage conductors during maintenance schedules, in order to ensure the safety of the maintenance staff. The fast earth switch, on the other hand, is used to protect the circuit-connected instrument voltage transformer from core saturation caused by direct current flowing through its primary as a consequence of remnant charge (stored online during isolation/switching off of the line). In such a situation, the use of a fast earth switch provides a parallel (low resistance) path to drain the residual static charge quickly, thereby protecting the instrument voltage transformer from the damages that may otherwise be caused. The basic construction of these earth switches is identical.

The earth switch is the smallest module of a gas insulated sub-station system. The module is made up of two parts: a fixed contact, which is located at the live bus conductor and which forms a part of the main gas insulated system; and a moving contact system mounted on the enclosure of the main module and aligned to the fixed contact. The moving contact system is an assembly of the moving contact, current transfer contacts, a mechanical system to convert the linear motion of the moving contact to rotary motion and an enclosure or the housing. An insulator is also used between the earth switch enclosure and the main enclosure to insulate the two systems electrically. The insulated connection of an earth switch helps in the diagnostics of gas insulated sub-station systems during and after commissioning by providing an alternate current path for calibration of the current transformer, measurement of circuit element contact resistances, etc.

Different operating mechanisms are used to achieve the desired operational speeds. The maintenance earth switch uses a simple mechanical linkage for operation in contrast with a regular circuit breaker mechanism used for the fast earth switch. The important make operation in a fast earth switch is accomplished in less than 50 ms to achieve the desired protection of the instrument voltage transformer during line isolation. The fast earth switch returns to its original (OFF) position before the line is restored. The lines are grounded for extended duration, by the slow maintenance earth switch after the residual charge has been drained by a fast earth switch. Earth switches are specified by their capacity to make, time to make and the pre-arcing duration. IEC-129 specifies their qualification requirements.

5.10 ACCESSORIES

Incomer and feeder connections are the main accessories of a sub-station. At the incomer, the supply is received from a higher level sub-station or from a ring main. The power is received and delivered through either the underground cables or the overhead lines, at a sub-station. If they are economically viable, underground cables are also employed for other similar power installations. In either case, interfaces are required to receive/deliver the power. Cable-to-gas and Air-to-gas terminations are employed as an interface to the two media in GIS installations. Conventional as well as the dry terminations are now available for such applications up to 170 kV voltage class. Beyond this voltage level, conventional terminations, with capacitive foil grading and liquid insulation, are employed. For air-to-gas termination, the use of composite insulator for bushings has been gaining importance because they are light-weight, and offer better mechanical and seismic performances. Figure 5.16 shows a gas-to-air bushing featuring a composite insulator.

An instrument voltage/potential transformer, used for metering and protection, forms a part of the GIS and is gas insulated. This equipment is directly mounted and connected to GIS, at times with an isolator/disconnector in series. Both the single-phase and the three-phase instrument voltage transformers (IVTs) are available for voltages up to 170 kV. Single-phase IVTs are common for system voltages higher than 145 kV. Figure 5.17 shows a single-phase instrument voltage transformer for use in GIS.

A gas insulated surge arrestor is a critical accessory required for a sub-station. This device protects the system from switching surges. Surge arrestors are commonly used for installation above 170 kV class, where an appreciable switching surge intensity is recorded. In exceptional cases, lower kV class sub-stations are also equipped with surge arrestors to provide additional safety and reliability. The conventional yard surge/lighting arrestors are used for gas insulated sub-station systems, where overhead lines are used to source/deliver the energy.

Fig. 5.16: Gas-to-air termination with composite insulator housing

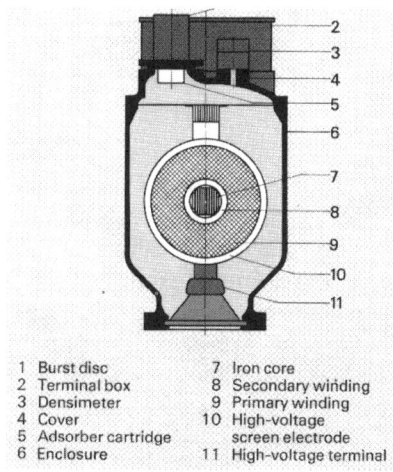

1 Burst disc	7 Iron core
2 Terminal box	8 Secondary winding
3 Densimeter	9 Primary winding
4 Cover	10 High-voltage
5 Adsorber cartridge	screen electrode
6 Enclosure	11 High-voltage terminal

Fig. 5.17: Gas insulated instrument voltage transformer

5.11 CONTROL PANEL

Both local and remote control panels are used in GIS. The local control panel (LCP) provides an access to the various controls and circuit parameters of an individual GIS bay. The local control panel facilitates the monitoring of gas pressures, status of the switchgear element and operating fluid pressures, of oil, SF6 and air. A dedicated local control panel for each bay is a common specification. The local control panel essentially features interlocks, operating buttons and a single line diagram. The panel has a swing panel and a clear glass door with padlocks. The operator can verify the status of the circuit through a glass panelled clear door, containing the mimicked single line diagram, indicators and push buttons. The circuit operations are possible only by the authentication and authorisation process based on physical issuance of the 'clear door key' by the concerned authority. The local control panel is capable of communicating digitally with the remote control panel, situated at another central location, through an IEEE-488 or RS-232 bus.

The number of remote control panels (RCPs) required depends on the choice of the utility. A single RCP is a must. A remote control panel, based on digital signal processing (DSP) and microcontroller (μc)/processor, can shrink the RCP to a single unit. A man machine interface (MMI), featuring bay-wise pages and many other similar functional integrations, is the best techno-commercial solution. Connectivity via information/data bus to the central load dispatch centre (LDC) or central controls enhances the utility of such an RCP. The RCP is sufficiently hardened against electrostatic and electromagnetic radiations (E and H-fields) to perform satisfactorily in harsh power switching and very fast transient over-voltage (VFTO) environments.

5.12 CONSTRUCTIONAL ASPECTS OF GIS

5.12.1 Enclosure

The enclosures for gas insulated sub-station equipment are fabricated using carbon steel and alloy steel, or are cast using aluminium. While corrosion-resistant stainless steel (SS-304 and SS-316) is the dominant enclosure material for isolated-phase equipment, cast aluminium enclosures are preferred as they are light-weight and entail low production costs.

Sharp protrusions and loosely/weakly held metal inclusions are checked and removed from the inner surfaces of the enclosures. While the inclusions are potential sources of conducting particles detrimental to the life of GIS, sharp protrusions form potential discharge locations. The inner surfaces of the enclosures are thus carefully examined and verified before the enclosures are permitted for use in GIS. Electro-polishing or anodising process is used to chemically blunt the sharp edges of the protrusions on the inner surface of the enclosure. In a few cases, inner surfaces are powder-coated to obtain a desirable surface finish and to bury inaccessible inclusions. The enclosures are sand/abrasive-grit blasted before powder coating to dislodge weakly held particles. In order to minimise/eliminate cast porosities in cast enclosures, vacuum/low-pressure die (LPD) casting techniques are preferred for GIS application. The enclosure, containing high-pressure gas are designed, manufactured and tested as per ASME pressure vessel code (Section 8) or as per IEC-517 recommendations.

The discharges in SF6 high voltage equipment are momentary and as the medium (SF6) is self-healing, the decomposed gas molecules re-combine in a majority of such cases. The full dielectric strength of the system is restored almost instantaneously. In some critical cases, where the discharge persists or re-strikes occur in a short span of time, a power follow-through current is likely to flow, causing a sustained discharge. Such discharges, dissipating an abnormally high amount of arc energy into the medium, anchor one of the arc roots at the metal enclosure. The dissipated energy raises the internal pressure of the system and the high current density at arc roots causes localised high temperatures that are sufficient to soften, melt and burn the enclosure metal at these locations. The burn-through of the enclosure is guarded in GIS and design provisions are made to prevent pressure rise and extended arcing.

Pressure rise, due to excessive dissipation of arc energy in the absence of burn-through, results in de-shaping of the enclosure. This is prevented by the use of spring-loaded pressure relief valves and rupture diaphragms. Rupture diaphragms are used as pressure fuses to prevent de-shaping of enclosures caused by excessive internal pressure rise. Brittle castings in aluminium or cast steel result in explosion instead of buckling and are tested for their ability to withstand higher pressure prior to use. The enclosures are tested for their reliable design by conducting a proof test, which is recommended at 2.3 times the operating pressure for fabricated enclosures, as per IEC-517. The cast enclosures are tested and certified for withstanding 3.5 times the operating pressure as per this standard.

5.12.2 Seals and Gaskets

Seals and gaskets are important components, which determine the insulation life and top-up frequency of a GIS. These critical components arrest leakage of the gaseous insulation in a pressurised installation, like a gas insulated sub-station system. O-rings with circular cross-sections and rectangular gaskets with rectangular cross-sections are the most common shapes of the pressure seals used in GIS. Single O-ring seals are common for indoor installation. Double O-ring seals or a single O-ring seal backed by a gasket is a common practice for outdoor installations. Elastomers/materials for GIS seals are carefully selected keeping in mind their application-related constraints. The following properties are considered before selecting the sealing materials for GIS:

- Resistance to decomposed SF6;
- Resistance to oil;
- Resistance to tension, compression and elongation;
- SF6 and moisture permeability;
- Environmental impact/ease of handling; and
- Service life.

Materials like EP rubber, nitrile rubber, silicon rubber and viton are common commercial elastomers. Nitrile and viton rubbers are used in GIS as they have the above properties. The compression distortion and the service temperature directly influence the service life of a sealing system. The compression distortion in the range of 10–12 per cent is considered optimal for a good sealing performance. A service temperature in the range of 20–35°C ensures a life expectancy of more than 100 years for this optimal compression. The life expectancy of such a sealing system can be calculated using empirical (like Arrhenius') equation.

Pre-loaded seals with PTFE base are used for rotary motion feed-through in GIS. Earth switches, disconnectors and circuit breakers are the GIS modules using such motion feed-through for rotary motion transfer. Proprietary seals like 'Dowty' with metal casing, are also used widely in GIS for gas circuit and gas pipeline joints for their features like quick and reliable joints.

5.12.3 Gas Circuit

The GIS enclosures are pressurised with SF6 gas at design pressures. The operating pressure for GIS varies from 0.1 MPa to 0.8 MPa. Different manufacturers recommend different operating pressures for their equipment. 0.1 MPa is a standard working pressure for vacuum interrupter-based medium voltage GIS systems. SF6 interrupter-based equipment, in medium voltage, uses pressures in the range of 0.25 MPa–0.45 MPa. The high voltage (>72.5 kV) gas insulated sub-station system modules operate at a gas pressure of 0.4 MPa; in these modules, the gas is principally used as an insulant. The GIS circuit breaker module is designed to operate at a gas pressure of 0.65 MPa or higher, as the gas is used in the module for both interruption as well as for insulation.

The enclosures are required to retain this insulant for the life of the equipment. Leaks, virtual or real, will allow leakage of the gas and reduction in the pressure or the density of the gas. Leak/loss of insulant directly causes a deterioration in the dielectric performance of the gaseous insulation. A one per cent (by volume) leak per annum per enclosure is permitted (IEC-517) for such systems. In order to ensure leak-tightness, the complete enclosure and seals are checked using the Mass Spectrometric Leak Detection (MSLD) system at manufacturing stage. An MSLD test, conducted with helium as the sniffing gas, can detect micro-leaks up to 5×10^{-10} mbarls^{-1} reliably. Simple pressure-drop with time has been widely used during assembly in the primary evaluation of a leak. Fluorine (a major constituent of SF6 gas) is a halogen; and therefore, a 'halogen sniffer' (Fig. 5.18) is reliably used to detect fine SF6 leaks (~5×10^{-4} mbarls^{-1}) on completed GIS assemblies at plant or at sites. These leak tests are also performed during the initial stages at the component level. Halogen leak detectors are sensitive to SF6 background and require an SF6-free ambience to operate satisfactorily.

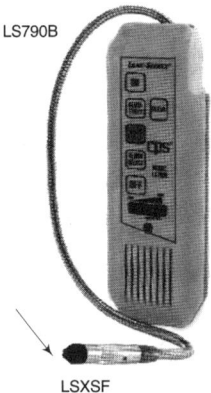

Fig. 5.18: Halogen leak detector (sniffer)

5.12.4 Expansion Joints

The copper and steel elements used in GIS, if assembled rigidly, are likely to change dimensions with a change in the ambient temperature. For an indoor sub-station, with a limited number of bays/sections, the expansion may not be such a critical issue. However, in major GIS systems, expansion and contraction of the metal-clad equipment with temperature is a critical issue that needs to be addressed seriously. Controlling sub-station temperature to contain the dimensional variation is a cost-effective and efficient solution for indoor GIS systems. In all the other cases of GIS layouts, the ambient temperature variations are considered during the design stage, thereby providing adequate means and implements for compensation for ambient temperature-related changes in the dimensions of the sub-station. Without such compensation, the sub-station equipment is likely to be stressed mechanically, causing damage to the enclosure and insulators. The sub-station life is likely to be reduced due to thermal cycling, fatigue, etc.

Expansion joint has been found to be an ideal solution for such duties. The addition of an expansion joint between two adjacent bays limits variation in dimension to the individual circuit, thereby maintaining the mechanical stability of the sub-station. The expansion joint also helps in the addition and removal of bays, if so required, during the life of a sub-station and makes such installations more flexible to changes in the circuit/configuration. These joints also provide margins for radial and axial non-alignment. Stainless steel flexible joints are the only alternatives for expansion joints in GIS, because of their flexibility, strength and thermal cycling capabilities. In indoor GIS, maintaining a constant sub-station temperature (minimum variation) controls the expansion and contraction of the metallic equipment. Care must be taken to add expansion joints in outdoor sub-stations, where the sub-station dimension changes due to a change in the ambient temperature during day and night, and retains robustness in the system. Figure 5.19 shows one such expansion joint used in a GIS.

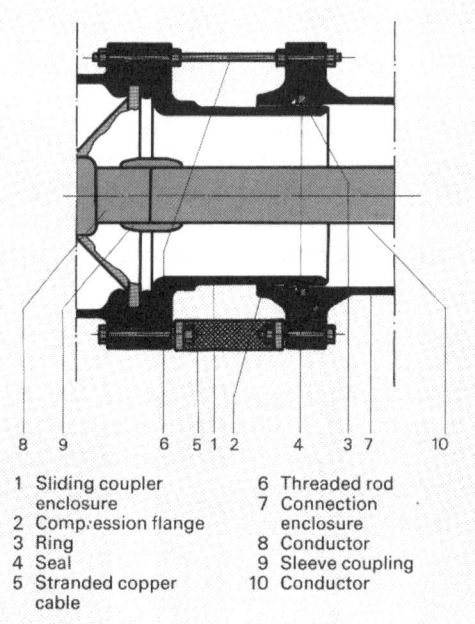

1 Sliding coupler enclosure
2 Compression flange
3 Ring
4 Seal
5 Stranded copper cable
6 Threaded rod
7 Connection enclosure
8 Conductor
9 Sleeve coupling
10 Conductor

Fig. 5.19: Expansion joint in GIS busbar

5.12.5 Current Transfer and Plug-in-Joints

In order to configure different circuit formations, gas insulated functional modules are connected in accordance with the desired circuit requirement or single line diagram (see Fig. 5.3). Various modules are connected suitably using bolted or plug-in joints. The use of copper flexible (stranded wire) connections is not permitted in GIS, since the strands are the potential sources of long metallic particles. A combination of solid bolted joints and plug-in joints is thus universal for GIS applications. Silver-plated interfaces are preferred for the bolted joints. The free sliding, plug-in-joints carry the full thermal current reliably and withstand large short time current efficiently. These joints are designed in a manner so as to contain the electrostatic and electrodynamic stresses to safe working limits. The geometry, material and the surface finish of these joints thus play a vital role in ensuring the satisfactory performance of a GIS. The following three types of plug-in-joints are generally used:

- Static
- Quasi-static
- Dynamic

The static joints are mostly plugged in only once at the time of the equipment assembly and are not disengaged throughout the life of the equipment. Quasi-static joints have flexibility for engagement and disengagement, which may be required as and when the equipment undergoes inspection, addition, etc. The dynamic flexible joints are used with functional elements like circuit breaker, disconnectors, etc., where the application demands dynamic current transfer during the normal operations of such equipments.

Static joints are designed with high insertion force and employ thick multi-lam contacts (> 0.25 mm thick). The quasi-static contacts use spring-loaded forged copper fingers housed in a casing. The entire assembly is plugged on to the tubular conductors to be bridged. The dynamic contacts are low insertion force, thin, multi-lam contact systems with a reduced current rating per contact. A sufficient number of contacts is assembled for this application to meet the thermal and short time current specifications of the equipment. A new variety of multi-lam contacts now available is simple to assemble and tolerant to wider dimensional variations and misalignments.

5.12.6 Support Structure

Gas insulated sub-station modules, when assembled to form a circuit configuration, assume the shape of a robust pipe structure with exceptionally good mechanical rigidity. In the initial stages of the installation, the main modules may require support for stability. These supports can be withdrawn after the installation is complete. Some permanent supports may also be necessary to support the terminal equipment. In isolated-phase construction, with vertical circuit breaker, the requirement to support the modules connected horizontally is always felt. The necessary support structure is thus designed for this configuration of GIS. Module assemblies mounted horizontally are suitably supported in GIS. Three-phase compact GIS configurations are self-supporting and the centre of gravity of the structure lies within the circuit breaker module. The circuit breaker module supports the rest of the bay equipment in such a configuration. Structures made of hollow rectangular sections or standard structural sections suffice for the purpose. In outdoor installations, sufficient care is taken to prevent corrosion of the support structure due to moisture and other pollutants.

5.13 INSTALLATION AND MAINTENANCE OF GIS

The weight and size of the GIS equipment do not change appreciably with the voltage class as the bulk of the current-carrying components and enclosures have identical dimensions for similar thermal and short time current. The additional insulation required for the next voltage class is achieved by increased gas density. Owing to these flexibilities, a few manufacturers offer the same equipment for two voltage classes (like 170/145 kV). Even when the GIS equipment is designed for an individual voltage class, the dimensions and weights of the equipments differ marginally. Table 5.1 shows the dimensions, weight and floor loading for three voltage classes of GIS.

1	Rated voltage	145 kV	170 kV	245 kV
2	Bay width, m	1.5	2.0	2.0
3	Bay depth, m	3.3	3.35	3.4
4	Bay height, m	3.2	3.4	3.4
5	Floor area, sq.m	4.95	6.7	6.8
6	Volume, m^3	15.84	22.78	23.12
7	Weight, kg	3800	5000	5700
8	Floor loading, kg/sq.m	765	750	840

Table 5.1 Size and Weight of GIS Equipment

It is clearly evident from Table 5.1 that the floor loading for GIS equipment is less than 1000 kg/sq.m and the floor area occupied by the bays is almost identical. The moderate floor loading of <1000 kg/sq.m suggests that the GIS equipment can be installed on any normal concrete floor with a safe bearing capacity of the above order. A 125–150 mm concrete slab, normally used for office buildings, meets this demand adequately. GIS systems have been installed indoors, at various floor levels, using their moderate floor loading requirements. The equipment/bay is anchored to the floor by using modular floor inserts and anchor bolts. Multi-point earth connections are recommended for GIS to reduce step and touch potentials. A 5-T overhead crane, installed before installation of the GIS equipment is employed to instal the GIS assemblies at an indoor site. A rail-mounted makeshift 'C-crane' is arranged for the installation of large outdoor GIS equipment.

GIS is a maintenance-free installation. Attempts to maintain GIS periodically/preventively have resulted in disasters, rendering the equipment useless. Good GIS installation over a period of time acquires better dielectric properties. Opening a clean GIS, at site, is likely to increase the otherwise low moisture content of the gas, insulator, etc., and add dangerous particles from the unclean site ambience. This may also additionally load the particle traps placed inside GIS. Maintaining gas pressure and density in the equipment along with monitoring of the performance and life of the actuators is a good maintenance practice for this equipment. The maintenance staff takes up periodic cleaning of particle traps, if so recommended.

5.14 LIFECYCLE COST OF GIS

The following costs are considered while planning a power sub-station:
- Cost of planning
- Cost of equipment
- Cost of protections and controls
- Cost of auxiliary cables
- Cost of civil works, including structures
- Cost of control room and building
- Maintenance cost

The total lifecycle cost (LCC) of a power sub-station with two types of equipment, air insulated and gas insulated, is discussed here.

The cost of planning an air insulated sub-station is based on previous data, survey reports for the site and the soil, and environmental conditions at site (including pollution, snow, rain, wind and seismic conditions). All these parameters help the planner specify the required electrical distances like creepage and mechanical strength for the porcelains and the base structures. Equipment foundations are designed to suit the soil load-bearing capacity. Planners also specify the cutting and filling required at site, control and power cable trenches and the earth mat design. Each of these principal activities is quite time-consuming which is why a lot of time is spent in planning an air insulated power sub-station. Since a gas insulated sub-station is smaller in size, it requires minimum land for its installation and consequently reduced site preparation. Since the equipment is immune to atmospheric variations and is installed indoors, it does not call for any special specifications even for the most stringent site. The low height of the equipment and its nested structure reduce the seismic hazards to a minimum.

Since the gas insulated sub-station is wired and dispatched as a single unit, it needs to be specified for one composite equipment and not for the constituent equipments of the circuit. This also eliminates the need for sourcing different equipment from different manufacturers and for large-scale paper work during the planning stage. The planning time is thus drastically reduced in case of gas insulated sub-stations as compared to air insulated sub-stations.

Since gas insulated equipment uses new efficient technology and is sophisticated in nature, its cost is almost twice that of the conventional equipment. The equipment cost also varies with the import content of the sub-station equipment and the duty structure prevailing at the time of its procurement.

Protections and controls for the two sub-stations are identical and thus the costs are also comparable. The auxiliary control cable lengths are evidently longer in case of an air insulated sub-station because of its larger size and substantial distance between the control room and the equipment.

Civil works, including site preparation, earth mat, trenches, foundations for equipment and the structures, supply lines for power, water and air, etc., are quite elaborate and expensive both in terms of time and finances in case of air insulated sub-stations as compared to gas insulated sub-stations. GIS systems require limited civil activity at site, eliminating the need for a majority of steel structures, fencing, and earth mats, and are serviced by low capacity power, water and air circuits. A saving of up to 75 per cent in the cost of civil works is possible with gas insulated sub-stations. It is possible to house equivalent GIS equipment in a building with dimensions similar to the control room of an air insulated sub-station. The cost of building for indoor GIS systems includes an EOT crane for unloading and installation of the sub-station equipment indoors.

As discussed earlier, the cost of maintaining the gas insulated sub-stations is practically nil because of their technical features and sealed construction. The air insulated sub-stations, on the other hand, need to be maintained periodically and require frequent replacement of oxidised and corroded parts. It is thus essential to recruit sub-station maintenance staff with an inventory of the consumables for the life of the sub-station. The salaries of the staff and the expenses on inventory are quantified for the life of the sub-station (30 years).

Fig. 5.20 compares lifecycle costs graphically for air and gas insulated sub-stations. Although it seems costly at first glance, gas insulated sub-station equipment is a cost-effective long-term solution for power sub-stations.

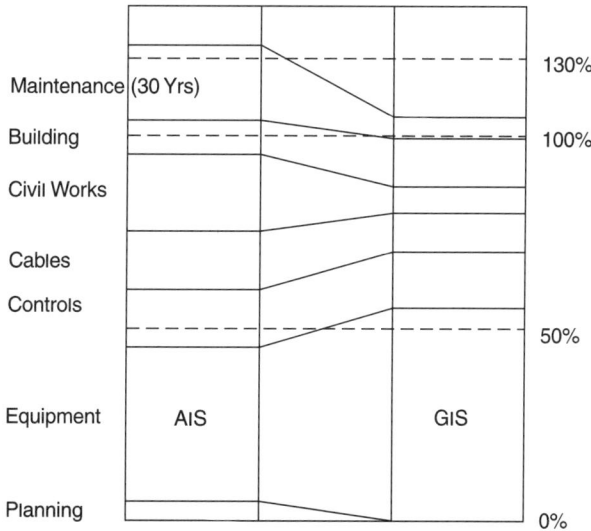

Fig. 5.20: Cost comparison of air and gas insulated sub-stations

5.15 CONCLUSION

Gas insulated sub-station systems offer a compact, cost-effective, reliable and maintenance-free alternative to the conventional air insulated sub-station systems. Their compact size offers a practical solution to vertically upgrade the existing sub-station and to meet the ever-increasing power demand in developing countries. The assured long life and freedom from frequent maintenance drills offered by these gas insulated sub-stations will help eradicate the conventional air insulated sub-stations in years to come.

References

1. International Electro-technical Commission (IEC) Standards; IEC-517, 186, 185, 129 and 694.
2. ASME, Pressure Vessel Code-IV.
3. S. A. Boggs, *et al.*, "Gas Insulated Sub-station Technology and Practice," *Proc. of International Symposium on GIS Technologies and Practices*, Toronto, Canada, 1985, Pergoman Press.

Chapter 6

AUTO-RECLOSER AND SECTIONALISER

Indrapal
Shubhra Chaturvedi

Electrical power for rural consumption is invariably supplied for economic reasons via an overhead line radial distribution network. Consequently, the system is exposed to interruption of supply by the effects of lightning, abnormal weather, birds, animals and wind-blown foliage.

Many years ago it was found that when a fault occurred on a rural system and the overhead line feeder circuit breaker tripped out, a subsequent foot patrol would often fail to find the cause of the fault. The reason for this was that the fault was transient in nature, i.e. once the supply was removed, the cause of the fault would burn out or fall clear. In the case of lightning or conductor clashing, which are momentary in nature, the line was allowed to re-energise safely without being noticed by the foot patrol. It became common practice to reclose the overhead line feeder circuit breaker to 'test' the line, when it tripped out on fault, and in more than half the instances, the circuit breaker was found to be closed and power was restored. Figure 6.1 shows an auto-recloser breaker and control unit at one of the site.

A second recloser was sometimes found to be necessary, but if this resulted in a further trip operation, the fault was deemed to be permanent and a foot patrol would be sent out to find the fault. The advantage of having a circuit breaker which would automatically reclose at least twice was realised, particularly for the circuit breaker which was located at a far away distance into the network.

Later with experience, it was seen that if such a recloser could be arranged to follow the instantaneous trips with a delayed trip, wherein the unit would remain closed long enough to cause fuses further down the line to blow and isolate the faulted section, even better continuity of supply be realised.

Fig. 6.1: An auto-recloser breaker and control unit at HPSEB in Himachal Pradesh

The power necessary for operation in the early pole-mounted auto-reclosers were derived from a weight suspended by a cable from the operating mechanism, but this had the disadvantage of a limited number of stored operations before a lock-out occurred. Today reclosers are used to protect power lines, transformers and other distribution equipment from being exposed to damaging levels of current. Auto-reclosers are now used in many countries including India for the rapid automatic restoration of supply following transient faults in the system. Studies of faults on overhead distribution lines have shown that most are transient and can be cleared without interrupting customer supply. This is especially true of areas that experience frequent lightning discharges or where power lines pass through dense woods having trees in the vicinity.

A recloser detects a fault and then opens for a pre-programmed time called dead time before closing automatically. This automatic close is referred to as an auto-reclose and multiple open and close operations can be utilised to clear transient faults. If the fault is transient and cleared when the line is de-energised, the next auto-reclose will restore supply. If the fault is permanent, the recloser will eventually open and not attempt to close as per the pre-set programme until manual intervention is done by the operator. This state is referred to as recloser lock-out. Manufacturers have standardised a maximum of four protection trip operations before lock-out occurs.

These auto-reclosers are normally designed to carry out one or two fast closing operations followed by one or two delayed closing operations (high set and low set). Experience has shown that over 80 per cent of the faults are of a transient nature and are cleared with these operations and supply can be restored to the system automatically. In the event of permanent faults, which are comparatively few in occurrence, the recloser is automatically locked after a maximum of four trippings and then it requires manual re-setting. All these operations are achieved automatically with the help of electronic control, which is fast replacing the conventional type of mechanical control. Figure 6.2 shows an auto-recloser breaker being assembled at BHEL, Bhopal.

Feeders with different electrical characteristics require different protection strategies. Distribution utilities worldwide have developed their own strategies to suit their particular system conditions. Independent configurations of the time to trip for each protection operation allow the utilities, preferred protection strategies to be implemented. Auto-recloser functions derive power

Fig. 6.2: An auto-recloser breaker being assembled at BHEL, Bhopal

for control unit from 11 kV incoming supply. These reclosers do not require any separate ac or dc control supply. They are totally self-powered, through an exclusive indoor type control transformer mounted inside the breaker itself. Vacuum auto-reclosers can provide trouble-free service for a long time without requiring any maintenance.

A significant advancement was made when solenoid operating mechanisms were introduced to use the power available in the overhead line itself for charging the mechanism spring, thereby avoiding the need for manual re-setting of the mechanism.

Early reclosers simply detected a level of current, opened after a fixed time, closed and repeated the sequence number of times as required. These devices were graded by number of operations, i.e. downstream reclosers had fewer auto-reclose operations than upstream reclosers. This method of grading subjected all the customers on the affected feeder to outages, but only the first recloser upstream of the fault locked-out and isolated downstream supply.

The next advancement was the introduction of current-dependent time to trip the auto recloser. The terminology widely used to describe current-dependent operating speed is IDMT (Inverse Definite Minimum Time) curves. In the first reclosers, IDMT application used electric and hydraulic mechanisms to provide a variable period from detection of the fault current to opening of the recloser. This time was inversely proportional to the magnitude of the fault current and allowed devices in series to be graded on current. These reclosers also incorporated mechanical means to provide a time multiplier to the curve, thus allowing better co-ordination of series devices with differing operating characteristics.

Subsequent generations of reclosers benefited from the introduction of electronic relays due to increased flexibility and accuracy of time to trip. The advent of microprocessor control has led to the current generation of reclosers with increasingly sophisticated protection capabilities. Numerical relays have ushered in a total revolution in the protection philosophy today.

6.1 TECHNICAL SPECIFICATIONS FOR AN AUTO-RECLOSER

The technical specifications for an auto-recloser are spelt out in Table 6.1.

1	Application standard	IS 7567, ANSI C37.60
2	Number of poles	3
3	Type	Outdoor polemounted Vacuum break type
4	Rated operational voltage	12 kV/36 kV
5	Rated maximum voltage	12 kV/36 kV
6	Impulse withstand voltage (2*50 micro-second wave crest)	
	(a) Main circuit to earth and between, poles	75 kV peak/170 kV peak
	(b) Across terminals in open position	75 kV peak/170 kV peak
7	One minute power frequency withstand voltage	
	(a) Main circuit to earth and between pole	28 kV rms/70 kV rms
	(b) Across terminals in open position	28 kV rms/70 kV rms
8	One minute power frequency withstands on auxiliary wiring to earth	2 kV rms
9	Rated frequency	50 Hz
10	Rated normal current	400 A/630 A
11	Rated symmetrical interrupting current (r.m.s. value)	6 kA/12.5 kA
12	Rated short time current for 3 secs	6 kA/12.5 kA
13	Rated short-circuit making current	15/31.5 kA
14	Operating force of handle	10–30 kg. 50 kg
15	Operating voltage	Between 85% to 110% of the rated control voltage
16	Mechanically latched operations	
	Closing	By solenoid or manually (electrical through push button)
	Tripping	By solenoid or manually (with shunt trip)
17	Closing time	Less than 80 ms at the rated control voltage
18	Opening time	Less than 60 ms at the rated control voltage
19	Bushing type CTS	
	(a) Accuracy class of CTs	5P10.
	(b) Ratio	200/1 with suitability for 200% continuous overloading, i.e. 400 A for 12 kV systems (in India only 12 kV auto-reclosers are being used today)
	(c) Rated burden	5 VA
20	Rated control supply voltage	250 V AC
21	Climatic conditions	
	(a) Maximum Ambient temperature	50°C
	(b) Altitude	1000 m above sea level
22	Weight of auto-recloser	350 kg approx.
23	Electrical life at rated current	10,000 operations
24	Mechanical life	30,000 operations

Table 6.1 Technical Specifications for an Auto-recloser

6.2 CONTROL TRANSFORMER

Auto-recloser is a self contained unit which does not require any external supply except 11 kV input. Control transformer auto-recloser breaker generates control voltage of 250 volts for the control unit.

This control transformer is single-phase double-wound epoxy resin, insulated and self-cooled type (see Fig. 6.2).

Rated Primary voltage	12 kV
Rated secondary voltage	250 V
Rated frequency	50 Hz
Rated capacity at rated voltage	2000 VA
Rated impulse withstand voltage on primary side	75 kV
Rated impulse withstand voltage on secondary side	5 kV
Rated power frequency withstand voltage on primary side	28 kV for 1 minute

Table 6.2 Rating and Characteristics of a Control Transformer

6.3 CONTROL UNIT FOR AUTO-RECLOSER FOR 12 kV SYSTEM

The control unit for an auto-recloser for a 12 kV system has the following characteristics.
- Wide phase current setting range of 20 to 200 per cent;
- Wide earth fault current setting range 5 to 50 per cent;
- Standard IDMT curves as per IEEE with time multiplier settings for phase and earth fault;
- Trip sequence settings in any combination of fast (high set) and IDMT (low set) trip timings;
- Trip test facility for ease of testing during installation and commissioning;.
- Programmable auto-reclosing timing and controls;
- Number of shots (reclosures): programmable from 1 to 4;
- Reclaim time: programmable up to 200 secs;
- Dead time 1 to 3 programmable up to 60 secs;
- Monitoring of breaker closing operations;
- Breaker closing is monitored and blocked if the closing voltage is unhealthy, thus improving the reliability of working;
- Digital fault zone indicator for easy tracking of fault;
- Trip counter to display the trip count;
- Steel cabinet suitable for outdoor applications;
- Meets IP-55 specification;
- Hinged door with lock for proper door closing;
- Padlock facility for security and for preventing unauthorised tampering;

- Cabinet suitable for pole mounting; and
- Pre-fabricated and connectable cables for fast and error-free wiring between the recloser control unit and breaker.

Numerical relays are the latest state-of-the-art relays with microprocessor-based logic. They have the following characteristics:

- Self-powered;
- Serial communication with modbus protocol;
- Event recording;
- Time tagging; and
- Programmable protection inhibit.

Figure 6.3 shows control unit for auto-recloser.

Fig. 6.3: Control unit for auto-recloser

6.4 WORKING OF AUTO-RECLOSER

Figure 6.4 shows the working of the auto-recloser when a permanent fault occurs. Then auto-recloser trips after a time delay as per the setting of the first trip on instantaneous or IDMT. It remains de-energised for a time as per the setting of the first dead time, DT1. Then it recloses and the supply is restored to the distribution system. If the fault persists, the recloser again trips for second time after a time delay as per the setting of the second trip on instantaneous or IDMT. It remains de-energised for a time as per the setting of the second dead time, DT2. Then it again recloses the second time and the supply is restored to the distribution system. The sequence is again repeated the third time and on the fourth trip when the fault still persists. The recloser goes into lock indicating that the fault is not a transient one but is of a permanent nature. As per the requirement, the setting in the auto-recloser can be done such that lock-out is achieved at the second or third trips also

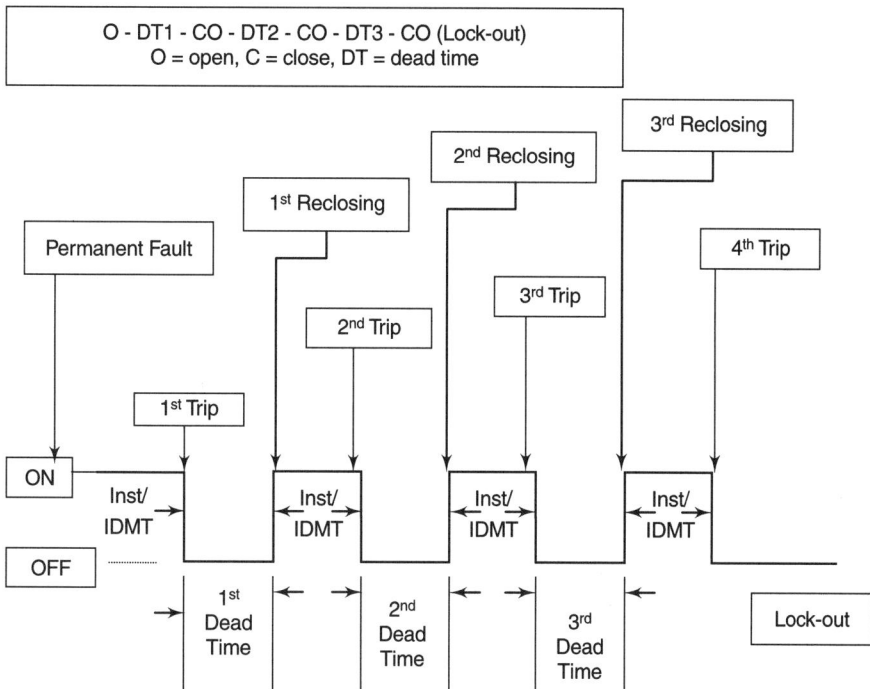

Fig. 6.4: Closing and tripping sequence of auto-recloser

6.5 NEED FOR AUTO-RECLOSER

Auto-reclosers are needed under various conditions. These are discussed below.

6.5.1 High Lightning Discharges

An electrical storm can generate extremely large faults with enormous induced voltages which can cause lot of damage. A fast tripping time reduces equipment risk and danger to human lives.

The use of instantaneous operation can allow fuse-saving strategies to be employed to prevent fuses on an upstream spur or transformer from blowing. This strategy is generally successful where the discriminating time lag is maintained between the fuse blowing time and auto-recloser tripping time.

Alternately, a fuse-burning strategy can be adopted. This allows a downstream fuse to clear a fault without the recloser operating. This strategy is most successful in rural areas with fuse-protected three-phase spurs. An instantaneous multiplier with a minimum time can also be selected as a modifying element to an IDMT curve to facilitate fast operation to clear high faults generated by lightning surges.

6.5.2 Large Concentration of Customers

Using reclosers on feeder junctions or spurs causes the smallest possible area to be left without HV supply. A fault causes the first upstream recloser to operate in order to isolate the faulted segment. The re-arrangement of normally open points together with SCADA capability allows the system reconfiguration to restore supply to the largest number of customers as quickly as possible, thereby reducing the customer time lost.

6.5.3 Remote Site/Difficult Access

Remotely controlled reclosers allow fast switching and network re-configuration to minimise the customer time lost. Where travelling times are significant, this can provide immediate benefits in terms of customer satisfaction and reduced call-out rates.

6.5.4 Bush Fire-prone Areas

Bush fires constitute a great danger in many parts of the world. Studies in Australia have determined that in order to minimise the risk of a fire, the primary protection must operate in less than one second. Reclosers can be closely co-ordinated in order to allow the maximum number of protection devices in series without exceeding this one-second rule. Some authorities set all their reclosers to operate only once to lock-out during periods of high fire danger, relying on current co-ordination to correctly isolate the fault. Remote control is then used to quickly restore supply once the fault has been located and cleared. The recloser that operates to clear the fault does not have to be attended by field personnel.

A study by an Australian electricity supply utility, Great Southern Energy, identified the contributing factors in the costs associated with line faults on rural networks. The factors that were found to influence costs are:

- Line length
- Faults per km—temporary and permanent
- Outages per km
- Restoration time

Of these factors, the restoration time and outages per km can be reduced through the effective use of reclosers.

Remotely operated reclosers allow quick detection of faults and quick analysis of the fault information. This reduces the time involved in generating switching sheets and dispatching line crews.

Since the distances in rural Australia are long, remotely controlled reclosers significantly reduce patrol and switching times. Coupled with the fault information provided by the event log or transmitted back to the control centre, the improvements in restoration time are substantial. The study determined that for a representative rural feeder, the initial outlay of $100,000 on recloser protection would help save around $120,000 annually.

This was based on the existing practice of extensive use of fuses for rural feeder protection and the assumption that 80 per cent of all the faults are temporary and could be cleared by recloser operation.

6.6 AUTOMATIC SECTIONALISER

The sectionaliser is essentially an isolating switch for the isolation of a faulty section after a fault has been cleared with auto-recloser type circuit breakers down the feeder lines. Sectionalisers are designed for use in combination with auto-reclosers. The sectionaliser comprises an automatic vacuum switch with its own control unit, a built-in control transformer (11 kV/250 VAC), suitable for an outdoor pole-mounted installation. It is totally maintenance-free because of the use of vacuum interrupters. The sectionaliser never interrupts fault currents and always opens with a time delay after the line has been de-energised.

The sectionaliser unit is of dead tank construction with six roof bushings with clamps for direct connection to the overhead lines. The sheet steel house, which is hermetically sealed, encloses a simple and reliable solenoid-operated vacuum circuit breaker. It has a few number of moving parts and eliminates the use of cumbersome hydraulic control equipment. In line with the fundamental requirements, the auto-sectionaliser is powered by the overhead lines at a primary voltage of 11 kV through an in-built control transformer of cast epoxy and enclosed in the same house as the circuit breaker. This has been possible due to low energy mechanisms usually associated with VCBs. For reliability, the tripping spring is charged during the closing operation itself.

Past experience has shown that porcelain bushings in sectionaliser may be damaged either in transit or by vandalism or by lightning stroke, thus requiring replacement at site. The design provides for easy replacement of bushings at site.

Mechanical ON/OFF indicators that have been provided can be seen from the ground level. An emergency manual operating device is also provided.

The features of the sectionaliser are detailed below.

1. Control Unit

   The control unit is housed in a weather-proof cabinet mounting on the same pole as that of the sectionaliser at a convenient height. It encloses the electrical control giving the necessary timing function for the operation of the sectionaliser.

2. Operations

   The sectionaliser always performs its isolating function in co-ordination with the operation of the auto-recloser in a sub-station. The operation can be briefly explained as follows:

   (a) The auto-recloser at the sub-station will always interrupt short-circuit or earth fault of the distribution line.

   (b) As no voltage condition is impressed on the control unit of the sectionalisers, they will open automatically after a time delay of two seconds in case of interruption of the circuit.

(c) The sectionaliser will close automatically one by one with a time interval of 10 seconds from the sub-station side to the end of the distribution line. If the faults are temporary and have been cleared by the auto-recloser, all the sectionalisers will remain closed.

(d) After the sectionaliser closes, the eliminating time (Y timer = 5 secs) of the control unit starts. If a fault occurs, and the recloser opens during the eliminating time, the sectionaliser will open and will be locked out.

(e) The sectionaliser will lock-out only if no voltage occurs during the eliminating time. It returns to its original position after the eliminating time has elapsed and the voltage is continuously present during the eliminating time.

(f) In case of a fault of permanent nature, the auto-recloser at the sub-station will interrupt again, when the sectionaliser that is faulty closes. At the same time, the sectionaliser of the faulty section will get locked out on the next opening, thus isolating the faulty section.

6.6.1 Control Relay Unit for Sectionaliser

The control relay unit for control of the sectionaliser is mounted on the same pole on which the sectionaliser is mounted. The unit can be mounted at a suitable height as per the requirement. Power supply for the control relay unit is fed by a control transformer located inside the sectionaliser. The primary winding of the transformer is connected to the 11 kV line. The principle of operation of the control relay unit is explained below.

The operating time of the control relay box is divided into two periods: a delay time (X-time) and an eliminating time (Y-time) (see Fig. 6.5).

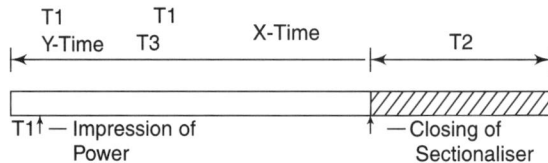

Fig. 6.5: Operating time of the control relay box of sectionaliser

At T1, the control relay unit starts the operation.

At T2, after the X-time period has elapsed from time T1, the sectionaliser is closed.

At T3, after the Y-time period has elapsed from time T2, the control relay unit completes its function.

Now, suppose 't' is a time period, from the closing of the sectionaliser to the tripping of the auto-recloser at the sub-station.

If time 't' is less than the Y-time of sectionaliser, then the control relay unit performs its eliminating function which will lock-out the sectionaliser. If time 't' is greater than the Y-time of the sectionaliser, then the control relay unit performs its normal closing function which will close the sectionaliser next. In this case, the setting of the X-time is always more than the setting of Y-time in order to ensure the proper functioning of the sectionaliser. Thus when there is no faulty section down the line, the sectionalisers are closed upon the transmission of power, because the timing of all control relay units have been adjusted to 't' greater than Y-time in every section. If there is a faulty section, then the sectionaliser feeding the faulty section will cause tripping of the auto-recloser at the sub-station within a short time 't' which will be lesser than the Y-time of the sectionaliser feeding the fault. Hence this sectionaliser will open and will be locked out.

Besides the X-time and the Y-time, the control relay unit has a Z-time of 1.5 to 3 secs (no-volt delay opening time) to prevent mal-performance due to instantaneous voltage drops caused by lightning and other factors. Since it is not required to trip on fault but to open on no-volt, the sudden low voltage due to a short-circuit or lightning may be interpreted as no-volt by the control unit. Hence a delay of Z-time is given before tripping.

X-time normally has a selectable range of 10, 20, 30, 40, 50 and 60 seconds and Y-time of 5 seconds. The working of sectionaliser and auto-recloser together is further explained in diagram form in Figs 6.6 and 6.7.

Figure 6.6 shows one of the arrangement for three sectionalisers and an auto-recloser being used together.

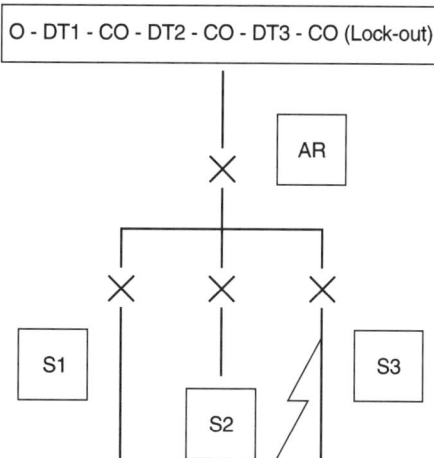

Fig. 6.6: Use of three sectionalisers and an auto-recloser together

6.7 USER'S BENEFITS

The main objectives when designing a sub-station automation system are reliability and economy of the electrical energy supply. Most of the benefits of the sectionaliser listed in this section result in cost savings while simultaneously increasing reliability.

6.12 Handbook of Switchgears

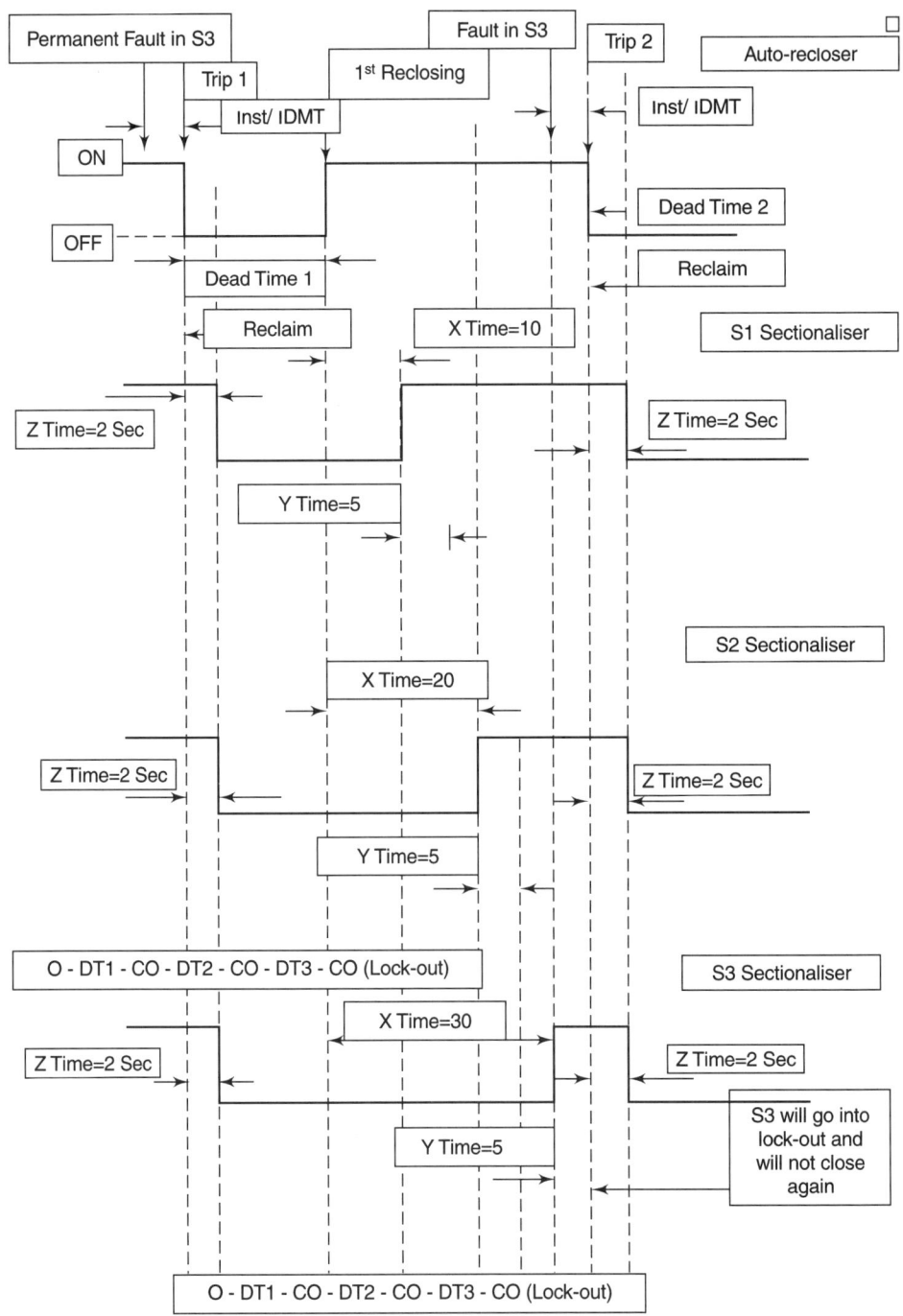

Fig. 6.7: Working of sectionaliser and auto-recloser together

- Lower maintenance costs
- Improved power quality
- Faster power restoration
- Reduced outage time
- More timely data for system planning
- Better knowledge of network behaviour and safe operation

6.8 A CASE STUDY OF AUTO-RECLOSER AT KPTCL

The auto-recloser at KPTCL at Vidhya Nagar, Bangalore displays the following features.

- The load is of about 270 A–300 A, 11 kV.
- Voltage varies from 8 kV to 11 kV.
- Load contains 28 no stone crushers of 30 HP, 415 V irrigation pumps and 6 numbers HT 11 kV industries. 50 per cent load is of the irrigation pump-set and 50 per cent of others as stated above.
- The auto-recloser gets its supply from the Yehalanka sub-station which has two transformers of 66 kV/11 kV, 20 MVA each. One of the transformers feeds 7 breakers while the other feeds 6 breakers. Out of these 7 breakers, one is an auto-recloser.
- The supply line of the auto-reclosure is about 30 km from Yehalanka to M. Hosaharlli.
- About 10,000 consumers are on the line including 50 per cent using irrigation pumps and 50 per cent, others.
- The energy supplied is approx. 6 MW.
- The revenue collected is approx. 20 lakhs per month.
- The occurrence of faults is approx.:
    - Transient faults—15 times a month
    - Permanent faults—6 times a month.
- The time taken to repair a permanent fault is approx. 2–3 hours.
- Charges for energy by M/s KPTCL:
    - Rs 3.25/unit by 11 KV HT industries
    - Rs 4.35/unit by commercial units
    - Rs 2.50/unit by floor mill, tanning industries
    - Rs 1.55/unit for combined light and heating
    - Rs 550/HP/year for irrigation pump-set flat rate
    - Rs 6.75/unit for temporary connections
    - 6 Permanent faults/month amounts to 72 faults/year
    - A 3 hrs fault will a cause loss of supply to consumers for 216 hrs/year.

- Loss of energy for revenue purpose = 216 hrs × 11 kV × 270 A

$$= 641520 \text{ KWh/year}$$

$$\cong 6.5 \text{ lakh units/year}$$

Even at an approximate cost of Rs 1.55/unit, an approximate loss of Rs 13 lakh/year will take place due to permanent faults. These losses can be minimised to half by using two sectionalisers and reduced further by increasing the number of sectionalisers.

Auto-reclosers can save losses in the case of temporary faults (80 per cent of the faults are temporary) while sectionalisers can save losses in the case of permanent faults. The cost of more than one auto-recloser and sectionaliser can be recovered within a year.

Chapter 7

CURRENT TRANSFORMERS AND VOLTAGE TRANSFORMERS

Vivek Johri
Manisha Singh

This chapter provides an overview of instrument transformers (current transformers and voltage transformers) which are required for the measurement of electrical parameters and for protection of equipment.

An ammeter can be designed most economically to read 1 A or 5 A at its full scale. Similarly, a voltmeter can be conveniently designed to operate at 110 V. These instruments are suitable for LT applications only. These small values of currents and voltages are not sufficient to measure power system parameters. The problem is accentuated when one deals with protective devices that are supposed to operate on transient current of say 40,000 A. The solution is to design some devices to extend the instrument/relay measuring range without subjecting them to rigorous high voltages and currents. It is here that instrument transformers come into picture.

An instrument transformer in which the secondary current is substantially proportional to the primary current and differs in phase from it by approximately zero degrees is called a current transformer (CT). A voltage transformer (VT) is an instrument transformer in which the secondary voltage is substantially proportional to the primary voltage and differs in phase from it by approximately zero degrees.

A CT is similar to a power transformer to some extent since both depend on the same fundamental mechanism of electromagnetic induction but there are considerable differences in their design and operation. CTs used for metering and indicating circuits are popularly termed as 'measuring CTs' and those used in conjunction with protective devices are termed as 'protection CTs'. For measuring CTs, the accuracy required is within normal working range of up to 125 per cent of the rated current. For over-current conditions beyond this, accuracy is not desired, rather there must be saturation in the core to relieve the connected instruments of the stresses due to over-current.

A accuracy is not required for currents below the rated value for protection CTs. But there must be accuracy at all higher values of current up to a maximum primary current equal to the maximum system fault level. The decision as to whether or not to use dual purpose CT for measuring and protection depends on various factors such as design, cost and space as also on the ability of the instrument to withstand short time over-current.

A voltage transformer, sometimes also called a potential transformer, is nothing more than an ordinary power transformer in which the error of transformation the (ratio and phase angle errors) are minimised. Unlike CTs, VTs have same accuracy in terms of both measurement and protective purposes. In some exceptional cases, VTs do comply with additional limits of accuracy depending upon how small or large the permissible errors over an extended voltage range are. For certain other forms of protection, VTs require a residual voltage winding to be connected in open delta.

7.1 THEORY OF INSTRUMENT TRANSFORMERS

Given below in brief is the electromagnetic theory on the basis of which current transformers and voltage transformers function. Detailed theoretical expressions are, however, not included.

7.1.1 Theory of Current Transformers

Connected burden and the degree of accuracy are defining factors for measuring CTs. To choose a CT with rated output considerable exceeding required output can result in increased errors. The required output should include the connecting lead burden alongwith instrument burden. As far as accuracy is concerned choosing an accuracy class higher than necessary is uneconomical because that would make the CT bulky. Unlike CTs, VTs operate at a fixed voltage and wide variations in flux are not present as in CT. Let us now discuss in detail the errors of instrument transformers.

In an ideal CT, the primary AT is exactly equal in magnitude to the secondary AT and is in precise phase opposition to it. But in actual CTs, errors with which we are concerned can best be considered through a study of the phasor diagram for a CT (see Fig. 7.1).

I_s — Secondary current

E_s — Secondary induced E.M.F.

V_s — Secondary terminal voltage

E_P — Primary induced E.M.F.

I_P — Primary current

$K_T I_s'$ — Reversed secondary current times turns ratio

(K_T = Turns ratio = secondary turns/primary turns)

I_O — Excitation current

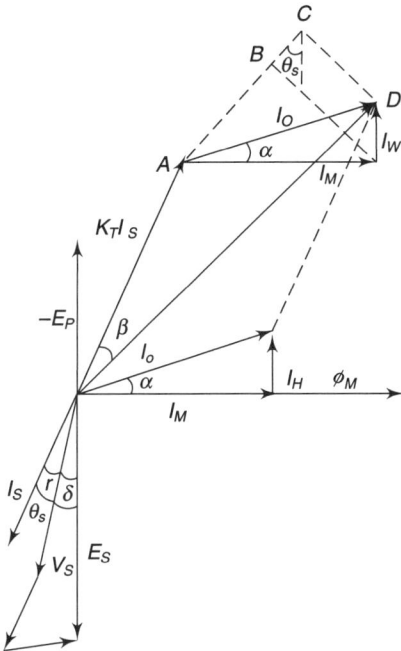

Fig. 7.1: Phasor diagram of a current transformer (lagging power factor)

I_M — Magnetising component of I_o required to produce flux

I_H — In phase component of I_o supplying core losses (eddy current and hysteresis)

$\varnothing_M$ — Main core flux

δ — Phase angle due to secondary winding

r — Phase angle due to burden

θ_s — Total secondary phase angle ($\theta_s = \delta + r$)

α — Loss angle due to core excitation

β — Phase angle between primary and reversal secondary current.

To start with, by taking flux as reference phase since it is common to both primary and secondary windings, the induced E.M.F. E_s and E_p lag behind the flux by 90° and can be drawn in. The magnitude of the phasors E_s and E_p is proportional to the secondary and primary turns. The excitation current I_O taken by primary, is made up of two components, viz. I_M and I_H. I_M is the reactive component supplying the hysteresis and eddy current losses in the core, this is in phase with the primary induced E.M.F. E_p. The secondary current I_s lags behind the secondary induced E.M.F. by an angle $\varnothing_s$. $\varnothing_s$ is made up of δ the angle produced by the burden connected to secondary winding is usually negligible and δ is zero. The secondary current is now transferred to the primary side by reversing I_s and multiplying by the turns ratio K_T. The resultant current flowing in the primary winding I_p is then the vector sum of $K_T I_l$ and I_O.

Flux Ø and exciting current I_O are determined by the secondary voltage required which is, in turn, determined by the burden connected to the secondary winding or for a given burden, by the current flowing in the secondary and primary windings. Hence with a CT, the flux density in the core is not constant but varies with the primary current. This is the basic difference between a CT and a VT where the voltage and flux density remain constant while the current varies with the load.

7.1.1.1 The Current Error (The Ratio Error)

The actual transformation ratio is not equal to the rated transformation ratio since the primary current is contributed by the magnetising and iron loss components. The error introduced due to this difference is turned as a current error. The current error in percentage is expressed by the following formula:

$$\text{Percentage current error} = K_N (I_S - I_P)/I_P \times 100 \qquad 7.1.1$$

where K_N = Nominal transformation ratio = $\dfrac{\text{Rated primary current}}{\text{Rated secondary current}}$

7.1.1.2 The Phase Error

For a perfect transformer, the angle between the primary and reversed secondary current vectors is zero. But for an actual transformer, there is always a difference in phase between the two due to the fact that primary current has to supply to the components of the exciting current. The angle between the above two phasors is termed as the 'phase displacement' or the 'phase error' of the CT. For sinusoidal current, it is said to be positive when secondary current vector leads the primary current vector. Since the phase error, β is very small, in practice $\tan \beta\beta \cong \beta$, it can be expressed by the following formula:

$$\beta = \frac{I_M \cos \theta_S = I_H \sin \theta_S}{K_T I_S + I_M \sin \theta_S + I_W \cos \theta_S} \qquad 7.1.2$$

or approximately, $\quad \beta = I_M/K_I I_S$

The phase error is usually expressed in minutes. The performance of the CT can be improved by reducing the current and phase error to a minimum. Referring to equations 7.1.1 and 7.1.2, this can be achieved by:

(i) Using a core of high permeability and low hysteresis loss magnetic material

(ii) Keeping the rated burden to the nearest possible value of the actual burden

(iii) Ensuring a minimum length of flux path and increasing area of the path, keeping joints to a minimum

(iv) Lowering the internal secondary burden, including its reactance as well as resistance, to the minimum possible value

(v) Effecting an increase in primary ampere turns, (i.e. the use of 'wound type' coil where short-circuit current allows such coil design)

(vi) Keeping the secondary burden to the minimum possible value

7.1.2 Theory of Voltage Transformers

An ideal VT is one in which when rated burden connected across the secondary, the ratio of voltage applied across the primary to the secondary terminal voltages is equal to the ratio of primary turns to secondary turns and furthermore the two terminal voltages are in precise phase opposition to each other. But in actual fact, the above relation doesn't hold good and errors are generated which can best be explained by the phasor diagram for a VT (see Fig. 7.2).

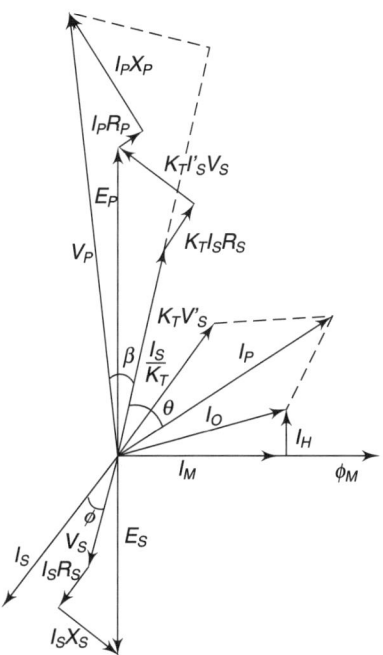

Fig. 7.2: Phasor diagram for voltage transformer (lagging power factor)

I_S — Secondary current
E_S — Secondary induced E.M.F.
V_S — Secondary terminal voltage
R_S — Resistance of secondary winding
X_S — Reactance of secondary winding
I_P — Primary current
E_P — Primary induced E.M.F.
V_P — Primary terminal voltage
R_P — Primary winding resistance

X_P — Primary winding reactance

I_O — No-load current

I_M — Magnetising current

I_H — No-load current loss component

$\emptyset_M$ — Flux

$\emptyset$ — Phase angle of burden

β — Phase angle error

θ — Angle between VS (reversed) and IP

K_T — Turns ratio = primary turns/secondary turns

Secondary terminal voltage V_S is generated from induced E.M.F. in secondary E_S after vectorially, removing the secondary winding's resistive and reactive drops. Secondary current I_S lags V_S by the phase angle of burden. The primary resistive and reactive drops ($I_P R_P$ and $I_P X_P$) are supplied by applied voltage V_P and are subtracted from V_P to derive primary induced E.M.F. E_P. E_P is in opposition to E_S. The angle between reversed secondary voltage $K_T V_S'$ and V_P is termed as the phase angle of transformer denoted by β. β is positive when $K_T V_T'$ is in advance of V_P.

7.1.2.1 The Voltage Error (The Ratio Error)

As it is defined for CTs, the voltage error for VTs is also defined as the error due to a difference in the actual transformation ratio and the rated transformation ratio. In percentage, it is expressed as

$$\text{Percentage voltage error} = \frac{K_N(V_S - V_P)}{V_P} \times 100 \qquad 7.1.3$$

where K_N = Rated transformation ratio = Rated primary voltage/Rated secondary voltage

7.1.2.2 The Phase Error

The phase difference between the primary voltage and the reversed secondary voltage vectors is the phase error. In order to keep the overall error within the specified limits of accuracy, the winding must be designed to have:

1. The internal resistance and reactance to an appropriate magnitude; and
2. Minimum magnetising and loss component of the exciting current required by the core.

7.2 TECHNICAL PARAMETERS OF INSTRUMENT TRANSFORMERS

Following are some of the commonly used terms for instrument transformers:

  (i) **Rated primary current:** The value of the primary current which appears in the designation of the transformers and on which the performance of the CT is based.

  (ii) **Rated secondary current**: The value of the secondary current which appears in the designation of the transformers and on which the performance of the CT is based.

(iii) **Rated burden:** The value of the load to be connected across the secondary of CT including connecting lead resistance expressed in V_A or ohms on which accuracy requirement is based.

(iv) **Rated frequency:** The value of the system frequency on which the instrument transformer operates.

(v) **Accuracy class:** The highest permissible percentage composite error at rated current/voltage prescribed for that accuracy class, for measuring CTs, standard accuracy classes as per IS2705 are 0.1, 0.2, 0.5, 1, 3, and 5, for protection CT, 5 P, 10 P, 15 P.

(vi) **Rated short-circuit:** The R.M.S. value of primary current which the CT will withstand for a rated time with its secondary winding short-circuited without suffering harmful effects.

(vii) **Rated voltage:** The R.M.S. value of the voltage used to designate the CT for a particular highest system voltage.

(viii) **Instrument security factor:** For measuring CTs, the ratio of instrument limit primary current to the rated primary (in the event of system fault current flowing through the primary winding of a CT, the safety of apparatus connected to transformer is the greatest when the value of ISF is small).

(ix) **Accuracy limit factor (A.L.F.):** For protection CTs, the ratio of rated accuracy limit primary current to the rated primary current; the standard values of A.L.F. as per IS2705 are 5, 10, 15, 20 and 30.

(x) **Rated accuracy limit primary current:** The highest value of primary current designed by the CT manufacturer, up to which the limits of composite error are complied with.

(xi) **Composite error:** Under steady state conditions, the RMS value of the difference between:

   (a) The instantaneous values of the primary current; and

   (b) The instantaneous values of the actual secondary current multiplied by the rated transformation ratio.

(xii) **Knee point voltage:** The sinusoidal voltage of rated frequency applied to the secondary terminals of CT, with all other winding being open-circuited, which when increased by 10 per cent, cause the exciting current to increase by 50 per cent. Minimum knee point voltage, V_K, is specified by the following formula:

$$V_K = K_I (R_{CT} + Z_s) \qquad 7.2.1$$

where

K — A parameter to be specified by the purchaser, depending on the system fault level and the characteristic of the relay intended to be used.

I — Rated relay current (1 A or 5 A).

R_{CT} — Resistance of CT secondary winding corrected to 75°C.

Z_S — Independence of secondary circuit (to be specified by the purchaser).

(xiii) **Rated insulation level:** The combination of voltage values (power frequency and lighting impulse, or where applicable, lighting and switching impulse), which characterise the insulation of a transformer with regard to its capability to withstand dielectric stress.

(xiv) **Rated primary voltage**: The value of primary voltage which appears in the designation of the transformers and on which its performance is based.

(xv) **Rated secondary voltage:** The value of the secondary voltage which appears in the designation of the transformers and on which its performance is based.

(xvi) **Voltage error or ratio error:** The error which a transformer introduces into the measurement of a voltage and which arises when a transformation ratio is not equal to the rated transformation ratio.

(xvii) **Rated voltage factor:** The voltage factor is determined by the minimum "operating voltage" which, in turn, is dependent on the system and the voltage transformer primary winding earthing conditions. For the standard rated voltage factor for different earthing conditions, refer IS:3156, Part I.

(xviii) **Knee point:** It is the point on the magnetisation curve at which a 10 per cent increase in flux density causes a 50 per cent increase in exciting ampere turns.

(xix) **Residual voltage:** The vector sum of three line-to-earth voltages.

7.3 DESIGN AND CONSTRUCTIONAL ASPECTS OF INSTRUMENT TRANSFORMERS

Various types of instrument transformers are currently in use on the basis of the varied applications of these devices. The construction aspects and sizing play an important role in the selection of proper types of CTs or VTs. The design principles of these transformers are different from those of power transformers because accuracy is of prime importance and these transformers are lightly loaded in comparison to power transformers. The design of CTs is a complex process and only some basic aspects and empirical assumptions are given below, which are useful from a user's point of view.

7.3.1 Types of Instrument Transformers

On the basis of their applications in the field, instrument transformers can be broadly classified into two types, viz. indoor and outdoor transformers.

7.3.1.1 Indoor Current Transformers

Current transformers designed for mounting inside metal cubicles are known as indoor CTs. Depending upon the method of insulation, these can further be classified as:
— Tape insulated
— Cast resin (epoxy, polyurethene or polycrete)

In terms of constructional aspects, these can be further classified into the following types:

(i) Bar Type: The CTs having a bar of suitable size and material used as primary winding are known as 'bar type CTs'. The bar may be of rectangular or circular cross-section.

(ii) **Slot/Window/Ring Type:** CTs having an opening in the centre to accommodate a primary conductor through it are known as 'ring type' (or 'slot/window type') CTs.

(iii) **Wound Type:** A CT having a primary winding of more than one full turn wound on the core is known as 'wound type CT'. The connecting primary terminals may be similar to those of a bar type CT or rectangular pads can be provided for this purpose.

7.3.1.2 Outdoor Current Transformers

These current transformers are designed for outdoor application. They use transformer oil or any other suitable liquid for insulation and cooling. A liquid-immersed CT which is sealed and does not communicate with the atmosphere is known as a hermetically sealed CT.

Outdoor oil-filled CTs are further classified as 'live tank type' or 'dead tank type' transformers.

(i) **Live tank type**: In this design of instrument transformers, the tank housing the cores is kept at the system voltage. A live tank CT is shown in Fig. 7.3. It can be noted that the bushing of this CT is prone to damages in transit as its centre of gravity is at a large height.

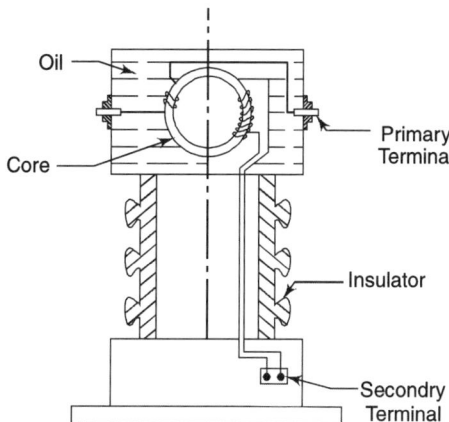

Fig. 7.3: Live tank CT

(ii) **Dead tank type:** In this design of instrument transformers, the tank housing the cores is kept at earth potential. Figure 7.4 shows a dead tank (single bushing) design which is mounting-wise similar to a live tank design but here the centre of gravity is low. Hence this type of CT is not damaged in transit. Figure 7.5 depicts dead tank CT (two-bushing) which is very compact in size and can be mounted on a steel structure near the outdoor circuit breakers.

A CT having more than one core and more than one secondary winding is known as a multi-core CT (for example, a CT having metering and protection cores). A CT in which more than one ratio is obtainable by reconnection or tapings in primary or secondary windings is known as a multi-ratio instrument transformer (e.g. a CT having a ratio of 800-400-200/1 A). In such transformers, tapings in primary windings must be avoided as far as permitted by the design.

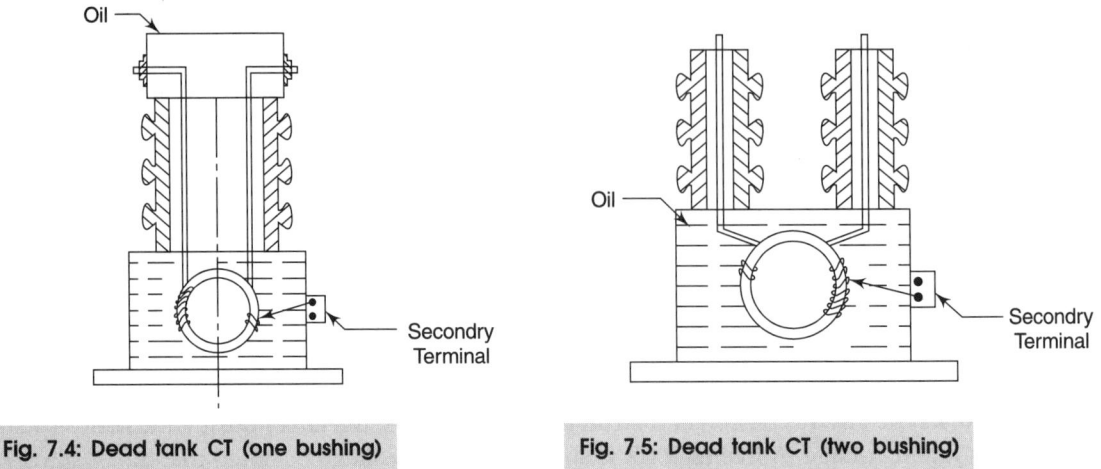

Fig. 7.4: Dead tank CT (one bushing)   Fig. 7.5: Dead tank CT (two bushing)

An instrument transformer intended to serve the dual purpose of measuring and protection is known as a 'dual purpose instrument transformer'.

A CT having a split metering core used for the measurement of current in a busbar is known as a 'split-core CT'. The spring action of a split core CT allows the operator to use this CT for enveloping a current-carrying low tension busbar, without stopping the current flow.

7.3.2 Types of Voltage Transformers

A liquid-immersed VT, which is sealed and does not communicate with atmospheric air, is known as a 'hermetically sealed VT'.

A VT that transforms the primary voltage to the secondary voltage entirely by electromagnetic phenomena is called an 'electromagnetic VT' (or 'conventional VT').

On the other hand, a VT comprising a capacitor divider unit and an electromagnetic unit is so designed and inter-connected that the secondary voltage of the electromagnetic unit is substantially proportional to and in phase with the primary voltage applied to the capacitor divider unit. Such an arrangement is known as 'capacitive voltage transformer' or CVT.

A VT having one magnetic core intended to serve the dual purpose of measuring and protection is called a 'dual purpose VT'.

A three-phase VT (or a group of three single-phase VTs) having windings connected in open delta so as to produce a voltage proportional to the residual voltage existing in three-phase system voltage is known as a 'residual VT'. A normal three-phase VT (not suitable for residual connection) has a three-limb core. But the three-phase residual VT must have a five-limb core to allow return paths for residual flux.

A single-phase VT, which has one end of the primary winding connected to earth is known as 'earthed VT'. Similarly, a three-phase star-connected VT having a star point brought out and earthed is also an 'earthed VT'.

A VT having two primary windings and two secondary windings connected in V-V connection is used only for the measurement of voltage and not for protection, there being no path for zero sequence voltages arising from earth faults. This type of VT is known as a 'V-connected VT'.

7.3.3 Design of Instrument Transformers

The E.M.F. induced by the primary winding of a transformer (E_P) is given by:

$$E_P = 4.44 \, f . \varnothing_M . N_P$$

Also, $\quad V_P = 4.44 \, f . A . B_M \, N_P$ (ideally, at no load) $\qquad$ 7.3.1

where $\quad E_P$ = Induced primary voltage

$\quad V_P$ = Voltage applied across primary

$\quad N_P$ = Number of primary turns

$\quad \varnothing_M$ = Main core flux

$\quad B_M$ = Maximum flux density

$\quad A$ = Area of cross-section

The E.M.F. induced in secondary winding (E_S) at no load is given by:

$$E_S = 4.44 \, f . \varnothing_M . N_S$$

or $V_S \, 4.44 \, f . B_M . A . N_S$ $\qquad$ 7.3.2

where $\quad E_S$ = Induced secondary voltage

$\quad V_S$ = Secondary voltage

$\quad N_S$ = Number of secondary turns

Dividing equation (7.3.2) by (7.3.1),

$$\frac{E_S}{E_P} = \frac{N_S}{N_P} = \frac{V_S}{V_P} = K_N \qquad 7.3.3$$

where K_N is transformation ratio. For ideal transformers, the transformation takes place without any losses in its various parts. Thus, for such transformers, the input and output can be approximately equal, i.e.:

Output $\quad V_A$ = Input V_A

or $\quad V_S I_S = V_P I_P$

or $\quad I_S/I_P = V_P/V_S = 1/K_N$ $\qquad$ 7.3.4

Hence the current in the transformers' windings is in the inverse ratio of the voltage across the windings. The above equations constitute the basis for the design of instrument transformers. The

7.3.4 Some Important Design Considerations

7.3.4.1 Selection of Secondary Current Rating

For a given distance between the burden (relay or instrument) and the CT, the adoption of 1 A as the rated secondary current reduces the lead burden to 1/25th of its value at 5 A secondary current. For any particular rated primary current and select value of primary ampere-turns, the lower the rated secondary current, the finer will be the adjustment to the ratio error by the use of turns correction. The disadvantage in selecting 1 A rated secondary current is that a more expensive, finer gauge wire is required. Also, 1 A secondary current would, all other things being equal, develop five times the peak voltage that of a 5 A secondary.

7.3.4.2 Selection of Primary Current Rating

The ratio error and phase error of a measuring CT are influenced by the number of available primary ampere-turns to a greater extent than by any other single factor, and this is equally true for the performance of a protective CT. Thus there is a minimum value of primary current for any particular application for which a ring type or similar CT may be used without making it unduly large or expensive. If the rated primary current is below this value, it is suggested that wound-primary construction be used but again there is a minimum limit determined this time by the maximum dynamic and thermal stresses for a given class of accuracy and burden. The lower the rated primary current, the larger would be the core required.

7.3.4.3 Selection of Rated Burden

The ratio error for any particular value of the flux density is lowest for an overall secondary power factor of unity. The difference between the actual and test values of the ratio error depends upon factors such as the flux density, the proportion of the total secondary circuit impedance supplied by the secondary winding, the core material and the type of core construction. In any application that demands high accuracy, the errors of CTs must be measured wherever possible with the working burden and not with the rated burden. The rated burden should be kept as near to the actual burden as possible to obtain better accuracy.

7.3.4.4 Selection of Class of Accuracy

The class of accuracy should be chosen carefully as it directly affects the size and cost of the CT. This effect applies more to lower ratio CTs.

7.3.5 Application of Instrument Transformers in Various Protection Schemes

7.3.5.1 Differential Protection

Differential protection is used for the protection of generators, generator-transformers, transformers, transmission lines, large motors, bus zones, etc. The principle of any differential relay is that it responds to vector difference between two or more similar electrical quantities. The protected zone is determined by the location of CTs or VTs secondary.

In Fig. 7.6, X is the winding of the machine to be protected. During healthy condition, the current entering in X is equal in phase and magnitude to the current leaving X. CT1 and CT2 are of such a ratio that during the normal conditions or for external faults, the secondary current of a CT is equal. The vector differential current $I_1 - I_2$ through the relay coil is therefore zero, i.e. $I_1 - I_2 = 0$ during normal conditions or for external faults.

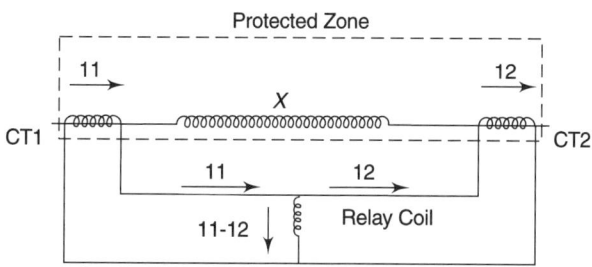

Fig. 7.6: Differential protection

When a fault occurs in the protected zone, the current entering the protected winding is no longer equal to the current leaving the winding as some current flows to the fault. The differential current $I_1 - I_2$ flows through the relay operating coil and the relay operates if the operating torque is more than the re-straining torque.

7.3.5.2 Percentage or Biased Differential Protection

In the above paragraph, we have discussed how differential protection of various machines is achieved. But in actual conditions, the differential relay has to overcome the trouble arising out of the differences in CT ratios for high values of external short-circuit currents. The differential relay is provided with an additional coil, i.e. a re-straining coil.

7.3.5.3 Restricted Earth Fault Protection

When the earth fault occurs at, say, point f of generator winding, voltage V_{AF} will be available which will drive fault current I_F through neutral to ground connection. This voltage V_{AF} will be relatively less when the fault point is nearer to terminal a. Hence I_F will also be reduced but if the relay setting is kept too low, it may respond during through faults. In practice, only 85 per cent of the generator winding is protected against a phase-to-earth fault and 15 per cent of the portion is left unprotected.

7.3.5.4 Interposing CT

An intermediate CT (ICT) is used for feeding protective or measuring devices which require operating current that is different from the main CT secondary current. For example, in case the main CT ratio is 50/5 A and the relay requires a 1 A rating, then an ICT of ratio 5/1 shall be used. It may be noted that the ICT's internal burden is to be fed by the output of the main CT.

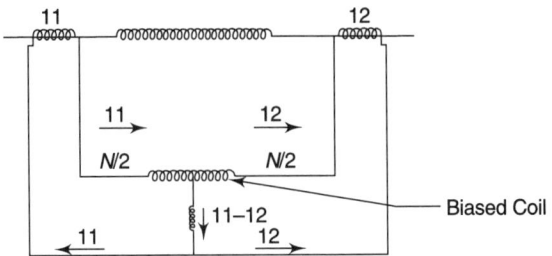

Fig. 7.7: Biased differential protection

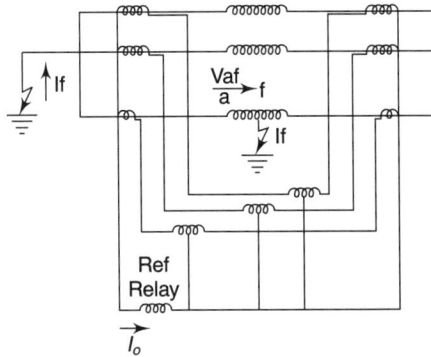

Fig. 7.8: Restricted earth fault protection

7.3.5.5 Core Balance CT

A core balance CT (or CBCT) is used in sensitive earth fault protection schemes. The CBCT is a ring type LTCT put around a three-core cable to detect zero sequence current (earth leakage current in the system). For single-core cables, all three cables must pass through a CBCT. The CBCT secondary feeds a sensitive earth fault relay. This arrangement can detect a primary earth leakage current of the order of 500 mA. It may be noted that the design of a CBCT is entirely different from that of a normal CT. Since the magnetising current and primary current are of the same order, the ratio of a CBCT is very significant, but one has to verify the operation of the relay at a specified current.

7.3.5.6 Summation CT

A summation CT seen in Fig. 7.9 is used for the vector addition of several feeder currents. It may be noted that the main CT ratio of feeder CTs 1 and 2 should be the same.

7.3.5.7 Current Balance CT

A special use of CTs is in a current balance scheme. It comprises a set of two numbers. The CTs used in this scheme per phase is known as a 'current balance CT'.

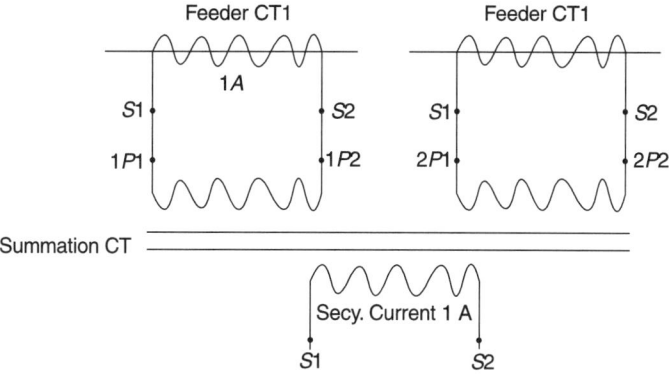

Fig. 7.9: Summation CT connections

Figure 7.10 shows the two circuit breakers CB1 and CB2, each feeding half the total load current (I). This arrangement is necessitated when a CB manufacturer does not have a CB of full impedance of paths A-CB1-B and A-CB2-B. Since the latter cannot be identical, the currents in the two paths will be:

$$I_1 \neq I_2 \neq I/2$$

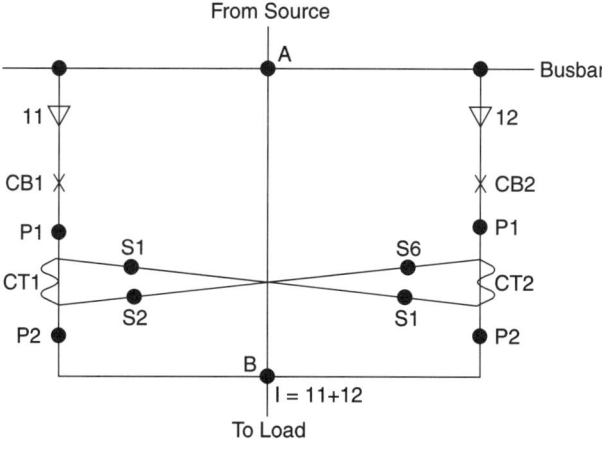

Fig. 7.10: Schematic diagram of current balance CT

In order to maintain the same $I/2$ in both branches, current balance CTs (CT1 and CT2) are used. Using cross-connected secondary, the primary of CT1 'sees' a reflected burden proportional to the impedance of branch 2 (and vice versa). This makes the apparent impedance of both branches equal and I_1 becomes approximately equal to I_2.

7.4 INSTRUMENT TRANSFORMER TESTING AND ANALYSIS OF TEST RESULTS

Instrument transformers are subjected to various tests in order to verify their capability to function properly under normal and transient conditions of the electrical system to which they are connected. The following tests are specified by standards:

(A) Voltage transformers

(i) **Type Tests**: These include the temperature rise test, lightning impulse test, high voltage power frequency withstand test, and error determination test.

(ii) **Routine Tests**: These include verification of terminal markings and polarity, power frequency voltages withstand on primary and secondary winding, partial discharge measurement and error determination test.

(iii) **Optional Tests**: These include the chopped lightning impulse test, capability and commissioning tests.

(B) Current Transformers

(i) **Type Tests:** These include short time current tests, temperature rise test, lightning impulse test, high voltage power frequency withstand test and error determination test.

(ii) **Routine Tests**: These include verification of terminal markings and polarity, power frequency voltage withstand tests on primary and secondary windings, over-voltage inter-turn test, partial discharge measurement and error determination test.

(iii) **Optional Tests**: These include the chopped lightning impulse test, measurement of dielectric dissipation factor and commissioning tests.

7.4.1 Type Testing of Instrument Transformers

We note that several tests on CTs and VTs are of a similar nature such as the temperature rise, impulse and HV withstand tests. The purpose of the temperature rise test is to verify the capacity of the transformer to carry rated continuous thermal current/voltage with the rated burden of the specified power factor connected. The maximum allowed temperature rises for various classes of insulation are specified in standards. For CTs and VTs meant to be installed inside sheet steel cubicles, the temperature rise test should be performed after mounting the device inside such cubicles.

The lightning impulse test along with the power frequency high voltage withstand test verifies the basic insulation level of the instrument transformers. The impulse test voltage is applied between one primary winding terminal and earth for VTs whereas in the case of CTs, primary winding terminals are short-circuited together and impulse voltage is applied between these terminals and earth. Secondary winding terminals are connected together and earthed for CTs and VTs. The transformer passes both the tests if no disruptive discharge or flashover occurs inside and along the insulation. Variations in the wave shapes of impulse voltage are also observed.

Similar connections are made for the HV test and absence of any disruptive discharge indicates suitability of the instrument transformer for rated HV withstand. For earthed type VTs, the primary test voltage is induced by energising the secondary winding. Sometimes, the frequency of exciting voltage is increased to prevent the drawing of excessive exciting current by the VT.

For conducting the short time current (STC) test on a CT, secondary winding is short-circuited and a current I is allowed to flow in the primary winding for a duration t such that:

$$I^2 t \geq I^2_{th} \cdot t_\tau \qquad 7.4.1$$

where I_{th} is the rated short time current and t_τ is the rated duration STC.

The over voltage inter-turn test on a CT is conducted by deliberately opening the CT secondary for one minute, while the primary is energised with full rated primary current. This test is intended to test only the inter-turn insulation. In service, CT secondary should not be left open-circuited otherwise excessive heating, dangerous over-voltage and permanent magnetisation of the core may occur. The partial discharge measurement test is an important routine test for assessing the quality of insulating material and the manufacturing process.

7.4.2 Applicability of Test Certificates

Depending upon various electrical installations' ratings the specifications of CTs and VTs vary widely. This gives rise to large variety of type testing requirements on a particular design of CT. Since type tests on CTs are of a destructive nature, it is not economical or practical to conduct type tests on all varieties. Hence, CT manufacturers typically conduct few tests on a particular design and offer the tested CT for various applications. The following method allows the CT user to verify the applicability of test certificates (TCs) for a particular application. The STC test verifies the capability of primary winding for carrying rated short time current (STC) for a rated duration. This test is significant for wound primary CTs. However, for slot type or bar type CTs, one should ensure that the reflected primary STC on the secondary should be within the thermal limits of the secondary winding.

After checking the drawing and design of tested CTs against the offered CT, the cross-sectional area of primary winding is noted. The cross-sectional area of the offered CT should satisfy the following condition:

$$\sigma_t \ (1 \ sec) \geq \sigma_o \ (1 \ sec) \qquad 7.4.2$$

where,

$\sigma_t \ (1 \ sec)$ = STC density in A/mm^2 for tested CT for a rated duration of 1 sec; and

$\sigma_o \ (1 \ sec)$ = STC density in A/mm^2 for offered CT for a rated duration of 1 sec.

In order to convert the STC density value from a three-second duration, the following relation may be used:

$$\sigma (1 \ sec) = \sqrt{3} \ (\sigma (3 \ sec)) \qquad 7.4.3$$

The equations 7.4.3 and 7.4.1 are derived from the Julean heating expression

$$I^2 RT = constant$$

For current I_1 to flow for T_1 duration and I_2 to flow for T_2 duration in a resistance R, we have

$$I_1^2 \, R \, T_1 = I_2^2 \, R \, T$$

or, $\quad I_1 = (\sqrt{T_2/T_1}) \cdot I_2 \quad$ 7.4.4

The condition 7.4.2 is not complete in itself, until, we do not check the capability of the CT to withstand the STC peak value also. Hence the following condition along with 7.4.2 should be met by the offered CT:

$$T_t \geq T_0 \quad \text{7.4.5}$$

where $\quad T_t$ = Dynamic ampere-turns for tested CT (2.55 × RMS value of the STC × no. of primary winding turns), and

T_0 = Dynamic ampere-turns of the offered CT, calculated as above.

7.4.3 Acceptance of Class PS CT Using Magnetisation Curves

CT cores are normally made of CRGO steel coils. The 'Hysteresis' and eddy current losses in the core give rise to magnetisation current which, in turn, causes ratio and phase angle errors. In ordr to match the magnetisation curves of CTs fitted at different locations in a differential or restricted earth fault protection schemes, one chooses class PS CTs. To understand this type of CT, we start with E.M.F. E_2 induced in CT secondary winding:

$$E_2 = 4.44 \cdot f \cdot N_2 \cdot \varnothing \, V \quad \text{7.4.6}$$

where, f is frequency in Hz, N_2 the number of secondary winding turns and $\varnothing$ is magnetic flux in Wb. This flux in the core is produced by the exciting current Ie. We have a non-linear relationship between Ie or H and $\varnothing$ or B as shown in Fig. 7.11, a typical B-H curve.

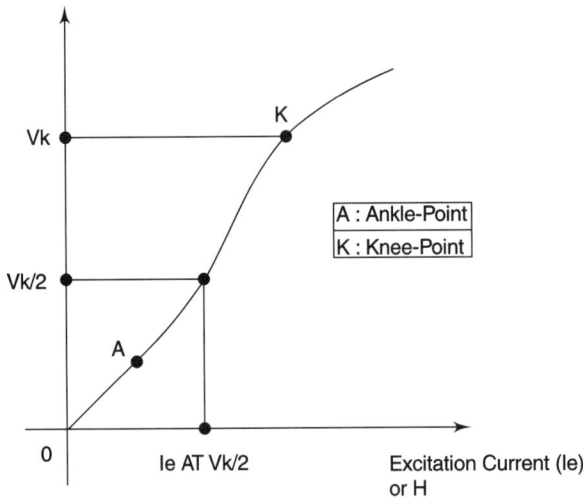

Fig. 7.11 Magnetisation curve

It may be noted that linearity is maintained in Fig. 7.11 between points A and K. These points are known as the 'ankle point' and 'knee point' respectively. The knee point voltage, V_K defined as that voltage, which when applied across secondary winding (with primary left open) and increased by 10 per cent, causes magnetic current Ie to increase by 50 per cent. In other words, if V_k causes Ie to develop, then 1.1 V_K shall cause a magnetic current of 1.5 Ie.

In differential and REF protection schemes, the specifying output, accuracy and ALF of the CT may not ensure compatibility of the magnetisation curves of CTs at the two ends. Instead, V_K and I_E are specified for such CTs, known as class PS (protection special) CTs. Specifying the absolute value of V_K (e.g. 300 V) generated by CT may not solve the problem since secondary voltage seen by the relay coil after voltage drops are significant. Thus, V_K is specified:

$$V_K \geq K.I.(RCT + 2.RL) \qquad 7.4.7$$

$$Ie \leq P \text{ mA at } (Vk/F) \text{ V} \qquad 7.4.8$$

where
K = constant governed by system parameters and relay characteristics (a typical value being 40),

I = rated relay current (1 or 5 A), (for delta connected CTs, this value is obtained by multiplying I_3 to CT secondary current),

R_{CT} = ac resistance of secondary winding in Ohms at 75°C,

R_L = one way cable ac resistance in Ohms at 75°C,

P = permissible magnetic current in mA, and

F = factor specified by relay manufacturer (typically 2 or 4).

The CT manufacturer specifies the R_{CT} value for each CT. The user has R_L value with him, enabling him to calculate V_K using formula (4.7). From the estimated magnetic curves of the CT, V_K is read for each CT primary tap. The following relation must be satisfied for the suitability of the CT:

$$V_K \text{ (obtained from curve)} \geq V_K \text{ (derived from formula)} \qquad 7.4.9$$

Similarly, at V_K/F volts, the value of Ie obtained from the curve should be less than the specified Ie value. The higher value of V_K signifies that the CT is of good quality whereas higher Ie signifies more errors.

7.5 STANDARDISATION AND COST CONSIDERATIONS OF CTS AND VTS

The installations wherein CTs and VTs are mounted to measure the normal and abnormal values of current and voltage, have great variations in key system parameters, including voltage class, insulation level, type of earthing, short-circuit level, rated current, rated burden, accuracy and ALF requirements, size and design requirements, type of protection schemes, etc. This leads to the use of numerous varieties of CTs and VTs. It is possible that even a frequent user of CTs may not come across two identical CTs over a period of many years. It is this limitation which makes it difficult to standardise CT and VT specifications, thereby resulting in seemingly endless lists. This problem could also result in numerous inventories and create unmanageable stockpiling of CTs and VTs if one attempts to stock these transformers. Part of the problem arises due to the tendency of the user to over-specify his requirement just 'to be on the safe side' thereby resulting in avoidable extra costs.

7.5.1 Recommendations for Standardisation and Cost Reduction

The following measures may be taken to address the problem of over-specification mentioned above. These measures are useful to some extent in solving the problem of over-specification and variety reduction.

7.5.1.1 Current Transformer Recommendations

These are detailed below.

(1) Specify one primary ratio for CTs up to 66 kV and consider replacement of the CT as and when the load increases. This is so because the cost of multiple ratio is substantial as requirement of the output from individual cores is desired at the lowest rating.

(2) Specify rated primary current higher than load requirement to fulfil the following relation:

$$Rated\ primary\ current \geq \frac{Rated\ STC\ for\ 1\ sec}{150} \qquad 7.5.1$$

This ensures a fairly economical design (please note that indicating instruments function well at the middle of the full scale).

(3) Out of the IS 2705 (part 1), 1992 list, following primary current should be chosen—10, 15, 20, 30, 50, 75 A, or their multiples or fractions.

(4) A rated secondary current of 1 A (and not 5 A) should be chosen if the total lead burden on a CT core exceeds 2 VA.

(5) The practice of specifying spare cores should be avoided.

(6) With modern static meters, the metering core burden of 5 VA, or at the most 10 VA, is adequate.

(7) The practice of specifying I_{SF} for the metering core for all ratios as 'less than 5' should be discontinued. If I_{SF} is less than 5 for the lowest ratio, it should be less than 10 for the higher ratio (assuming 1:2 ratios). Since better STC withstand instruments are available, the minimum I_{SF} requirement should be increased.

(8) Relay burdens in ohms at various settings should be added and the CT voltage, V should be sufficient to pass the required current A (multiple of relay setting) through the total ohmic impedance of the relay. Hence, burden and ALF should be chosen among the following standard values so that they satisfy the relation given in equation 7.5.2.

Burden in VA: 2.5, 5.0, 7.5, 10, 15 and 30,

A.L.F.: 5, 10, 15, 20 and 30,

$$V = \frac{Burden\ in\ VA \times A.L.F}{Rated\ current} \qquad 7.5.2$$

(9) Accuracy classes for protection core are 5 P, 10 P and 15 P. In order to correctly design CT (having toroidal core) operating at a minimum of 200 AT, accuracy class 5 P is automatically obtained. Hence the accuracy of 5 P should be generally specified, as specifying 10 P does not offer any price advantage.

(10) For class PS CTs, the value of the knee point voltage V_K is specified as

$$V_K \geq K.I. \ (R_{CT} + 2R_L) \ V \qquad 7.5.3$$

where symbols have the same meaning as in equation 7.4.7.

The values R_{CT} and V_K should not be specified by the user but only K and R_L may be specified. The manufacturer should choose suitable V_K and R_{CT}.

(11) The constant K, calculated as two times the reflected fault current, should be specified realistically. For example, a value of 6 is sufficient for the motor feeder against a value of 20 for other feeders.

(12) If fault current I_F is less than 150 times the rated primary current I_P, the short-circuit duration t should be such that:

$$I_F = \frac{150 \ I_p \ A}{\sqrt{t}} \qquad 7.5.4$$

7.5.1.2 Voltage Transformer Recommendations

These are as follows.

(1) Normally VTs are specified as having the following two sets of secondary windings:
   — Metering winding (having accuracy class of 0.1, 0.2, 0.5, 1 or 3), and
   — Protecting winding (having accuracy class of 3 P or 6 P).

   Sometimes some instruments need accuracy class of 0.5 while for others it is 1 to 3. In such cases, two separate metering cores with lesser burden having a different accuracy class may be specified. This will result in considerably low cost VTs.

(2) VT burdens are normally over-specified. Instruments typically need 2 VA VT burdens, with a maximum of 5 VA. It is recommended that the instrument burden should be calculated accurately and the lead burden kept at the minimum value by re-location/lead length reduction/increase in lead cross-section.

(3) Open delta burden is zero in normal operating conditions and hence this should not be taken into account while determining the accuracy of the main metering winding.

(4) Specifying non-simultaneous loading conditions on individual winding will reduce cost of VT.

(5) HT fuses routinely specified on VTs have high rating and may not protect the VTs from overload/short-circuit on secondary side. For this, suitable protective devices in the secondary side should be used instead of HT fuses.

(6) CT/VT combined sets for distribution metering may be formed from two independent CTs and three star-star VTs.

7.5.1.3 Special Instrument Transformers

The recommendations for special instrument transformers are detailed below.

1. The use of interposing CT (ICT) at the load end helps to minimise the VA burden of the lead by using a 1 A secondary main CT.
2. For suppressing zero sequence components from the out-of-balance differential current, the use of ICT on one end of the main CT is recommended. The ICT secondary of three phases can then be connected in closed delta.
3. The summation CT is used for the vector addition of current of several feeders. If the main CT ratios are different, then primary winding turns of summation CT should be altered with the objective of obtaining current on the secondary side, which is proportional to the summated load current.
4. In the case of core balance current transformers (CBCTs), the choice of the CBCT ratio should be left to the manufacturer to facilitate the best possible results. The following information should be provided by the user:
   (i) Cable size or minimum ID required for CBCT;
   (ii) Minimum earth leakage primary current to be detected; and
   (iii) Specifications of the relay to be used, including relay burden and relay setting proposed to be used.

7.5.2 A Typical Instrument Transformer Standardisation Model

In order to solve the daunting problem of standardising CTs and VTs, an attempt is being made to standardise CTs and VTs for indoor medium voltage applications. The model for CT is discussed below:

A basic indoor wound/slot type CT with two cores is chosen with the following rating. The third core is optional.

	Core 1 Protection	Core 2 Metering	*Core 3 Class PS (Optional)
Rated burden	15 VA	15 VA	(1) V_K formula such that the worked out V_K is less than 800 V (2) $Ie \leq 30$ mA at $V_K/2$
Accuracy class	5 P	1.0	—
Accuracy limit factor	10	—	—

*Applicable for CT secondary current of 1 A and only $1/\sqrt{3}$ A voltage class and insulation level that is any of the three depicted in Table 7.2.

Table 7.1 Rating for Indoor Wound/Slot Type CT

Sl. No.	Nominal System Voltage kV (rms)	Highest System Voltage kV (rms)	Power Frequency Withstand Voltage kV (rms)	Lightning Impulse Withstand Voltage kV$_p$
1.	3.3	3.6	10	40
2.	6.6	7.2	20	60
3.	11	12	28	75

Table 7.2 Basic Insulation Levels for CT'S

The I_{SF} for the metering core should be < 5 for the lowest ratio and < 10 for the higher ratio. CT dimensions for wound type and slot type CTs should be consonant with slot size and should be clearly mentioned. The STC rating, duration of STC and CT ratios are specified in Table 7.3 involving a total of 312 types of CTs.

S.No. and Type	Ratio (A/A)	Duration of STC	Code no. for Various STC Ratings					
			13.1 kA	18.4 kA	26.3 kA	31.5 kA	40 kA	44 kA
1 (Wound)	50/1-1	1 sec	C1	C2	C3	C4	C5	C6
		3 sec	C7	C8	C9	C10	C11	C12
2 (Wound)	75/1-1		...	...	...	...	...	...
3 (Wound)	100/1-1		...	...	...	...	...	...
4 (Wound)	150/1-1		...					
.	.							
8 (Wound)	400/1-1		...					
9 (Slot)	500/1-1	NA	Same code for all					
10 (Slot)	600/1-1	NA	Same code for all					
.	.							
15 (Slot)	2500/1-1	NA	Same code for all					
16 (Wound)	50/5-1	...	...					
.	.							
32 (Slot)	2500/5-1	...	...					
33 (Wound)	50/5-5	...	...					
.	.							
48 (Slot)	2500/5-5	...	...					

Table 7.3 List of CT Code Numbers

Each of the 312 codes mentioned in Table 7.3 is reproduced in Table 7.4. Here, the CT user is asking for price addition/deletion for various factors from the manufacturer.

For intermediate ratios, the price of the nearest lower ratio shall be considered.

Use of above CT selection and pricing tables solves the problem of CT costing and standardisation. This obviates the need for the user to obtain cost data from the manufacturer every time. The standardisation model for VT can be formulated along similar lines.

CT Code	Basic Price	Extra Price for				Reduction in Basic Price for	
		Accuracy 5P 20 agnst. 5P 10	Accuracy 0.5 agnst. 1.0	Addition third core (Class PS)	For 2 nos. Class PS in Place of Met+ Prot Cores	Double Ratio 100-50/1-1 Instead of 50/1-1	Deletion of Any Core	Output 10 VA Against 15 kA
C1							
C2							
.							
.							
.							
C312							

Table 7.4 Prices for Various CTs

7.6 MODERN CURRENT SENSORS (ROGOWSKI COIL)

Till recently, there was no alternative to the conventional electromagnetic type current transformer for industrial grade current measurement. This involves windings, ferromagnetic core and inherent non-linearity expressed in the form of saturation at higher currents. But modern trends ssuggest the emergence of new smart current sensors (Rogowski coils), which do not involve any ferromagnetic circuit and thus do away with the problem of saturation.

The term 'Rogowski coil' or 'Rogowski torus' means a conductor winding coiled on a toroidal former made of non-ferromagnetic material (e.g. plastic), thereby ensuring characteristics of excellent linearity. A Rogowski coil is particularly suitable for use in measuring the magnitude of sinusoidal current at the mains frequency. The Rogowski coil is looped around the bus bar-carrying current to be measured. The shape of the loop is not important, hence it is ideal for awkward shaped busbars and other components of a modern switchgear. This makes the Rogowski coil an ideal measurement device in the case of gas insulated switchgear (GIS).

The voltage induced in a Rogowski Coil is proportional to the rate of change of current enclosed by the coil loop. It is therefore necessary to integrate the coil voltage in order to produce an output voltage that is proportional to the current being measured. This explains why such a simple idea (first used in ac sensors in 1912) could not be put into common practice earlier as the electronic integrator design was not so advanced earlier.

7.6.1 Theory of Rogowski Coil

If a uniformly wound coil (N turns/m) on a non-magnetic former of constant cross-sectional area (A sq.m) is formed into a closed loop then the voltage E induced in the coil is given by the equation—

$$E = \mu_0 N A \frac{di}{dt} = H \frac{di}{dt} \qquad 7.6.1$$

where H (Vs/A) is the coil sensitivity, and

I is the current to be measured passing through the loop.

This is based on Faraday's law which states, "The total E.M.F. induced in a closed circuit is propotional to the rate of change of total magnetic flux linking the circuit".

An integrator with a low-pass filter network in parallel with the integrating capacitor is used along with the Rogowski coil. The filter reduces the gain at frequencies below the transducer bandwidth (typically 0.5 Hz) so as to reduce the low frequency noise and dc voltage offset drift. The resistor R_d [(approximately equal to ($\sqrt{L/C}$))] provides appropriate damping for the coil.

For frequencies within the bandwidth of the transducer $E = E'$, the integrator behaviour is given by:

$$V_{out} = (1/C_1 R_0) \cdot \int E \cdot dt = -R_{sh} \cdot I \qquad 7.6.2$$

where $R_{sh} = H/(C_1 R_0)$ is the transducer sensitivity. We see from equation 7.6.2 that the output voltage (typically a few milli-volts) is proportional to the current being measured.

7.6.2 Design Considerations of Rogowski Coil and Integrator

The absence of an iron core virtually eliminates circuit loading and saturation problems and the coil is suitable for almost unlimited over-current. Modern Rogowski coils have an extremely wide measurement range from 30 A to more than 100 kA at full scale, with sensitivities ranging from 0.01 mV/A to 100 mV/A.

A simple flexible Rogowski coil is bent into a closed path to completely capture the flux passing through its aperture. The wire loop may be configured as a single turn, a simple helix, a toroid, or other configuration used to form a sensor. One feature of this configuration is the coaxial routing of the coil end back to the beginning. This allows the coil ends to be temporarily separated to allow installation around a primary conductor. If this coaxial return was not incorporated, the sensor would essentially become a one-turn loop around the conductor and would become sensitive to stray magnetic fields.

The maximum current range is affected by both the frequency and magnitude of the measured current. This limitation exists due to the nature of technology and the finite gain bandwidth and output swing of the integrator electronics. Most probes provide a di/dt capability from 250 A/sec to 250 A/μ-sec. Variables such as the number of windings, cross-sectional area, amplifier gain, zero drift and shielding affect the di/dt range. The lower limit of di/dt is determined by the integrator design.

Accurate measurement is dependent upon a uniform coil cross-section. Bending flexible coils into a closed path deforms the circular cross-section into an oval, thus decreasing the turns area. Exceeding the minimum bend radius of the flexible form can cause some turns to permanently shift or even break. The interruption in turns at the ends of the coil or 'gap' can also cause non-uniformities that contribute to position sensitivity errors. This can occur where there are large gradients in the fields impinging on different parts of the sensor, such as measuring currents on a small conductor relative to the measurement head. Additional windings near the probe gap reduce this problem.

Two main difficulties encountered in measuring current by means of a conventional type of Rogowski coil pertain to the temperature sensitivity of key parameters and imperfect axial symmetry. These factors give rise to large errors in measurement. A particular design employs a printed circuit board (PCB) having a circular cut-out, the coil being formed by rectilinear metal deposits on each of the two faces of the PCB and extending along radii such that the geometrical projections thereof intersect at the centre of the cut-out. Deposits of copper on the PCB constitute the coil. The deposit surface is chemically stabilised by a deposit of gold, thereby controlling the internal resistance of the Rogowski coil.

The commercially available sensors have a specified accuracy of +/– 1 per cent of the calibrated value. Practically, accuracy of +/– 0.5 per cent can be obtained at the centre of the loop. As the busbar shifts away from the centre of the Rogowski loop, an accuracy of only +/–2 per cent to +/–3 per cent is expected. A linearity value of 0.05 per cent of full scale can be achieved. This shows the superiority of these sensors over conventional CTs. Large external currents and high voltage transients (>100 $V\mu$ sec) in the vicinity of the Rogowski coil also cause errors in measurement. The design of the switchgear should thus be such that these external effects are minimised.

A fundamental disadvantage of the conventional inverting integrator is that with an abrupt change in di/dt, such as when a switching device is turned on/off, the sensor output shows a transient pre-shoot which can be followed by oscillations. Hence, it is necessary to include a filter to reduce the magnitude of the pre-shoot and oscillations. This filter, however, significantly reduces the high frequency bandwidth of the sensor.

7.7 VOLTAGE SENSORS

Conventional voltage transformers depend on the magnetic circuit of the core for transferring primary voltage to the secondary. The magnetic circuit is inherently non-linear in nature, which causes errors in steady state measurements (ratio and phase angle errors). Also, the transient response of the VT is unstable. Moreover, modern digital measuring/protective equipments have high input impedance and thus impose a very small VA burden on the VT. This situation calls for the development of a compact, linear voltage sensing device, which has a stable transient response.

The voltage sensor discussed below is a device that fulfils these conditions. It is a simple resistive voltage divider or potentiometer consisting of HV and LV arms. The problems of power loss, voltage drop, transient stability, and insulation failure normally associated with potentiometer type dividers are addressed while designing the voltage sensor. The resistor is highly stable under steady state and transient conditions of the network.

The HV arm is a special high power resistor made of porcelain tube coated with special resistive strips. Metal paint is applied at both ends of the resistance for external connection. A layer of glazed silica covers the resistive strips to provide protection. The porcelain tube of the HV arm has suitable dimensions for withstanding rated one-minute power frequency and impulse voltages. The resistor is housed into a metal chamber for safety and for blocking electrical interference.

The resistance value chosen for the HV arm is of the order of 100 mega ohms in order to keep the current flowing through the small divider. This small current (about 60 µA for an 11 kV system) does not load the circuit under measurement. Moreover, the power loss at the rated voltage is a few watts. The resistor has a power rating of 120 W.

The LV arm resistor is a normal high voltage resistor housed into an insulation chamber at the bottom of the sensor. Suitable over-voltage protection is provided for the LV arm in the form of a surge diverter. The sensor is totally encapsulated with epoxy resin. The LV arm of the sensor consists of high voltage non-inductive resistor strips having a porcelain base. The LV arm is housed into an insulation chamber at the bottom of the sensor with necessary over-voltage protection. The resistance value of the LV arm is 64 kΩ, with a power rating of 5 W. This arrangement generates sufficient output voltage for operating commercially available electronic relays.

The voltage sensor must be tested for the following:

(a) One-minute power frequency withstand voltage test
(b) Impulse withstand voltage test
(c) Partial discharge test at 2 PC sensitivity
(d) Voltage ratio test for different voltage levels
(e) Temperature rise test at continuous operating voltage

7.8 FIELD PERFORMANCE OF INSTRUMENT TRANSFORMERS

An instrument transformer that is well designed and tested at the manufacturing stage and found to be of high quality, may still develop problems in field use. A major problem occurring in instrument transformers is that of insulation failure. System parameters like basic insulation level and the type of earthing are studied to decide the suitable insulation of instrument transformers. The failure to specify correct insulation requirements to the manufacturer causes such failures.

Sometimes in the case of VTs, ferro-resonance causes insulation failure. Take the case of an isolated neutral system having a set of three VTs connected in star and neutral point earthed. Such configurations are common in busbars of medium voltage switchgear meant for running power plant auxiliaries. Here, if there is no loading on the bus (or the bus is lightly loaded), the switching on of the incomer causes resonance to set up between the VT reactance and the stray capacitances of the busbar. This phenomenon is known as ferro-resonance. The circulating in-rush current causes very high voltage to appear at VT terminals, leading to insulation failure. In order to avoid this type of mal-operation, the VTs should be provided with one set of tertiary winding which should be connected in open delta. Across this open delta, a resistor of good thermal capacity having a value of 100 ohms or 200 ohms can be connected. During conditions leading to ferro-resonance, the resistor dissipates the surge energy which is then attenuated to a safe value.

For three single-phase earthed VTs connected in star earth, the connection is isolated at site for conducting the field HV test. Usually the testing staff forget to re-connect the star point to earth and commission the VT. Since these earth terminals are lightly insulated, an insulation failure occurs when the energised VT senses a neutral shift due to earth fault.

The opening of the CT secondary, while primary winding is energised, is also a serious condition. This causes dangerously high voltage to appear across opened joints which, in turn, leads to excessive heating and permanent magnetisation of the core. In order to avoid this, the CT secondary wire size should be of a larger cross-section than routine connections. Also, the terminal block joints should be of adequate size and the sliding contacts should be operated carefully. Some CT cores that are accidentally magnetised can be de-magnetised and re-tested by the manufacturer by subjecting the core to repeated magnetisation loops of decreasing magnitude.

Insulations of window type or slot type CTs tend to fail inside the window, the reasons for which are an unequal air gap around the rectangular busbar passing through the window. To avoid this failure, an equipotential shield should be embedded inside the resin at the location of the window. A lead should then be used to connect the shield with the busbar (see Fig. 7.12). This arrangement reduces the electric stress in the air gap to zero, as the gap now has parallel plates on both sides having the same potential.

Fig. 7.12: Side view of window type CT showing equipotential shield connections

A core balance CT (CBCT) is put around a three-core power cable to detect zero sequence current in the cable. The metallic armour is earthen at the switchgear end. Hence, the CBCT is in effect sensing the earth fault current of the main cables minus the return current through the armour. In order to avoid this error, the armour at the switchgear end should be earthed after passing the earth conductor through the CBCT opening. This nullifies the effect of the armour current.

Chapter 8

SURGE PROTECTION

Vivek Johri
M. Mohana Rao

An electrical surge is a transient over-voltage occurring in electrical networks. It is a voltage wave characterised by steeply rising voltage (wave front) followed by slowly decaying voltage (wave tail). A standard 1.2/50 µS impulse voltage wave is an example of a surge voltage. The electrical network and equipment connected with it are subjected to external and internal surges.

The most common cause of external surge is atmospheric lightning which induces a voltage surge in transmission lines. This lightning surge travels towards both the source side and the load side on the transmission line because of distributed line inductances and stray capacitances. The surge travels at the speed of light. At the ends of the transmission line, the surge-impedance changes and the wave is reflected back. This back-and-forth travel of the surge wave continues until the energy of the surge is attenuated by line resistance. This causes the voltage of the line to rise to several times the rated voltage.

Another type of surge voltage wave is generated by the switching action of certain types of circuit breakers. These internally generated surges are known as switching surges. Their mode of travel/reflection at the end of cable attenuation is similar to the lightning surge described above. The basic arc-quenching techniques of SF6 and vacuum circuit breakers give rise to current chopping and multiple re-ignition. These phenomena are responsible for the generation of switching surges, especially when the connected load is inductive in nature. This can be explained by the expression $L.di/dt$ for voltage induced across circuit breaker contacts. Surge protection devices are used to protect the equipment connected in the system.

8.1 SURGE PROTECTION DEVICES

Surge protection devices limit the over-voltages in electrical systems to the specified protection level, principally lower than the withstand voltage of the equipment. These devices are used to protect the conventional as well as gas insulated sub-stations from temporary over-voltages (TOV), switching impulses (SI), lightning impulses (LI) and to a certain extent, very fast transient over-voltages (VFTOs).

The impressed over-voltages have different shapes and amplitudes depending on the sub-station configuration, rating of sub-station, physical location of the sub-station, type of protection, etc. The maximum amplitude of over-voltages which occur rarely, is very high as compared to the normal system operating voltage. The amplitude of lightning over-voltages is enormous (~1000s of kV). Strengthening a sub-station against these over-voltages is uneconomical on the basis of probabilistic methods and insulation co-ordination. Necessary protection and insulation levels for the sub-station are designed. The calculation is based on input parameters such as voltage-time characteristics of insulation, tower footing impedance, sub-station configuration, etc. Surge protection devices limit the incident over-voltages and help in economising the insulation requirements for basic equipment and the sub-station. Surge protection devices have been used worldwide in power systems because of their excellent surge limiting/clamping capabilities and for the protection of system insulation.

Historical surge protection devices were the air insulated spark gaps. Because of their high response time, susceptibility to ambient conditions and incapability to handle phase-to-ground faults during operation, these surge protection devices could not protect the sub-station and the equipment effectively from over-voltages. The next generation surge protection devices used thyrite piles with a series spark gap. Series gap limited the flow of the power frequency through a thyrite pile. The advent of silicon carbide (SiC) elements was the next development in this series. These blocks carry a considerable amount of leakage current during normal system operation. These protection devices are in wide use today. In EHV and UHV systems, parallel resistor blocks are necessary to maintain uniform voltage distribution across series gap in SiC surge protection devices. Moreover, the response time of the series gap and high residual voltage of the SiC arrester blocks limit the protection characteristics of these surge arresters. In the process of innovation for highly non-uniform V-I characteristics, metal oxide-based surge arresters have been identified as a complete solution to the protection of electrical power equipment and the sub-station.

Lightning arresters provided at the ends of the line clamp the surge and bring it to a level which can be withstood by the equipment at the switchyard/sub-station. Hence the effect of lightning surges is minimal on the electrical equipment installed inside the plant/sub-station. In contrast, the effect of switching surges is immediately felt on the equipment in the absence of any lightning arrester in between.

Normally equipment like oil-filled transformers, switchgear, cables, transmission/distribution line, etc. can withstand the switching surges without any problem because their insulation is designed to withstand large over-voltages. Rotating machines, on the other hand, have little space in the stator slots, resulting in less insulation against over-voltage. Similarly, dry type transformers and electric arc furnaces also cannot withstand large switching surges. For such equipment, surge protection devices are required.

Another aspect to remember is that the circuit breaker generates the switching surge and the latter travels towards both the source side and the load side. But surge arresters are typically provided on the load side only, because source side equipment are sufficiently insulated to withstand the surge.

Broadly speaking, the devices used for dealing with surges can be classified as shown in Table 8.1.

Source of Surge	Type of Surge	Surge Protection Device	Mounting
External	Lightning surge	Lightning arresters	Outside sub-station plant
Internal	Switching surge	Surge suppressor/ arresters	Load side of circuit breaker

Table 8.1 Devices Used for Dealing with Surges

Some of the other devices used for the purpose of avoiding the ill-effects of surges are surge capacitor, spark gap, and surge suppression reactor. This chapter deals with surge protection devices meant for mitigating the effects of switching surges only. The two main types of surge protection devices used for reducing the magnitude of the surge voltage are:

(i) Metal oxide (e.g. ZnO) type surge arrester;

(ii) Capacitance-resistance (C-R) type surge suppressor.

The surge protection device based on metal oxide blocks is called a metal oxide arrester, gapless surge arrester or ZnO surge arrester. These blocks show quick responses of the order of a nano-second to surge voltages and absorb the incoming surge without appreciable time delay. The metal oxide surge arrester does not require series gap and is hence referred to as a gap-less metal oxide arrester. The metal oxide type surge arrester is a non-linear resistance. It consists of discs of metal oxide arranged in a stack inside a suitable insulator. The top end is meant for connection to a phase conductor on the downstream side of the breaker. The bottom end of the stack is earthed. Three such surge arresters are needed for a three-phase system.

Normally the device offers very large resistance (ideally infinite) to earth. When a switching surge appears, the resistance of the surge arrester decreases thus causing a large discharge current to flow to earth. The switching surge is thus diverted to the earth and its energy is absorbed by the surge arrester. Once the surge disappears from the system, the resistive properties of the surge arrester improve and it again acts like an open circuit. Thus the system voltage is not subjected to an earth fault.

The C-R type surge suppressor, on the other hand, is a device for balancing the load inductance and thus altering the surge impedance offered to the switching surge wave. It is a simple series capacitance-resistance combination connected between line and earth for each phase. The device changes the impedance seen by the surge and reduces the magnitude of the surge produced during switching operation. Thus we see a basic difference between the two types of surge protection devices. The ZnO type diverts the surge to earth and absorbs the energy of the surge.

On the other hand, the C-R type suppresses the surge before it is produced. This is the reason why the ZnO type device is called a 'surge arrester', while the C-R type device is called a surge suppressor.

8.1.1 Metal Oxide Type Surge Arrester

The mechanical design of this type of surge arrester comprises discs of metal oxide. The most commonly used metal is zinc. Hence this type of arrester is also known as zinc oxide (ZnO) type surge arrester. The discs are arranged in a cylindrical stack. The stack is put inside a polymer/porcelain housing. The stack is held in place by spring loaded cups at the top and bottom. After electrodes are fitted at either end, the remaining space inside the housing is filled with fibre-reinforced composite material. Mastic seal is used as an inner lining in the polymer/ceramic lining. The entire assembly forms a solid component, having threaded studs on either end for connections to line and earth respectively. The design of metal oxide surge arrester is a compromise between the energy absorption capacity of the discs and reduction of leakage current during the normal operation of the electrical system. The normal operation is defined as a condition when no surge is present and the arrester is subjected to normal system voltage only.

Metal oxide arrester blocks with a highly non-uniform current voltage (I-V) characteristic is applicable for over-voltage suppression. The non-linear resistance of the block is an inherent bulk property and consists of mainly zinc oxide (90–95 per cent) with relatively small amounts of several additives of other metal oxides (5–10 per cent) like alumina, antimony trioxide, bismuth oxide, cobalt oxide, zirconium, etc. On a macroscopic scale, the additives are almost homogeneously distributed throughout the arrester block. Figure 8.1 shows the view of a metal oxide block. The micro-structure of the metal oxide block represents a network of series and parallel arrangements of highly doped zinc oxide (ZnO) grains separated by inter-granular junctions. The non-linear behaviour is the superimposition of non-linear characteristics of the individual junctions. The current-carrying capacity of the arrester block is proportional to the total cross-section of the block. The I-V characteristics of a non-linear resistor are expressed as

$$\frac{I}{I_{ref}} = \left(\frac{U}{U_{ref}}\right)^a \qquad 8.1$$

Fig. 8.1: Metal oxide surge arrester block

$$\alpha = \ln\left(\frac{I}{I_{ref}} \Big/ \frac{U}{U_{ref}}\right) \qquad 8.2$$

where I_{ref} and U_{ref} are the reference current and voltage respectively of the surge arrester block. For effective voltage protection, the a value of the arrester block should be greater than 20. For the metal oxide block, α is in the range of 30 to 40 and for SiC elements, it is only in the range of 3 to 4. Figure 8.2 shows the I-V characteristics of a metal oxide block. From this figure, it is clear that, for a normal system, the voltage current increases linearly and in the range of fraction of milliamperes. Beyond the reference current (transition point), only a small change in voltage is observed with an increase of current in kA. These non-linear characteristics of the metal oxide block provide protection against over-voltages.

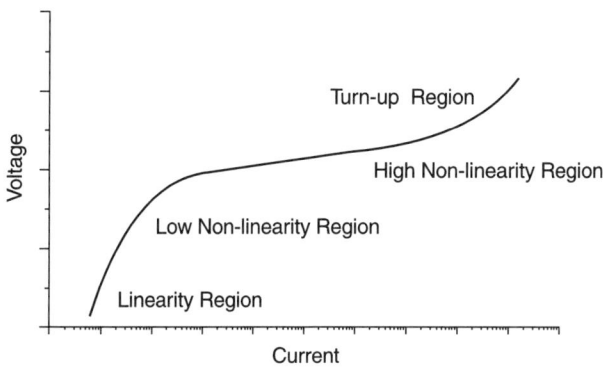

Fig. 8.2: I-V Characteristics of metal oxide block

The I-V characteristics of the metal oxide arrester blocks are highly dependent on the operating temperature. The current at a given voltage increases with the temperature (i.e., the block has a negative temperature coefficient). Since the metal oxide block carry continuous leakage current, the thermal design of the arrester housing plays an important role in the effective dissipation of heat during the post-current discharge. There is a critical temperature depending on the voltage rating of the block, beyond which joule heat generated in the block cannot be dissipated. The heat generation in the ZnO block is proportional to the product of the applied voltage U (effective value) and the peak current I flowing in the blocks).

$$P = k \cdot U \cdot I \qquad 8.3$$

where k is 0.4 to 0.5 in the current range of 0.1 to 1.0 mA. The current magnitude depends on the block temperature and applied voltage. Heat dissipation outside the arrester is proportional to the difference between the block temperature and the ambient temperature T_o.

$$Q = \lambda(T - T_o) \qquad 8.4$$

The value of λ depends on the dimensions of the arrester housing and the thermal design of the surge arrester. When high currents are discharged through metal oxide blocks, the temperature of the block increases. If the temperature is less than the critical temperature, the block may come to its original state for power frequency follow up currents, otherwise the block temperature may continue to increase and finally lead to thermal runaway.

The residual voltage of the block (voltage across the block during discharge current) does not depend on temperature. This insensitivity to temperature also ensures a very good homogeneity of current flow during discharges and allows the parallel connection of blocks to increase absorption capability.

Metal oxide blocks can absorb energy of more than 7 kW-s/kV continuous rating, without damage. The performance of ZnO blocks for fast rising over-voltages is observed to be remarkable since the block can absorb energy more than 12 kW-s/ kV in a single event. The metal oxide surge arrester is basically a non-linear resistance. Accordingly, its principal ratings are:

1. *Voltage Ratings*:
    (a) Rated voltage and continuous operating voltage (COV); and
    (b) Basic insulation level expressed in temporary over-voltage withstand/lightning impulse withstand/power frequency voltage withstand capacities.
2. *Thermal Ratings*:
    (a) Discharge capability in kJ/kV; and
    (b) Current rating expressed as nominal discharge current.
3. *Impedance Ratings*:
    (a) Resistance (expressed as residual voltages attained as a result of the application of various current impulses. The types of current impulses are lightning, steep and long duration impulses; and
    (b) Impedance distribution expressed as resistive and capacitive components of continuous current at COV.

Following the rule of thumb for voltage ratings of a surge protection device, we can assume that the maximum COV is 85 per cent of the nominal voltage and a 1 sec temporary over-voltage is 120 per cent of the nominal voltage. For quantifying various other ratings described above. Table 8.2 gives the typical ratings of a zinc oxide arrester used for protecting motor feeders/dry type transformer feeders in a 6.6 kV switchgear.

Rating	Value
Rated Voltage	7.2 kV rms
Maximum Continuous Operating Voltage (COV)	5.62 kV rms
Nominal Discharge Current	10 kA
Line Discharge Class	3
Discharge Capability	
(a) Repetitive	3 kJ/kV
(b) Single Shot	9 kJ/kV
Long Duration Current Impulse (2400 µS)	500 A
Rated Frequency	50 Hz
Maximum Resistive Component of Continuous Current at COV	0.5 kA_p

(Contd.)

(Contd.)

Rating	Value
Maximum Capacitive Component of Continuous Current at COV	1.5 kA$_p$
Reference Current	3 mA$_p$
Reference Voltage	7.56 kV$_p$ to 8.64 kV$_p$
Maximum Residual Voltage at Current Impulse of 8/20 µS Duration and having a value of:	
(a) 5 kA	18.8 kV
(b) 10 kA	19.9 kV
(c) 20 kA	21.9 kV
Maximum Switching Impulse Residual Voltage at Current Impulse value of:	
(a) 1000 A$_p$	15.9 kV
(b) 500 A$_p$	15 kV
Maximum Steep Current Impulse Residual Voltage at Current Impulse value of:	
(a) 10 kA$_p$	21.9 kV
(b) 1 kA$_p$	16 kV
High Current Impulse Withstand (4/10 µS)	65 kA
Temporary Over-voltage Withstand Capacity for a duration of:	
(a) 0.1 S	11.71 kV
(b) 1 S	11.18 kV
(c) 10 S	10.68 kV
(d) 100 S	10.18 kV
Basic Insulation Level:	
Lightning Impulse Withstand Voltage (Dry)	60 kV$_p$
Power Frequency Withstand Voltage (Wet)	20 kV rms
Maximum Radio Interference Voltage	45 µ V

Table 8.2 Ratings of a Zinc Oxide Surge Arrester 7.2 kV, 10 kA

The continuous operating voltage should be more than the line-to-earth voltage of a system (which is equal to the line-to-line voltage divided by $\sqrt{3}$). At this voltage, the arrester operates normally with a small discharge current. If the voltage rises above this value, the resistance of the arrester drops and large discharge current passes through it.

8.1.2 Capacitance Resistance (C-R) Type Surge Suppressor

This is a simple surge protection device, which has a power capacitor and resistor in series. Similar to the metal oxide type of surge arrester, it is also connected between line and earth on the feeder side of the switchgear panel. The capacitor used has all the ratings of a normal power capacitor used for the designated voltage class. The suppressor works in the following ways:

1. The C-R surge suppressor reduces the effective surge impedance as seen by the breaker on the load side. This is so because typical loads are inductive in nature and the C-R surge suppressor balances the inductance with its capacitance. This results in the generation of

lower over-voltage during current chopping as the effective value of L in the expression $L.(di/dt)$ is reduced.

2. The energy of the switching surge is dissipated fast by the resistor R connected in series with the capacitor. It reduces multiple travelling of the surge wave and suppresses voltage escalation.
3. This device lowers the rising rate of the voltage appearing between the contacts after interruption and suppresses the occurrence of re-ignition.
4. The device suppresses the high frequency current at re-ignition because the current waveform takes the shape of an exponential curve attenuated with the C-R time constant.
5. The device suppresses virtual current chopping. This type of chopping is caused by multiple re-ignition. The instance of multiple re-ignition is made to cease by using the C-R type surge suppressor.
6. C-R surge suppressors reduce over-voltage even when the breaker is performing a make operation.

The values of C and R are determined by the non-oscillatory conditions of an equivalent circuit through which a high frequency current flows when re-ignition occurs. The capacitance value must satisfy the following energy equation, among other things:

$$(1/2) LI^2 = (1/2) CV^2 \qquad 8.5$$

where the symbols have their usual meanings, L being the load inductance and C the capacitance of the surge suppressor.

Empirical relations exist for the calculation of C and R. When the results obtained by empirical relations are tallied with experimental results, it is seen that the following values of C and R give satisfactory results of suppressed over-voltages for medium voltage applications:

$\qquad$ C – 0.04 µF to 0.3 µF

$\qquad$ R – 20 Ω to 1000 Ω

Of the above values of C and R, the most commonly used values adopted for the protection of motor controlled by medium voltage switchgear are:

$\qquad$ C – 0.1 µF

$\qquad$ R – 100 Ω

The other ratings of this type of surge suppressor are nominal system voltage, highest system voltage, rated frequency, power frequency voltage withstand and impulse voltage withstand test.

The C-R type surge suppressor is subjected to the following test.

1. **Type tests**: These include measurement of capacitance, thermal stability test, short-circuit discharge test and impulse test.
2. **Routine tests**: These include measurement of capacitance, tan delta test, and HV power frequency voltage withstand test.

8.1.3 Application of Surge Protection Devices

Surge protection devices are used in switchgear to clamp transient over-voltage at levels below the allowable BIL of the equipment being fed and the switchgear feeding it. The switching surge generated by the breaker is propagated along the cable until it reaches the slot entry of the motor, where it encounters a change in the surge impedance. This causes reflections and refractions of the wave.

Depending on the slot dimensions and the position of the conductors in slots, there is an influence of the mutual capacitance and inductance between parts of the winding on the surge wave shape. The wave shape of the surge is also affected by the eddy current losses in the iron core. These factors play an important role in the selection of surge protection devices.

The application considerations of the two types of surge protection devices described above are discussed below.

8.1.3.1 Metal Oxide Surge Arrester

The metal oxide type surge arrester offers smooth turn-on and turn-off at the lower surge current levels. In order to apply such arresters in switchgear, one must consider factors such as ambient temperature, the lead length of the conductor connecting the arrester to the equipment being protected and the phase spacing between the arresters.

As the temperature of the device increases, the leakage current also increases. This increase in the leakage current causes the temperature of the device to exceed beyond the allowable limit for metal oxide discs. The internal resistance of the surge arrester drops and the turn-on voltage of the device is lowered. If the temperature exceeds the thermal capability of the arrester, thermal runaway occurs and the arrester turns-on at the system operating voltage and will not turn-off, thus resulting in a permanent line-to-earth fault.

The length of the lead connecting the arrester to the protected equipment must be kept to a minimum. Similarly, the lead length between the arrester and earth must be kept to a minimum. Since the lead resistance offered to the travelling surge wave adversely affects the discharge voltage of the surge arrester, the arrester discharge voltage is decreased by approximately 5 kV per metre length of lead. This factor is more prominent in higher voltages, where the BIL co-ordination margin is minimum.

Due to the above condition of minimum lead length, the most appropriate place to mount the surge protection device is at the terminal of the protected equipment (e.g. the motor). However, due to practical mounting considerations, it is mounted at the feeder side of the switchgear panel.

While the arresters are being mounted, adequate phase-to-phase spacing must be maintained inside the switchgear. If any heat shrinkable sleeve is used, it must not appreciably bridge the outer creepage distance of the arrester.

8.1.3.2 C-R Type Surge Suppressor

The C-R type surge suppressor is effective in suppressing prospective surges which could be generated during the breaking and making operations of the circuit breaker. Its multi-pronged action is described in section 8.1.2 above. Table 8.3 gives a typical test result of over-voltage measurements with different ratings of H.T. motor fitted with different types of surge suppressors.

Motor Rating →	3 kV, 37 KW		6 kV, 150 KW	
Switching instant → Protective Device ↓	When motor is running at normal speed	Just after the motor has started	When motor is running at normal speed	Just after the motor has started
Surge Capacitor (0.1 µF)	1.7 p.u.	4.0 p.u.	3.2 p.u.	3.7 p.u.
C-R Surge Suppressor (0.1 µF, 100 Ω)	1.4 p.u.	2.0 p.u.	2.1 p.u.	1.1 p.u.
C-R Surge Suppressor (0.05 µF, 600 Ω)	1.3 p.u.	2.1 p.u.	1.9 p.u.	1.1 p.u.

Table 8.3 Typical Test Result of Over-voltage Measurements

The test results show that the C-R surge suppressor is superior in performance of surge protection the surge capacitor. With C-R surge suppressors, the instances of higher over-voltages during switching immediately after the starting of the motor are minimised. This is because of the energy dissipation feature of the series resistor R. The maximum over-voltage experienced by the motor is around 2.1 p.u., which is not harmful for motor insulation. The surge generated in making current is usually less than that in interrupting the current.

8.2 OVER-VOLTAGES IN THE ELECTRICAL SYSTEM

Surge protection devices are used in electrical sub-stations to limit the over-voltage level to the protection level. The insulation co-ordination of conventional/gas insulated sub-stations indicates the specification of BIL of the station equipment, air clearances of both phase-to-ground and phase-to-phase, rating and location of surge protection devices, etc.

8.2.1 Over-voltages in Sub-stations

The basic inputs for designing a metal oxide surge arrester are as follows:
1. Determine the maximum phase to neutral operating voltage and the maximum per unit over-voltage that can occur in the power system;
2. Determine the duration for which over-voltage conditions can persist.

By using the voltage-time characteristic, one can determine the rating of the arrester such that the duration of over-voltage is within the withstand capability of the arrester selected. The main type of over-voltages in sub-stations are detailed below.

8.2.1.1 Temporary Over-voltages

Temporary over-voltages determine the continuous operating voltage of the arrester and thus the protection level of the arrester. For grounded neutral systems, the temporary over-voltages during fault conditions may not exceed 1.4 p.u. In an isolated neutral system, the over-voltages are upto √3 p.u. An arrester with $U_c = 1.05\ U_m/\sqrt{3}$ is called a 61 per cent arrester, and may be suitable for grounded neutral systems. The rated voltage of the arrester is generally 1.2 times the continuous operating voltage U_c. For an isolated neutral system, the rated voltage of the arrester U_r is equal to maximum system voltage U_m. Thus, depending on voltage-time characteristics, the rating of the surge arrester may be more than 61 per cent. Further, in many operating systems, TOVs may persist for longer periods of time, the peak value of the rated voltage V_{rp} should be less than the maximum value of the reference voltage V_{ref} of the arrester. For better protection, the ratio of the above two parameters is in the range of 0.9 to 0.98.

8.2.1.2 Switching Impulse (SI) Over-voltages

The switching over-voltages may be divided into the following three groups depending on the method of over-voltage control:

1. The over-voltages without any control may be upto 3.5 p.u.;
2. The over-voltage level is upto 2.5 p.u. with closing resistors; and
3. The over-voltage level is upto 2.0 p.u. with control switching.

The sub-station protection against switching surges is ensured, if the system is well protected from lightning over-voltages.

8.2.1.3 Lightning Impulse (LI) Over-voltages

The time period of the impulse wave depends upon type of the lightning stroke, i.e. direct/indirect. The steep wave of a less than 1 µs rise time can occur only through a back flashover (in direct stroke) at a relatively short distance from the sub-station. When a lightning stroke hits the earth wire or tower, a rise in voltage may occur because of the tower footing impedance and flashing over of insulator string may take place. This flashover is known as a back flashover. During the designing the GIS (gas insulated sub-station) insulation, multiple strokes should be considered. The reason for this is that after the circuit breaker opening due to the first stroke, the GIS may be stressed and an over-voltage wave generated by the second discharge. The second wave then meets an open end with the possibility of higher over-voltages due to a change in the reflection conditions.

8.3 EXPERIMENTAL STUDY IN USE OF COMBINATIONS OF SURGE PROTECTION DEVICES

In order to study the effects of different combinations of surge protection devices, an experimental study was undertaken, which is discussed below. Table 8.4 gives the test results of over-voltages using different combinations of various surge protection devices. The motor effect was simulated using IEC equivalent circuit and the surge protection devices were installed first at motor terminals and then at C.B. terminals.

Surge Protection Devices Mounted at	Surge Voltage Measured at	Without Protection	Surge Voltage in p.u. with Protection				
			C(0.5 µF)	R-C (50 Ω & 0.5 µF)	ZnO	ZnO & C Parallel	ZnO & C Series
Experiment no.→		(1)	(2)	(3)	(4)	(5)	(6)
Motor terminals	Motor terminal	3.89	1.79	1.56	1.11	1.51	1.2
	C.B. Terminals	4.89	2.79	2.56	2.10	2.52	2.19
C.B. terminals	Motor terminal	3.89	2.66	1.56	1.11	1.50	1.11
	C.B. terminals	4.89	3.65	2.56	2.10	2.50	2.10

Table 8.4 Test Results of Over-voltages Using Different Combinations of Surge Protection Devices

In experiment no. (1) in the table, no protection was used, it gave rise to a maximum surge of 10.5 kV, i.e. 3.89 p.u. at motor terminals. A high frequency voltage of 2.5 kHz was imposed on the power frequency voltage. Next, in experiment no. (2), a surge capacitor of 0.5 µF was used. This prolonged the rise time to 0.6 mS at surge level of 1.79 p.u. The use of capacitor decreased the high frequency value of surge to 0.31 kHz. In the third experiment no. (3), R-C series circuit of 50 Ω and 0.5 µF were used. This increased the rise time to 0.5 mS with a surge strength of 1.56 p.u. There are no oscillations, and all the travelling waves phenomena were obviated by the circuit.

In experiment no. (4), the use of ZnO type surge arrester did not help to increase the rise time but decreased the surge level to a superior value of 1.11 p.u. This is so because the voltage peak was clamped to the threshold voltage level of the arrester. The parallel combination of ZnO and C combined the advantages of both the devices and resulted in a reduction of surge strength to 1.51 p.u. Lastly, the series connection had the effect of dividing the voltage among the two devices. This resulted in the ZnO arrester being operated in delay and hence the surge value was more as compared to that obtained when only ZnO arrester was used. Series combination is not useful because of discrepancy induces by delayed operation. The surge voltages measured at the C.B. terminals were more than the motor terminal values by about 1 p.u. This is because the motor voltage is added by reflection to the system voltage.

Afterwards, the experiments were repeated with surge protection devices installed near motor terminals. In this condition, the travelling wave phenomena took place in the cable and the voltage surges at motor terminals were increased by reflection. The stress on the source side is thus higher in experiments wherein only capacitors were used for protection.

The conclusion drawn from the above experiments is thus that as far as surge level considerations are concerned, the use of ZnO arrester alone is the best. But this type of arrester reduces the rise time of the surge wave, which may result in stressing of the motor insulation. It follows that for older motors, R-C protection can be used because of its advantages of cost, simplicity and surge level. But for new frequently switched H.T. motors, a combination of ZnO and C in parallel can be used. The parallel combination offers the advantages of both methods of protection with the only limiting factors being the paucity of space available inside switchgear panels and increased cost.

8.4 SURGE PROTECTION DEVICES FOR GAS INSULATED SUB-STATIONS

Gas insulated sub-station is a sealed-for-life electrical installation. Hence, it requires certain special considerations with respect to design and selection of surge protection devices. These considerations are discussed below.

8.4.1 Over-voltages in GIS

GIS is exposed to Very Fast Transient Over-voltages (VFTO) along with other over-voltages like temporary over-voltages, switching impulse over-voltages and lightning impulse over-voltages. In a gas insulated sub-station, the VFTOs are generated mainly due to switching operations. These transient voltages have a rise time of about 3–10 ns with a peak magnitude of about 2.8 p.u. for the most onerous condition of switching. The number of VFTOs in each switching operation may be in the range of 20–100 depending upon the operating speed of the switch. Each VFTO may appear for a time period of 2 to 10 μs depending upon the switching operation. The operation of the surge arrester for these transient over-voltages is a topic of research.

The voltage strength versus the time of the GIS insulation is nearly constant from the normal system frequency to the frequency range of lightning over-voltages, whereas the strength curve of the air insulation of the conventional sub-station increases with frequency. Thus, the insulation co-ordination of a gas insulated sub-station is different from that of conventional sub-stations. At the same time, the insulation breakdown in GIS is more severe than in a conventional sub-station. From the point of view of over-voltage protection and insulation co-ordination, lightning over-voltages are the most difficult to tackle because of high frequency and possible increase in voltage due to reflection at the transition points in GIS. The switching over-voltages are associated with relatively low frequencies and hence switching over-voltages are almost constant throughout the GIS. The necessary measure to protect the GIS against lightning over-voltages is highly dependent on the design of the connected transmission line. A capacitor at the entrance of GIS reduces the steepness and the crest value of the over-voltages and thus reduces the risk of failure.

8.4.2 Why Gas Insulated Metal Oxide Arrester (GISA) for GIS?

The modular concept of gas insulated sub-stations offers more flexibility in the design of switchgear and sub-stations in terms of an efficient use of space. The layout of gas insulated switchgear consists of modules such as circuit breakers, isolators, earth switch, and current and voltage transformers enclosed in a compressed insulating gas (e.g. SF6).

The following types of protection may be used for GIS application:

1. Conventional air insulated surge arrester (SiC/spark gap);
2. Conventional metal enclosed surge arrester (SiC/spark gap);
3. Air insulated metal oxide surge arrester; and
4. Gas insulated metal oxide surge arrester.

The response time of the series gap and high residual voltage of the SiC arrester blocks for discharge currents limit the protection characteristics of the conventional air insulated and metal enclosed surge arresters. The active part of an enclosed surge arrester is connected with extremely short leads. However, overhead line arresters (air insulated metal oxide surge arresters) are normally connected via long leads to line and to earth. The connection leads between the overhead line and the arrester terminal creates an inductance and may influence the performance of the surge arrester. The residual voltage of the arrester varies only moderately with the discharge current amplitude, but the steep current will generate voltage additional to the residual voltage due to the inductance of arrester leads. Further, the arrester connection leads of the arrester causes a delay in the response time of the arrester.

(a) 36 kV (b) 145 kV

Fig. 8.3: Gas insulated surge arresters

In practice, conventional surge arresters are installed outside the GIS preferably at the transformer end and the protection is limited by the size of the sub-station, electrical distance between GIS and the arrester, among other things. However, the gas filled surge arrester forms an integral part of GIS. The main advantages of a gas filled surge arrester over a conventional porcelain/silicon-clad surge arrester are:

1. Quick response for steep discharge current because of small length of connecting leads
2. No connecting lead voltage drops like in conventional metal oxide surge arrester
3. Lower protection level and hence reduction in the cost of insulation
4. Immunity from external environment like pollution, snow, rain, etc.
5. Reduction in over-voltages seen by the connected transformer/equipment
6. Flexibility to locate the arrester at an optimum position in GIS

The gas filled surge arrester comprises mainly the stack assembly, HT shield for uniform voltage distribution under normal operating conditions and the housing of the arrester. The stacks developed are modular in design and effective in reducing the assembly time. The development of gas filled arrester for better performance is achieved by controlling the following parameters:

1. Overall inductance of the gas filled arrester;
2. Uniform voltage distribution across the metal oxide blocks;
3. Height of the surge arrester; and
4. Thermal design of the gas filled arrester.

The uniform voltage distribution across the blocks under ac voltage is achieved by using an HT shield. The shield, which is designed for uniform voltage distribution across blocks by simulating the surge arrester with equivalent electrical circuit, takes into account the stray capacitance from the blocks to enclosure, height of the stack, capacitance of the blocks, capacitance between stacks, capacitance between shield and second stack, etc. The analysis has been carried out using PSPICE/EMTP software-based programme. The voltage variation across blocks should be within ±8 per cent for better performance. In a gas insulated surge arrester, leakage current flows through stray capacitance between the elements and grounded metal enclosure. The ZnO element behaves as an insulator of capacitive nature for the normal power frequency voltage. If all the elements are in a single stack (without shield), there is a possibility of 35 per cent over stress on the topmost element of the stacked ZnO elements (depending upon the diameter of the enclosure). This may lead to a thermal runaway problem after the discharge current flows through surge arrester blocks. Figure 8.3 shows the view of gas filled surge arresters for 36 kV GIS and 145 kV GIS.

8.4.3 Location of Arrester

In GIS, surge arresters are strategically located for its efficient protection since the damage caused by a flashover in such a sub-station is more severe than in conventional air insulated sub-stations. Because of lower surge impedance of the GIS elements, the arrester gives a greater protection length as compared to conventional sub-stations. The crest value as well as steepness of lightning over-voltages in GIS may vary from position to position in the station. When the lightning over-voltage enters the low surge impedance GIS, a fraction of wave is reflected. The wave transmitted into the GIS is reflected at the transformer and voltage at the transformer increases. Further, a reflection with voltage rise occurs at a relatively high surge impedance of the overhead line. In order to study possible over-voltage levels for different configurations of gas insulated sub-stations, BHEL has developed a PSPICE model. The modelling of sub-station components and behaviour of the metal oxide surge arrester during lightning are key aspects of the above model. The modelling of the surge arrester is based on experimental results. The surge arrester behaviour in full current range can be represented in the form of $I = kV^n$. This model is valid for TOV, SI and LI over-voltages with reasonable accuracy. Alternatively, a high frequency model is used to represent the surge arrester for very fast transient over-voltages.

The following procedure is adopted to identify the location and quantity of surge arresters required for the protection of gas insulated sub-station/conventional air insulated sub-station from lightning over-voltages:

1. Estimate over-voltage levels at different positions of the sub-station for a lightning impulse input, without metal oxide surge arrester. The peak value of lightning over-voltage is in the range of withstand level of transmission lines (above 3 p.u. depending upon the system voltage).

2. If the maximum over-voltage in the sub-station is greater than 3.0 p.u. (the protection level of surge arrester is 2.2 – 2.5 p.u.), provide one surge arrester at the terminal of the transformer.
3. Repeat the above analysis and identify the positions of over-voltages above 3 p.u. Provide the surge arrester at the entrance of GIS.
4. Repeat the above analysis and identify the positions of over-voltages above 3 p.u. Provide the surge arrester internal to GIS particularly near the open switch. If the length of the gas insulated system is in tens of meters, more than one surge arrester may be required inside the GIS.

8.4.4 Performance Evaluation

The performance of the gas filled surge arrester is evaluated in two stages. In first stage, metal oxide blocks are evaluated while in the second stage, full assembly of the surge arrester is evaluated.

8.4.4.1 Evaluation of Individual Surge Arrester Block

The following tests are conducted to verify the performance of an individual surge arrester block:
1. *Residual voltage test*: The zinc oxide elements are tested for the protective level at a nominal discharge current of 10 kA with 8/20 micro-second wave.
2. *Reference voltage test*: The power frequency reference voltage test is carried out on all the blocks in the transition area between the leakage current and the conduction current region normally in few mA.
3. *Leakage current at MCOV*: The leakage current at the rated voltage and maximum continuous operating voltage (MCOV) are measured and the blocks satisfying the specified leakage current levels are selected only for the surge arrester assembly.
4. *Energy withstand capability test*: The surge discharge capability of metal oxide blocks is defined in terms of high current (4/10 µS), classification current (8/20 µS) and rectangular wave (2 mS) discharge. These discharge currents verify the material uniformity, current density (area of the block) and temperature rise of the block within tolerable levels.

8.4.4.2 Evaluation of Gas Insulated Surge Arrester

The following tests are conducted to verify the assembly of blocks, uniform voltage distribution across blocks, energy dissipation capability of the housing and verification of protection level of the gas filled arrester:
1. *AC withstand test*: The shield and FRP tube assembly without blocks are to be tested for partial discharge at normal voltage and ac withstand voltage of 275 kV (rms) for one minute, to verify its suitability to withstand over-voltages as per IEC-99-1 standards [1].
2. *Lightning impulse withstand test*: The performance of a gas insulated surge arrester mainly depends upon the effectiveness of the HT shield. The shield and FRP tube assembly without blocks are to be separately tested for lightning impulse (LI) of 650 kVp of 1/50 µs wave shape, to verify its suitability to withstand over-voltages as per the above standards. Figure 8.4(a) shows the view of a 145 kV gas insulated surge arrester for the LI test.

3. *Leakage current measurement*: The metal oxide arrester carries a small leakage current of the order of 600 µA under the normal system voltage. The zinc oxide arrester element is a negative resistance element. The temperature of the element increases with an increase in the resistive component of the leakage current. Further, an increase in temperature results in higher leakage currents. This is a cumulative process. Thus, for a surge arrester to be acceptable, the value of the leakage current should be in limits even at elevated voltages. Figure 8.4(b) shows the view of an experimental set-up for leakage current measurement. The leakage current is measured up to the rated voltage of the arrester. The continuous leakage current measurement has been carried out both at the maximum continuous operating voltage (MCOV) and elevated voltage (1.2XMCOV) for 30 minutes.

(a) Lightning Impulse Test

(b) Leakage Current Test

Fig. 8.4: Test set-ups for evaluation of GISA

No significant change in leakage current has been reported during the above measurement. The assembled arrester is also tested for reference voltage which is greater than the rated voltage. Figure 8.5 shows the variation of leakage current through the blocks for different voltages upto rated voltage.

4. *Residual voltage test*: The protection level is normally evaluated in terms of the values of the residual voltage at the nominal discharge current of 10 kA, 8/20 us wave shape. Figure 8.6 shows the I-V characteristics of a 145 kV gas filled surge arrester in the high current zone. From this graph, it is clear that the voltage across the surge arrester is almost constant beyond a 1 kA discharge current. Figure 8.7 shows the waveforms of residual voltage measurement across the 145 kV gas insulated surge arrester for a discharge current of 16 kA at a test station. The residual voltage of the surge arrester along with bushing is at an acceptable level as per standards. Alternatively, the residual voltage of the complete arrester assembly may be determined as the measured value of the residual voltage of the section multiplied by the ratio of the rated voltage of the complete arrester to the rated voltage of the tested section.

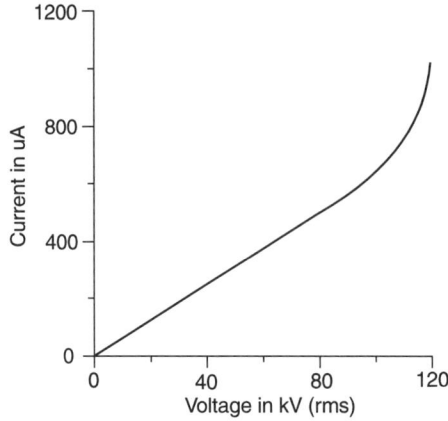

Fig. 8.5: V-I characteristics of 145 kV surge arrester in low current zone

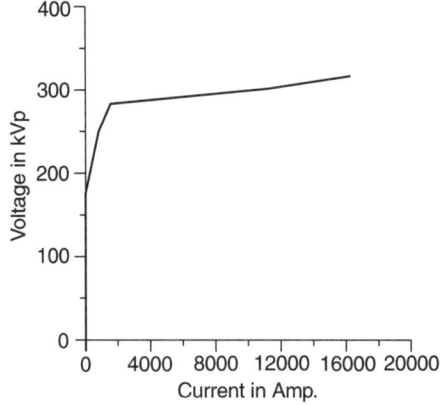

Fig. 8.6: I-V characteristics of 145 kV surge arrester in high current zone

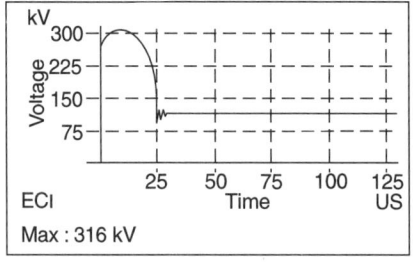

Fig. 8.7: Residual voltage profile for 16 kA, 8/20 μs impulse

8.5 PERFORMANCE MONITORING OF SURGE PROTECTION DEVICES

The performance monitoring of surge arresters may be classified into two groups. The first group, pertains to online measurement when the arrester is connected to the system and energised with the service voltage. The second group concerns off-line measurement when the arrester is disconnected from the system and energised with a separate voltage source, at a site or in a laboratory. The major disadvantage of off-line measurement is that the arrester has to be disconnected from the system. The number of operations of a surge arrester during service is monitored continuously using surge counters. Clamp-on meters are used for the watt loss measurement of the surge arresters.

8.5.1 Surge Counters

Surge counters operate for impulse currents above a certain amplitude, or above certain combinations of current amplitude and time duration. The counter does not provide specific information about the condition of the arrester. For safety reasons, the surge counter should be installed beyond easy reach of personnel. It should be located where it can be read from the ground level with the arrester in service. The installation should be done without considerably lengthning the earth connection or reducing its cross-section. The arrester should be equipped with an insulated earth terminal, and a conductor between the arrester and counter should be insulated from earth.

8.5.2 Measurement of the Leakage Current

The leakage current mainly comprises the capacitive current and the resistive part of the current. The resistive part of the current is only a fraction of the total leakage current, under normal working voltage. The capacitive and resistive components of leakage current differ in phase by 90°. Therefore, a large increase in the resistive current of the non-linear metal oxide resistors is needed to observe a significant change in the total leakage current level. In addition, the leakage current is sensitive to the installation since the capacitive current depends upon the stray capacitances of the system.

Online measurements of the leakage current are extensively used in practice by means of conventional mA meters built into the portable instruments, showing the r.m.s., mean or peak value of the leakage current. The low sensitivity to changes in the resistive current level makes the measurement of the leakage current suitable as a diagnostic indicator only in those rare cases when the resistive current is in the same range as the capacitive current.

8.5.3 Measurement of the Resistive Component of Leakage Current

The surge arrester performance is evaluated by measuring the resistive part of the leakage current, which is proportional to the watt loss of the arrester. The increase in watt loss is a direct measure of deterioration. If the resistive component is at a considerable level, then the leakage current waveform is not purely sinusoidal and rich with third harmonics. The output of the measurement generates a data of amplitudes of different frequency components. Measurement of these frequency components up to the ninth harmonic give good feedback about the performance. These harmonics increase with an increase in the resistive leakage current. The CT clamp type measurement is recommended because disconnection of ground wire is not required for the above measurement. Figure 8.8 shows the view of a CT clamp type meter used for watt loss measurement.

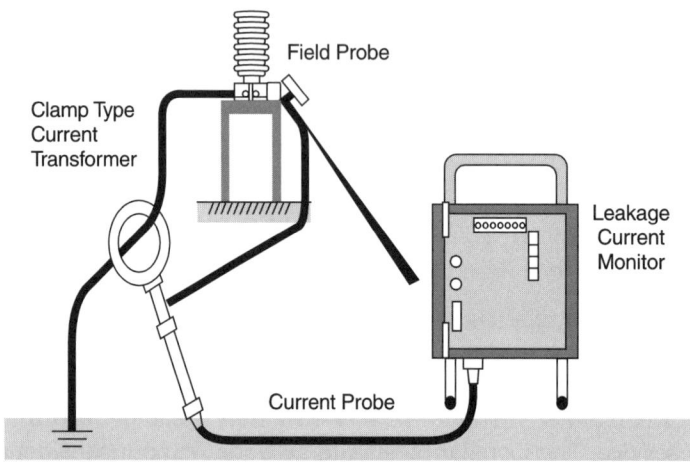

Fig. 8.8: Leakage current measurement using clamp-on meter

References

1. IEC 99-1,"Lightning Arresters, Part 1: Non-linear Resistor Type Arresters for ac Systems", 1970.
2. Gandhare, W.Z. and M.S. Agarwal, "Computer Simulation of Surges Caused by Direct Switching of High-voltage Motors", *Fourth International Seminar on Switchgear and Controlgear*, 1996, IEEMA, IIT, Mumbai.

Chapter 9

PROTECTIVE RELAYS

M.K. Srivastava
Neelam Bhogal

All electric power systems constitute certain basic components such as generating stations, transformers, transmission lines and motors. But we all know that for proper and efficient functioning, the power system incorporates many other important components. Protective relays are one of those vital constituents of the power system.

The function of protective relay is to affect disconnection of any faulty section of the power system from service. A fault is said to have occurred in a section when it suffers a short-circuit or when it starts behaving in an abnormal manner due to any other reason. This may cause damage to the equipment or otherwise endanger the effective and healthy operation of the system. A protective relay senses the abnormal condition and the task of isolation of the faulty section is achieved through a circuit breaking device which is capable of disconnecting the faulty element.

With the evolution of technology, there has been a continuous development in the design of protective relays. The use of microprocessors for achieving various relay functions is one of the most notable advancements that has taken place. The use of computers has imparted enormous flexibility in designing the required protective features in the relay.

Protective relays need the fundamental power system quantities, i.e. current and/or voltage as an input to provide the protection function. These input(s) are given to the relay(s) either directly or through instrument transformers, where necessary. Electrical relays are of many types and it is difficult to give a precise definition of a protective relay which covers all types. The following definition is, however, generally accepted:

"An electrical relay is a device designed to produce sudden, pre-determined changes in one or more electrical output circuits, when certain conditions are fulfilled in the electrical input circuits controlling the device."

Hence, we can say that a protective relay is an electrically operated device designed to sense the identified circuit parameters and to initiate disconnection, with or without a warning signal, of the intended part of an electrical section in case of any abnormal condition in the installation and with minimum interruption to overall system.

In order to ensure the smooth functioning of any modern power system, the provision of protective relays is a must. However, it must always be kept in mind that as in all good engineering practices, here also economics plays an important part. An optimal balance needs to be established between the cost of relaying and the required system reliability.

Like all other constituents of the power system, protective relaying should also be evaluated on the basis of its contribution to the best economically possible service to the customers. The contribution of protective relays is to help the rest of the power system to function as efficiently and effectively as possible in the event of abnormal conditions. By prompt sensing and removal of the faulty section, the effects of the following undesirable situations, which adversely effect the overall economy of power system operation, are minimised:

1. The cost of restoring the damaged section
2. The chances of the fault spreading and subsequent damage
3. The down-time of the equipment
4. The loss in revenue and the strained public relations due to equipment outage

Although all varieties of the protective relays aim to achieve the same functions as described above, these can be grouped into the following three distinct generations:

1. First generation   :   Electromechanical relays
2. Second generation :   Static relays
3. Third generation   :   Numerical relays

9.1 ELECTROMECHANICAL RELAYS

9.1.1 Operating Principle of Electromechanical Relays

Electromechanical relays work on the principle of a mechanical force generated due to the current flow in a coil wound on a magnetic core. This force results in the operation of a contact arrangement which is used for relaying the operated condition to the desired circuit in order to achieve the required function. Since the mechanical force is generated due to an electric current flow, the term 'electromechanical relay' is used.

As mentioned above, the mechanical movement of the operating mechanism is imparted to a contact assembly to open or to close the contacts. When we say that the relay operates, we mean that it changes the state of its contacts, i.e. from Open to Close or from Close to Open. A majority of the relays are provided with a 'control spring' or are re-strained by gravity, so that they assume a given position when in a de-energised state. The contacts which are closed in this condition are termed as 'normally closed' and the ones which are open are termed as 'normally open'.

Electromechanical relays operate basically on any one of the two fundamentally different operating principles, viz. electromagnetic attraction, and electromagnetic induction.

The first variety of relays, i.e. electromagnetic attraction type relays operate as a result of a plunger being drawn into an electromagnet (solenoid), or an armature being attracted to the poles of an electromagnet. The electromagnetic induction relays operate on the induction motor principle wherein a torque is developed due to electromagnetic induction in a rotor. This operating principle, however, applies only to relays actuated by alternating currents.

Following are the different types of electromechanical relays:

1. Attracted armature relays
2. Moving coil relays
3. Induction relays
4. Thermal relays
5. Motor operated relays
6. Mechanical relays

The above relays are discussed in detail below.

9.1.2 Attracted Armature Relays

These relays comprise an iron-cored electromagnet which attracts a movable armature that is hinged, pivoted or otherwise supported so as to achieve motion in the magnetic field. The motion is controlled by an opposing force generally due to gravity or a spring.

As the armature movement starts, the air gap reduces and the flux density rises. This is inevitable since it is only by moving into the strongest field that the armature is enabled to do any work. As the flux density increases, the magnetic pull increases rapidly, so that in order to re-set the relay after an operation, the current must be reduced by an appreciable amount. The ratio of re-setting to operating current level is known as the 'returning ratio' or 'drop-off to pick-up ratio'.

The effect of a ratio below unity is to give the device a snap action which is, in general, beneficial as it facilitates positive action and good contact operation. However, a low re-setting value may not always be acceptable. By preventing the armature from closing the magnetic circuit completely, the effect can be reduced. It can be further offset by spring control. This is illustrated in Fig. 9.1. The variation of magnetic pull on the armature with armature position is shown by curve A. As the armature operates, the restraining spring is compressed and the restoring force increases, as shown by the straight line B. The unbalance operating force in the closed position is therefore AB; if a constant restraining force had been applied, corresponding to the horizontal line, the unbalanced quantity would have been AC.

In order to permit the relay to re-set, the current must be reduced below I_2, equivalent to point B in the case of the spring-controlled relay and I_3 in the case of gravity control. The latter has a lower re-setting value, other conditions being equal. It is clear that in case the slope of curve B is increased, i.e. if the spring rate is increased, the re-setting value of the relay will be raised. This will mean a reduction of the output force AB and reduced contact pressure. In fact, by using a sufficiently strong spring, the relay can be given a proportional movement instead of a snap

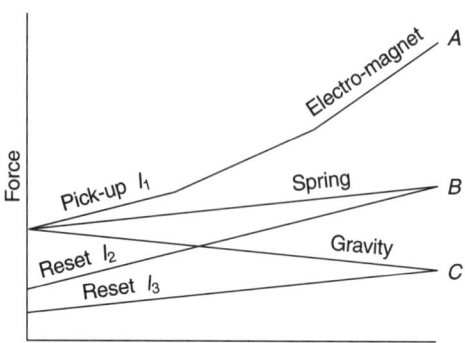

Fig. 9.1: Armature travel

action, but will then have a correspondingly low contact making capacity. Typical relays may have a drop-out value of 25 per cent of the pick-up value or alternatively 90 per cent or even higher for special designs. in the former case, the relay will have high speed snap action and will be capable of doing a considerable amount of work in closing contacts, whereas the later relay with a high drop-off value will tend to be relatively slow at current values marginally above setting and may have a reduced contact rating.

9.1.2.1 AC Relays

Owing to the inherent characteristics of ac quantities of passing through zero at every half cycle, the relays energised by ac flux tend to vibrate, as at zero points the armature gets released slightly from the magnetic pole. Certain design modifications have been worked out to overcome this effect. The provision of shaded loop is one solution. For this, the magnet is split into two sections and one section is surrounded by a low resistance copper band. The eddy currents induced in this loop cause a phase delay in the flux passing through the loop as compared with that in the other half of the pole. With this arrangement, the flux does not come to zero on the entire face of the magnet pole at any instant and a hold on force is always available.

Another method of arresting the vibrations is to use a rectified ac. This method is adopted where a shaded pole has not been provided. In this case, the coil inductance maintains the current during the zero points of the ac cycle.

Fast relay operations are achieved by reducing the armature weight and the total contact travel. For example, reed relays, described in Section 9.1.2.3, can operate in one milli-second.

9.1.2.2 Solenoid Relays

The operating principle in these relays is similar to the attracted armature relays except that the coil in which the plunger is attracted, is longer. Steel plates are provided outside the coil for completing the magnetic circuit. A longer stroke is achieved in this design. This design is adopted where a large amount of work has to be done.

9.1.2.3 Reed Relays

These are low inertia attracted armature relays designed to achieve low operating times, generally of the order of one milli-second. The design uses stiff steel fingers which are mounted such that their tips overlap and are normally separated by a small gap. When a magnetic field is applied, the two fingers are attracted to each other. The fingers are fitted with suitable contact making materials so as to be capable of closing a circuit.

9.1.3 Moving Coil Relays

The main types of relays in this category are discussed below.

9.1.3.1 Moving Coil Permanent Magnet Relays (Rotary Type)

In this design, a permanent magnet is used to achieve a radial field in which a light weight pivoted coil is suspended. The coil is free to rotate at an angle (known as operating angle) which can vary depending upon the requirement. The restoring force is achieved through a restoring spring.

The magnetic field is usually arranged to be uniform over the operating arc, so with a given current the operating torque will be constant and independent of the coil position. It is therefore possible in a long travel relay to provide a calibrated scale by giving a range of setting adjustment. Alternatively, the movement of the relay coil may be limited to a small arc, but adjustment may be provided for the initial wind-up of the spring so that a similar range of setting values is possible.

The coil can be wound without former or on an insulating former. Usually, however the coil is wound on an aluminium or copper former which acts as a short-circuited winding and thus provides a dynamic drag in opposition to the motion of the coil, because of the induced currents which are produced by such a motion. This provides an efficient damping system which can effectively prevent over-swinging and, if made sufficiently strong, can provide a useful time delay.

As an alternative or supplement to the above action further damping can be obtained by providing a low resistance shunt path to the moving coil, whereby the back E.M.F. in the latter can cause a circulation of current in the local circuit of coil and shunt, thereby producing a damping force.

9.1.3.2 Dynamometer Type Relays

This type of relay is generally similar to the rotary type moving coil relay but has an energised field which may be either iron-cored to provide a radial field as with the permanent magnet types, or may be entirely air-cored. While the permanent magnet type designs are essentially dc energised relays, which can only be applied to ac schemes by the use of rectifiers, the dynamometer pattern is a universal dc or ac relay. It is, however, much less sensitive, owing to the need to provide the field system flux electrically, and is therefore used only in special applications.

9.1.4 Induction Relays

The basic principle of induction motors is applied to relays designed to operate on the induction principle. The moving conductor is placed in the two magnetic fields, displaced both in time and phase, and produces the required torque. The two fields are derived from a single quantity by energising two electromagnets with the required phase shift. Another arrangement can be that of energising two magnets by separate sources. In both the cases, the torque generated is given by:

$$T = K \phi_1 \phi_2 \sin \alpha \qquad 9.1$$

where T = torque

ϕ_1, ϕ_2 = flux produced in the two electromagnets

and α = angle between ϕ_1 and ϕ_2

9.1.4.1 Single Quantity Relays

In this arrangement, a C-shaped electromagnet is used. This design is generally applied to over-current and over-voltage relays. The current generates a flux across the air gap which passes through an aluminium disc placed in the air gap. The pole faces of the electromagnet are not solid but are divided into auxiliary poles. One of these poles is surrounded by a solid copper loop. The induced current circulating in this loop causes a phase displacement between the flux emerging from the shaded pole and that in an adjacent pole. The effect is to produce a laterally moving field which, while cutting across the relay disc, produces a dragging force on the latter because of the currents induced in the disc. The driving torque is theoretically proportional to the square of the current value over the linear range of the electromagnet, but becomes modified from this simple relation at high current values owing to saturation of the magnetic circuit.

The restraining force is achieved by a spiral spring, the force of which must be overcome by the driving torque before any operation can begin; this determines the setting or minimum operating current of the relay. The disc is further controlled by a permanent magnet which produces an eddy current braking torque, this torque being proportional to the speed at which the disc rotates.

The following equation represents the relay response:
(considering disc inertia to be negligible)

$$K_1 I^2 = S + \frac{K_2 d}{t} \qquad 9.2$$

where S = spring torque

d = distance moved

t = time

K_1, K_2 = constants.

Also:

$$K_1 I_0^2 = S$$

where I_0 = setting current. Hence:

$$K_1 (I^2 - I_0^2) = \frac{K^2 d}{t}$$

that is

$$t = \frac{Kd}{(I^2 - I_0^2)} \qquad 9.3$$

The spring torque S makes the characteristics asymptotic to the setting value in place of zero. For current setting adjustment, the taps provided on the operating coil are used.

The relay is normally provided with various setting ranges for adjustment and the setting is done by inserting a single pin plug in the appropriate position on a 'plug selection board' provided on the relay. It is possible to achieve much wider setting ranges but that is not recommended due to the fact that it results in poor coil utilisation corresponding to the higher current taps. For example, for a 10 to 1 range, only one-tenth of the coil turns would be in circuit for the highest setting and the coil rating as a multiple of the setting would be low for this tap. This is significant in case of earth fault relays for which high multiples of the setting current may be relevant for all settings.

Tapping the coil adversely affects its accuracy. Although the relay operates with a given value of coil ampere-turns, in practice all turns are not equally effective. In addition to the useful flux, leakage flux is produced in the electromagnet which alters the flux density and hence the saturation level in part of the magnet iron circuit. This effect varies with the position of the active winding with the result that the curve shape is also affected.

Characteristic Time/Current Curves

The majority of over-current relays follow standard characteristics as per in BS 142 (shown in Fig. 9.2 curve A). Other characteristics are also possible and have certain advantages in application. In particular, steeper curves which have been designated as 'very inverse' and 'extremely inverse'

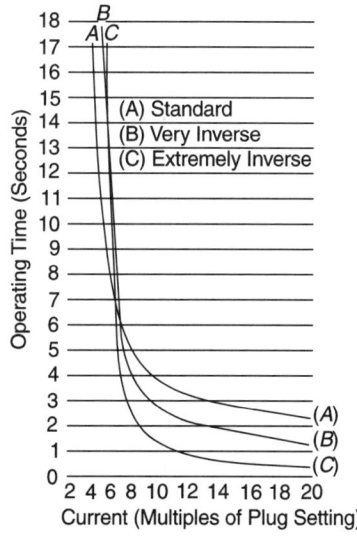

Fig. 9.2: Operating curves in over-current relays

are also shown in Fig. 9.2 (curves B and C). These curves are obtained by controlling the degree of saturation which occurs in the electromagnet. Such relays have lower values of working torque.

9.1.4.2 Double Quantity Relays

In this type of induction relays, two separate input quantities are utilised to produce interacting fluxes. The electromagnet system may be disc type or cup type.

The upper portion of the electromagnet of the disc type element is of a low power factor. When these circuits are energised by a voltage, the resulting flux lags the applied voltage by a large angle. This angle can be effectively made 90° by compensation arrangements and in this condition, the torque developed is proportional to the power.

9.1.5 Thermal Relays

The design of these relays is based on thermal effects. One of the most commonly used properties is the expanding property of metals when these are heated. The sensing arrangement consists of two layers of different metals (known as bi-metal) welded together. The bi-metal strip bends when heated as the expansion of the two metals is different. These bi-metal relays are mainly used for thermal overload protection. The operating time depends upon the type of thermo-sensitive element used.

9.1.6 Motor Operated Relays

A fractional horse power electric motor is employed in this design which drives a contact making arrangement through suitable gearing arrangement. These relays may be used to provide time delays and for performing an operational sequence.

9.1.7 Mechanical Relays

Relays which are designed to respond to gas pressure, liquid flow, liquid level, etc. are classified as mechanical relays. The most popular and widely used relay in this category is the Buchholz relay, which is used to protect oil-immersed transformers when the pressure exceeds a permissible limit. This relay consists of either one or two floats contained in a closed housing, which is located in the pipe running from the transformer tank to the conservator. Any fault in the transformer results in decomposition of oil which, in turn, results in gas generation. This gas passes up the pipe towards the conservator and is trapped in the relay. In case of a heavy fault, a bulk decomposition of the oil takes place. In a two-float relay, one float is meant for sensing the slow accumulation of gas due to mild or incipient faults while the other float gets deflected by the oil surge caused by a major fault. The floats control the contacts to raise an alarm in the first case, and in the second case, isolate the transformer by tripping the associated circuit breaking device.

9.2 STATIC RELAYS

9.2.1 Development of Static Relays

Protective relays using static components are now well established and widely accepted. When these relays were introduced about 25 years ago, many apprehensions regarding their reliability were raised but subsequently these relays gained wide acceptance for their range, flexibility and reliability.

The initial designs were based on silicon transistors but owing to the rapid pace of development in the field of electronics during the last one-and-a-half decades these were replaced by medium and large scale integrated circuits and then by microprocessors.

The ever-increasing size and complexity of the power system, demands relay designs which offer improved sensitivity and selectivity, reliable performance in all types of climatic conditions and more flexible characteristics. All these have been made possible by using the latest components and techniques. One of the added advantages has been the reduction in size which results in panel space saving.

The static relays have been designed to replace almost all the functions which were being achieved earlier by electromechanical relays. However, auxiliary relays of electromechanical design are still being used in static design for output contact arrangement. These simple design features of electromechanical relays are, therefore, going to co-exist with static relays.

Following are some of the new aspects that have emerged with the introduction of static relays:

(a) A wide variety of electronic components are used and it is rare that all these are produced by the relay manufacturers themselves. Hence, it has become imperative to introduce stringent quality control checks for both the components as well as the final relay assembly in order to ensure the reliable operation of the relays.

(b) Testing becomes essential at all production stages and automatic test set becomes imperative to cope with the large volume of production.

(c) Highly sophisticated isolation and filter circuits are required to be built into the relay design to take care of electromagnetic interference and transient switching disturbances in the power system.

(d) Highly reliable power supply circuits are required.

The following two philosophies are followed for static relay designs:

1. Discrete static modules intended to replace one or more identified function of electromechanical relays are used.

2. Many inter-connected modules are used to achieve complex multi-function schemes where the scheme logic is achieved through digital components or microprocessors and outputs such as indication, trip and alarm signals are given. This facilitates optimal exploitation of the inherent speed of the measuring elements and the digital logic circuits. The total scheme consists of many plug-in modules in a rack-mounted case. A common power supply module is fitted to provide power to the complete arrangement.

9.2.2 Relay Circuits Using Analogue Techniques

The power system input quantities such as current, voltage, phase angle and power are analogue quantities. These are usually compared singly or in some combination with a reference 'setting' level and a digital (one zero) decision is given as a result of this measurement.

As the development of static relays progressed, certain circuit modules were applied repeatedly in various designs. This is similar, to some extent, to the way in which a small number of basic electromechanical elements is used for differing applications in electromechanical relays.

9.2.2.1 Over-current Relays

The following basic circuits are used in over-current relays:

1. ac to dc converter
2. Level detectors
3. Timers

Each of these circuits forms a part of the time delayed over-current relay shown in the block diagram in Fig. 9.3. In an ac to dc converter, ac is converted to a dc voltage by means of a current transformer of suitable ratio, a bridge rectifier and a resistive shunt load. This voltage is compared with a set level by a level detector, which gives a start command to the timer when this level is exceeded. This timer provides either a fixed time in case of definite time relays or a time inversely proportional to the magnitude of the input current in which case a curve shaping circuit is required. The timer usually charges a capacitor such that when the charge reaches the level set on the corresponding level detector, the later gives a signal to the output switching circuit. Instantaneous operation for short-circuits can be obtained by means of a third level detector which bypasses the time delay circuit.

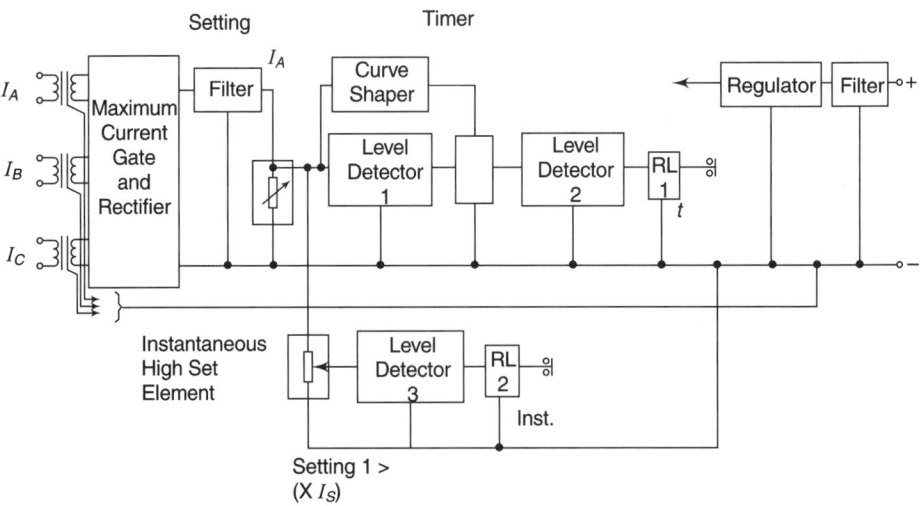

Fig. 9.3: Block diagram of a time delayed over-current relay

Although early designs used transistors, analogue measuring circuits now universally use the integrated circuit operational amplifiers. This is a high gain dc coupled amplifier which amplifies the voltage difference between its input terminals. The operational amplifier is a vital component of the static circuitry.

An ideal operational amplifier is characterised by:

1. Infinite voltage gain
2. Infinite input impedance
3. Zero output impedance

The use of the operational amplifier to measure an analogue quantity imposes a negligible burden on the measured quantity and its output can drive substantial loads. The operational amplifier can be used in a number of basic relaying circuits such as level detector, polarity detector and integrator.

9.2.3 Relay Circuits Using Digital Techniques

The output of all static relays is digital, i.e. they perform some kind of switching (on/off) function. Therefore, the analogue input signals are converted to digital signals at some stage.

With the rapid development of digital integrated circuits, logic gates, microprocessors and microcomputers, relaying circuits are increasingly using digital techniques. Analogue signals such as voltage and current can be converted into digital signals by means of voltage and frequency converters or analogue to digital converter (ADCs). The resultant digital signal can be processed either by discrete logic or by a microprocessor. Generally, discrete logic can perform a large number of relatively simple operations in parallel, whereas a microprocessor can perform highly complex logical tasks but is not ideally suited for parallel operations.

9.2.4 Microprocessors

The invention of microprocessors has ushered in revolutionary changes in protective relay designs. Since they offer enormous flexibility, microprocessors are widely used in protection schemes.

9.2.5 Housing of Static Relays

One of the main advantages of the use of static components in protective relay applications is their small overall size. In the beginning, these relays were housed in similar cases meant to house electromechanical relays. But later, a comprehensive analysis of the housing aspects of static relays was undertaken by leading manufacturers. The conclusions that emerged from this analysis are:

(a) It should be possible to provide a complete protection scheme unit within one housing.

(b) The design should be such that the relay circuitry is immune to external interferences and harmful surges. Also, external wirings should be minimised by providing internal connections as far as possible.

(c) The design should provide access for maintenance and testing.

On the basis of the above considerations, the relay manufacturers have come up with a modular relay housing system which has gained wide acceptability. For example, standard 19-inch racks which are built as open frame structures and are either free standing or wall mounted. Static relays are generally designed in modular construction with individual modules plugged into sockets by means of printed circuit board edge connectors. The connections between modules are made by using back-plane wiring. Current monitoring and test points are brought out to the front plate of an individual module. This facilitates fast checking of any particular circuitry and trouble shooting.

With the help of a card extender, the components of any module can be accessed for testing, while the connections remain intact. Card replacement can be done quickly in the modular concept, and servicing and repair also become convenient. In order to offer this flexibility and maintain compatibility with rack-mounted schemes, many manufacturers have introduced a range of 'building block' relays. These relays can either be panel-mounted singly or mechanically joined together in groups before being mounted in panels or on racks. These individual 'building blocks' are plug-in type modules housed in metal cases with all the inputs and outputs wired to suitable terminal blocks.

As the relays can be mechanically joined in the form of a composite assembly before they are mounted on a rack or in a panel, it is possible to make inter-connections between modules and test-specific schemes as per customers' requirements in the relay manufacturer's works itself.

9.2.6 Quality Control and Testing

The issue of quality control in case of relays using semiconductor components calls for increased attention as compared to the case of electromagnetic relays, mainly due to the following reasons:

1. The manufacture of components is undertaken by semiconductor device manufacturers. This calls for extensive quality control tests on electronic components before these are used in the relay so as to ensure the quality and reliability.
2. Rapid advancements have taken place in semiconductor technology and improvements in the quality control and checking systems have to keep pace with these advancements.

The need for better quality control techniques of components and the introduction of improved test methodologies has grown as the complexity of the electronic circuits has increased from linear operational amplifiers to digital gates and logic, followed by large scale digital integrated circuits, memories and microprocessors.

The ever-increasing production volume and the complexity of electronic circuits used in static relays put a continuous pressure on conventional methods of testing and led to the invention and introduction of automatic test equipment which are programmable and flexible, thereby ensuring effective testing and hence quality.

The main aims of improvements in the testing methods and quality control techniques are:

1. Minimising testing time thus reducing the overall production cost
2. Maximising reliability during service

The testing time will be minimum when the relay is tested after it is completely assembled. If the relay passes the tests, it indicates that all the constituents of the relay are OK. But if the relay fails to pass, it would be difficult to diagnose and rectify the failures. Moreover, it is likely that a fault in one component or sub-assembly may cause the consequential failure of other components or assemblies, thus magnifying the failure and complicating the task of diagnosing it. It is a general rule that the cost of rectifying a failure increases by a factor of ten for every stage of the assembly.

Another philosophy highlight the need for conducting a test at every stage of the production. Although this increases the cost due to repeated testing, it will also enhance the reliability of the product.

9.2.7 Component Testing

In order to eliminate early life failures, it is essential to carry out inspection as per the applicable quality plans and testing of critical parameters of the incoming components. Analogue devices such as semiconductor transistors and operational amplifiers are subjected to burn-in at an elevated temperature and the critical parameters are also monitored during the burn-in test.

The identified parameters are re-checked after the burn-in test also and compared with the values measured before the start of the test. The devices exhibiting drift beyond a certain value of the parameters are rejected as possible early life failures. A drift limit of up to 10 per cent is generally accepted.

In some cases, the devices are loaded into test jigs and are tested by a programmable tester which identifies the parameters that are out of limit.

In the case of digital circuits, the dynamic burn-in test is normally conducted at 70°C for 72 hours or the static burn-in test 15% conducted at 85°C for about 200 hours, with parametric tests preceding and following the burn-in test.

9.3 NUMERICAL RELAYS

The first protection devices based on microprocessors were employed in 1985. The widespread acceptance of numerical technology by the customer and the experiences of the user helped in developing the second generation numerical relays in 1990.

Modern power system protection devices are built with integrated functions. Multi-functions like protection, control, monitoring and measuring are available today in numeric power system protection devices. Also, the communication capability of these devices facilitates remote control, monitoring and data transfer.

Traditionally, electromechanical and static protection relays offered single-function, single characteristics, whereas modern numeric protection offers multi-function and multiple characteristics. Some protections also offer adaptable characteristics, which dynamically change the protection characteristic under different system conditions by monitoring the input parameters.

The measuring principles and techniques of conventional relays (electromechanical and static) are fewer than those of the numerical technique, which can differ in many aspects like the type of protection algorithm used, sampling, signal processing, hardware selection, software discipline, etc.

9.14 Handbook of Switchgears

First generation numerical relays were mainly designed to meet the static relay protection characteristic, whereas modern numeric protection devices are capable of providing complete protection with added functions like control and monitoring. Numerical protection devices offer several advantages in terms of protection, reliability, and trouble shooting and fault information. Numerical protection devices are available for generation, transmission and distribution systems.

The following sections cover relay hardware, relay software, multiple protection characteristics, adaptive protection characteristics, data storage, instrumentation feature, self-check feature, communication capability, additional functions, size and cost-effectiveness.

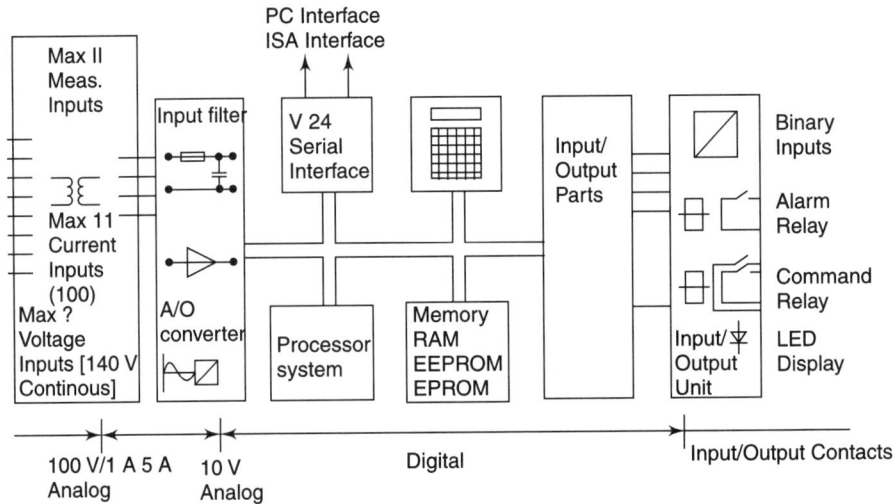

Fig. 9.4: Block diagram of numerical relay

9.3.1 Relay Hardware

Numerical relays use a specialised digital signal processor (DSP) as the computational hardware, along with associated software tools. The relaying voltage and currents are passed through an isolation transformer. The voltage inputs of the relay are scaled down from the nominal voltage to a low level determined by the ADC (analogue-to-digital converter) input range. The current inputs to the relay are scaled down from nominal 5/1 A and are converted into equivalent voltages. These scaled signals are filtered by using a low-pass filter to prevent a liasing of the high-frequency components into the fundamental frequency component. The filtered signals are multiplexed using an analogue multiplexer and amplified, if needed, by using a programmable gain amplifier. The multiplexed analogue signal is sampled and converted into digital data using the ADC. The more complex relay uses multi-processor architecture wherein the digital signal processor executes complex algorithm calculations and the host processor performs all the other tasks.

Communication between the processors is provided by the dual-ported memory. Flash memory is used for storing the programme and RAM (random access memory) is used for temporary storage of variables, target information and oscillography. Contact inputs and outputs, user interface (keyboard and liquid crystal display) and the serial communication ports (RS-232 and RS-485) are interfaced to the host processor.

The digital signal processor executes a variety of signal processing algorithms to estimate several parameters of the digitised voltage and current signals, and transfers them to dual-ported memory. The host processor receives these parameters from the dual-ported memory and performs relay logic and other timing functions to generate appropriate trip or alarm output signals. The host processor, running under a multi-tasking operating system, also performs several other tasks including communications, set-point updates, target updates and user interface.

Providing back-up for critical components enhances the reliability of the relay. Certain digital multi-function relays use redundant power supplies. Both power supplies are continuously running in a hot stand-by configuration and if one supply fails, the other continues to provide the required uninterrupted power to the relay. The relay sends an alarm indication about the power supply failure to alert maintenance personnel.

The analogue signal inputs (voltage and current), contact status inputs and communication circuits are conditioned and protected to withstand the harsh electrical and environmental conditions of the sub-station and power plant. The design of the relay input, output and power supply circuits must incorporate filtering to reduce EMI (electromagnetic interference). The primary method of reducing unwanted induced ac voltage is to bypass these voltages to ground with capacitors. Other components such as varistors, chokes and ferrite beads, are also applied to suppress surge voltages and EMI.

9.3.2 Relay Software

The software provided in a numerical relay is commonly organised in a series of tasks operating in real time. The main component is the real time operating system (RTOS) whose function is to ensure that other tasks are executed as and when required on a priority basis. The software provided will vary on the basis of the following relay-specific functions.

1. System services software: This controls the low level IO for the relay (i.e. drivers for the relay hardware, boot-up sequence, etc.).
2. HMI interface software: This is a high level software for communication with the user via front panel controls that are connected through a data link to another computer running suitable software, storage and setting data.
3. Application software: This is a software that defines the protection function of the relay.
4. Auxiliary function: This software implements other features offered in the relay.

The reliability of software in the digital relay is critical to the overall reliability of the product. The majority of software problems in digital relays can be attributed to design and implementation errors. Careful planning and designing before coding result in a more reliable product. A software quality assurance plan must be carried out throughout the product development programme. Test plans, documentation, detailed software validation and audit programmes can greatly reduce software errors.

Software verification and testing of multi-function relays provide greater reliability. Following are some of the key tests to be conducted at various design phases of digital multi-function relays:

1. Relay algorithm simulation testing
2. Static functional testing
3. Dynamic functional testing
4. Environmental and hardware-related tests
5. Site installation and testing

9.3.3 Multiple Protection Characteristics

In simple Inverse definite minimum time lag (IDMTL) over current and earth fault relays, multiple characteristics like NI, VI, EI, LTI and DTL characteristics are available in relay and the required characteristics can be selected at site.

9.3.4 Adaptive Protection Characteristics

Numerical relays can adapt themselves to different system conditions by monitoring the operating quantities from the digital inputs of the relay. Some examples of these adaptions are

1. In motor protection relay, different settings can be adapted for the starting condition and running condition. Over-current protection, earth fault protection, unbalance protection and stall protection cannot be set at a very sensitive level during start-up because the starting current will influence these settings. However, sensitive settings can be adapted for these protections during the running conditions of the motor. The relay switchover from one group setting to the other during the running condition.
2. In transformer protection relay, the third slope is called the bias slope limit, which is used to afford greater stability and this can be achieved by monitoring the through fault current. Once the through fault current is above the set value, the relay bias automatically increases to facilitate greater stability for through fault condition.
3. In distance relays, for different system configurations like parallel feeder in, and parallel out and grounded, different settings can be adapted for this condition.
4. In IDMTL relays, depending upon the system condition like the number of incomers and outgoing feeders, different settings can be selected and better protection can be achieved.

9.3.5 Data Storage

In numerical relay data storage is done in three formats, i.e. fault record, event record and waveform record.

Fault record gives the complete information about the fault, i.e. fault current magnitude, type of fault, fault phase details, etc. with time stamping. Also in motor protection relays, additional information pertaining to positive sequence, negative sequence, equivalent current [(square root $(I_1^2 + KI_2^2)$] and difference current (between phases) is available. Similarly in distance relays fault location is also possible. The fault record gives all information about the fault to the operating personnel.

Event record stores inside the relay all the changes taking place in the system like protection element, pick-up, and drop-off operation, output and input energisation and setting change, etc. All the events are time-stamped.

Waveform record stores the analogue and digital channel values during the relay operation or on demand. This record gives the nature of fault waveforms (transient) and instantaneous values, etc. It is also possible to calculate the relay operating time and breaker operating time. These records can also be converted into the 'comtrade' format, with which data can be played back into the relay using the digital test system. The records can also be manually triggered and viewed to verify the phase relationship during commissioning. These records can also be fed to the harmonic analyser software to find out the harmonic content in the fault quantities.

9.3.6 Instrumentation

Numerical relays are provided with a metering function and separate panel mounted meters can be eliminated. Some relays can also give the energy meter function. The relay provides metering facility, where the operating personnel can view online different parameters and these metering features can be used as a valuable tool during the commissioning period when primary injection is carried out.

This is perhaps the most obvious and simplest function to implement, as it involves the least additional processor time. The values that the relay must measure to perform its protection function have already been acquired and processed. It is therefore a simple task to display them on the front panel, and/or transmit them as required to a remote computer/HMI station. A number of extra quantities can be derived from the measured quantities, depending upon the input signals available. These may include:

(a) Sequence quantities (positive, negative, zero)
(b) Power, reactive power and power factor
(c) Energy (kWh, kvarh)
(d) Maximum demand in a period (kW, kvar; average and peak values)
(e) Harmonic quantities
(f) Frequency
(g) Temperatures/RTD status
(h) Motor start information (start time, total number of starts/re-accelerations, total running time)
(i) Distance to fault

The accuracy of the measured values can only be as good as the accuracy of the transducers used (VTs, CTs, A/D converter, etc.). As CTs and VTs used for protection functions may have a different accuracy specification than those used for metering functions, such data cannot be sufficiently accurate for tariff purposes. However, it is sufficiently accurate for an operator to assess the system conditions and take appropriate decisions.

In motor protection relay, the instrumentation mode can also display motor status, time to trip (during abnormal condition), time to start (if the protection is locked out), the last starting time, last starting current, positive, negative, difference current values, etc.

9.3.7 Self-check Feature

Self-diagnostics is one of the most important features of numerical relays; it was not available in either electromechanical or static relay design. The ability to detect and correct a failure before the protection system has to operate is in contrast to the traditional protection system wherein a relay failure remains undetected until it fails to operate correctly during an event or until the next maintenance test. The most important self-diagnostic functions implemented in digital multi-function relays are detailed below.

9.3.7.1 Data Acquisition System Testing

Power supply voltages and ground are connected to the analogue input channels of the multi-plexer and checked against warning and failure thresholds. This also verifies the analogue data acquisition system including multiplexer, programmable gain amplifier and ADC. The ADC's conversion time is also checked to see if it is within the specification.

9.3.7.2 Memory Testing

The flash ROM contents are checked by calculating the check-sum and comparing it with the pre-computed and stored check-sum. The check-sum is calculated as the modulo-256 sum of all the bytes. The RAM is tested by writing and reading a test pattern.

9.3.7.3 Set-point Testing

Set-points are stored in the serial EEPROM and a copy of these set-points is also stored in the RAM for executing relay logic. Whenever any set-point is changed, the check-sum of the set-points is calculated from the contents of the EEPROM. This check-sum is then compared with the calculated check-sum of the set-points stored in the RAM every time a set-point task is executed.

9.3.7.4 Watchdog Timer

The relay hardware design includes a watchdog timer re-set circuit to take the processor through an orderly re-set should the programme get lost due to hardware/software glitches.

The relay hardware and software are continuously monitored by the relay itself and if any abnormality is detected, the relay shuts down safely and indicates the nature of failure. Operating personnel can immediately take corrective action, if the relay fails in service.

9.3.8 Communication

Communication makes the relay more intelligent and the operating personnel can set the relay and also download the fault information. It is also possible to upload the revised software to the relay at site without sending the relay back to the manufacturer. Thus improvements in numerical relays can be upgraded at site without despatching the relay to the manufacturer which reduces the relay downtime.

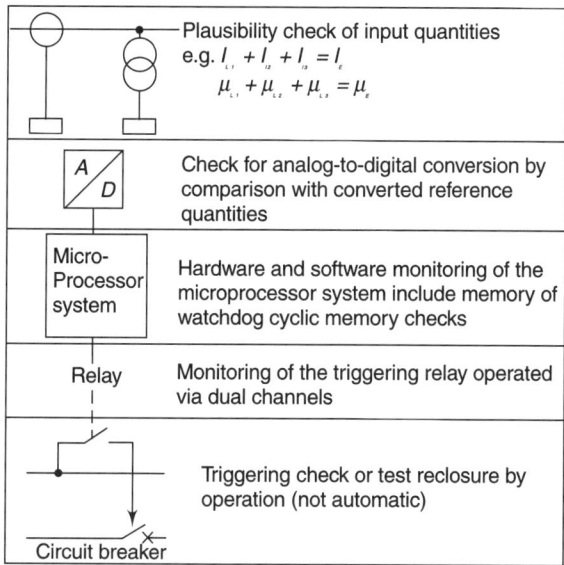

Fig. 9.5: Self-monitoring system

Numerical relays carry out the processing of input quantities using a digital technique. The processed data can be accessed through the relay communication port. Although a separate high-end communication system is available today for numerical relay data communication and control, a simple communication system is also possible for remote control and monitoring using telephone line as a communication medium. Figure 9.6 and Fig. 9.7 shows the arrangement for remote communication using telephone line.

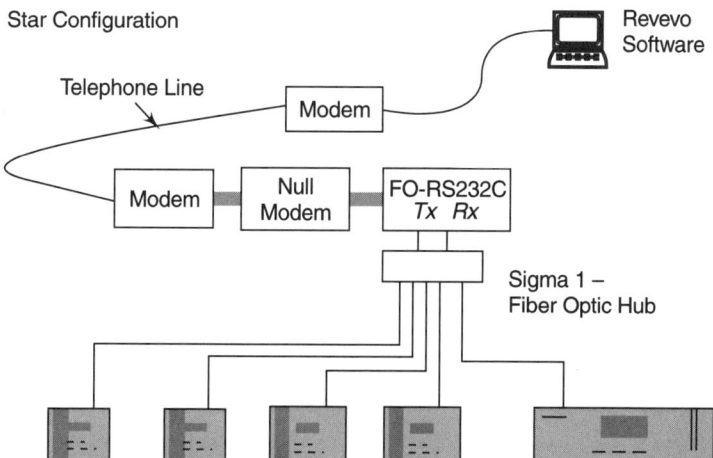

Fig. 9.6: Arrangement for remote communication through telephone line-star configuration

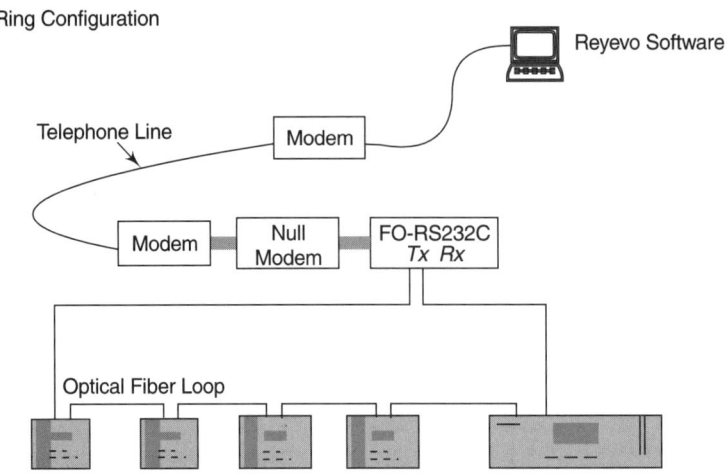

Fig. 9.7: Arrangement for remote communication through telephone line-ring configuration

The relays can be connected in an optical loop (Fig. 9.7) or by using fibre optic hub for star configuration (Fig. 9.6). The relays are provided with optical ports TX (Transmit) and RX (Receive). The advantage of optical communication is that, it is completely free from electromagnetic interference and power system induced transients. The optical signal has to be converted into a digital signal for communication and fibre optic to digital (RS232) converter is used for this purpose. The optical to RS232 converter output is connected to a telephone modem via a null terminal converter. The null terminal converter is required, since both the telephone modems and RS232 converter are data communication equipment (DCE). The null terminal connector switches various control lines. The modem performs modulation and demodulation of digital data at the sending and receiving ends respectively. Since the telephone line was designed for speech communication, it is necessary to convert the binary data into a form, that is compatible with a speech channel at the sending end of the line and to re-convert this signal back into its binary form at the receiver.

The advantage of this type of remote communication is that it is very cost-effective. The following functions can be achieved by using remote communication:

1. Change of settings
2. Change of group settings
3. Control of breaker (closing and opening)
4. Instrumentation viewing
5. Event recorder downloading
6. waveform recorder and fault record down loading

9.3.9 Additional Functionality

Modern numerical relays have other additional functions like:

1. Circuit breaker fail
2. Loss of load, conductor broken
3. Trip circuit supervision
4. Circuit breaker conditions monitoring including programmable digital output and inputs for various logic build generally for blocking
5. Auxiliary relays like Buchholz relay alarm/trip, and winding temperature alarm/trip contacts, which can be connected to relay as logic input and details of relay operation can be seen in the event log sheet
6. The relay user can also make custom built logic by using internal elements and inputs, i.e. VT/CT supervision

9.3.10 Size of Numerical Relays

The size of the numerical relays is much less as compared to that of electromechanical and static relays due to the multi-functionality approach of the former. For example, we have the following protections in numerical relays in-built into one relay.

Motor Protection Features Available:

1. Thermal protection
2. Unbalance protection
3. Under-current protection
4. Over-current and earth fault protection, and temperature protection (RTD)
5. Number of starts
6. Hour run meter
7. Ammeter

All these functions can be built into one relay, which will be very compact in size and also reduced weight. A typical picture of a numerical relay is shown in Fig. 9.8. The size of this relay is 103 mm × 177 mm × 250 mm.

9.3.11 Cost of Numerical Relays

Modern numerical protection devices have multiple protection elements and this technique makes the cost of the relay comparable with that of electromechanical and static relays (discrete). The cost of the microprocessor and digital hardware is falling day by day and the end-user can get a cost benefit in numerical relays in terms of greater functionality at a reduced price. The following cost reduction benefit the consumer.

- The number of protective relays and auxiliary relays are reduced.
- CT and PT of lower VA burden are sufficient. Thus the cost of CT and PT is reduced.

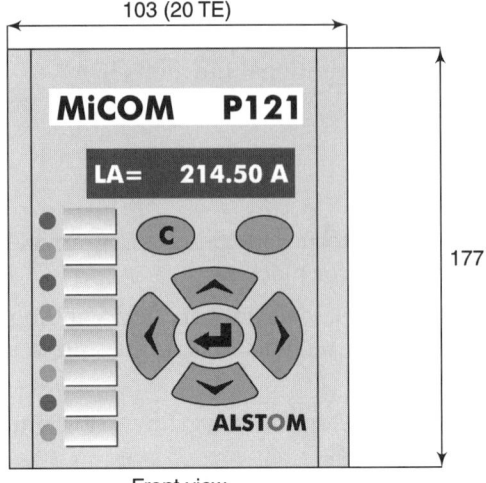

Fig. 9.8: Typical O/C numerical relay

- The panel space required is very less. Thus the size of instrument panels is reduced, thereby resulting in further cost reduction.
- The wiring of relay is much less as compared to electromechanical relays/static relays. This adds to savings in wiring material and labour.

Thus numerical relays should be used for complex requirement and multi-functional uses.

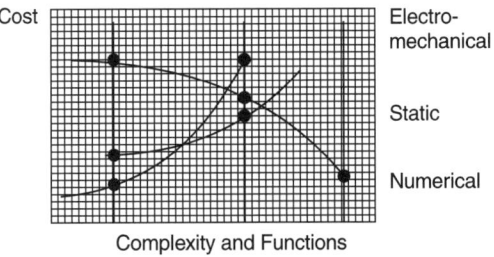

Fig. 9.9: Cost comparison of numerical relay with electromechanical and static relay

9.3.12 HV and Environment Testing

Numerical relays are tested as per IEC 255-5 for dielectric, impulse voltage and insulation resistance. It should be suitable for 2 kV dielectric voltage, 5 kV impulse voltage and insulation resistance > than 1000 M-ohms.

9.3.12.1 Electrical Environment

Relay is tested for high frequency disturbance, fast transient, electrostatic discharge, radio frequency and impulse as per IEC 1000-4 and ANSIC 37.90.

9.3.12.2 Atmospheric Environment

A relay is generally suitable for temperature as per IEC 255-6 for storage at –25°C to +70°C, and for operation at –25°C to +55°C. A relay is generally suitable for humidity as per IEC 68-2-3, for enclosure protection IP 52 as per IEC 529, for siesmic, shocks and bumps as per IEC 255-21 for Class 1 and for Class for vibration test as per IEC 255-21.

9.4 CONCLUSION

Modern numerical protection devices offer innumerable features as compared to static and electromechanical protection system. These are given below.

- Protection is enhanced due to their complex, multiple protection characteristic.
- Self-check feature improves the protection system reliability.
- Communication capability makes the numerical protection system more intelligent and provides valuable information to the user.

Hence for complex, reliable, multi-functional requirement numerical relays shall be used.

References

1. The Art and Science of Protective Relaying by C. Russel Mason.
2. Protective Relays Application Guide: GEC Alstom T&D, 1987.
3. Network Protection and Automation Guide: Alstom, 2002.
4. Reyrolle Product Technical Manual.
5. SEL Technical Papers on Distance Protection with M3425 Technical Manual.
6. Alstom Relay Manual.

Chapter 10

APPLICATION OF MEDIUM VOLTAGE SWITCHGEAR

N.N. Shenoi
Md. Azam Khan

Medium voltage switchgears (commonly known as MV switchgears) play a significant role in the modern electrical networks right from generating stations, various transmission sub-stations at different voltages, distribution sub-stations and load centres. Besides the supply network, switchgears are necessary in industrial works, industrial projects, and domestic and commercial buildings for controlling various electrical equipments.

Switchgears consist of switching devices (like circuit breakers, load break switches, contactors) along with protective equipment, metering, instrumentation and control devices to perform the switching, protection and control functions.

The requirement of circuit breakers for different applications varies depending upon the location, rating and local requirements. When a switchgear is to be applied in an electrical power system, certain considerations should be kept in mind regarding location—whether indoors or outdoors, system parameters—system earthing, frequency and insulation level, ratings— both normal rating as well as short time rating, ambient conditions, etc. These requirements are detailed below.

10.1 VARIOUS CONSIDERATIONS

10.1.1 Location Considerations

The switchgears for various applications can be located indoors or outdoors and accordingly these are classified as indoor switchgears or outdoor switchgears.

10.1.1.1 Indoor Switchgears

An indoor switchgear is one which is exclusively intended for installation within a building or other enclosures, wherein it is protected from wind, rain, snow or abnormal dust deposits, abnormal condensations, ice and hoary frost.

Indoor switchgears are normally of a metal-clad design. The various components forming the switchgears are arranged in compartments separated by earthed metal partitions. Thus we have breaker compartment, CT/PT compartment, cable termination compartment, busbar compartment, surge suppressor compartment, LT busbar compartment, instrument panel chamber, etc.

Depending upon the design, certain devices are combined in one compartment, e.g. some manufacturers combine CTs, PTs and the cable termination arrangement in one compartment. In some other design, surge suppressors or PTs are mounted in the breaker compartment (on the truck carriage base).

Normally one set of busbars is used per switchgear and sometimes multiple busbars are used to ensure reliability of the system in which case each busbar system must be accommodated in a separate compartment.

Metal-clad switchgears are built and tested in the factory with the complete assembly of the compartments and the entire unit is transported in one package. Metal-clad switchgears can be installed as a single independent unit or in board formation. When these are used for individual control, a separate compartment for incoming cable and outgoing cables has to be arranged. In the board formation, several switchgears are joined together in a row for different applications thus forming the complete board called a switchboard. A complete switchboard has switchgears for different applications, e.g. incoming from generator, station supplies from grid, outgoing feeders for different applications, potential transformer panel for bus voltage/feeder voltage measurement and supply to voltage dependent measurement/protection devices. Regardless of whether switchgear is manufactured as a single unit or in switchboard formation, the end panels should be provided with suitable covers to prevent accessibility to HT and LT busbar/connections.

10.1.1.2 Outdoor Switchgears

Outdoor switchgears are intended for installation in open space which is directly subjected to rain, dust and the environmental effects of the location.

These can be mounted inside a metal enclosure in the form of kiosks for taking HT connections or mounted on a structure (porcelain-clad switchgears) or poles (pole-mounted) depending upon the application. When metal-clad switchgear is to be installed outdoors, the enclosure has to be weather-proof.

10.1.2 Rating Considerations

Circuit breakers are rated at normal rated voltage and maximum operating voltage. This maximum operating voltage should not be exceeded by the power system to which the circuit breaker is applied.

The circuit breaker rated current is the continuous current that it can carry without exceeding the temperature rise. This is essential for the life of the insulation of the main power conducting parts.

The breaking capacity and the short time rating of the circuit breaker should be selected based upon the fault level of the system and location/application, i.e. whether it is located at the generating point/source or it is far away or located after other electrical equipment like transformers.

Circuit breakers are normally rated for 50 or 60 Hz frequency. In special applications where rated frequency is 200 Hz or above, the interrupting capability of the circuit breaker will be reduced.

10.1.3 Ambient Considerations

MV switchgears are designed to operate successfully at the ratings specified on the rating plate under standard ambient conditions. Standard ambient conditions include a temperature of 40°C and altitude up to 1000 meters. When these conditions change, the need for derating the switchgear arises. Also the use of surge suppressors should be considered for all such high altitude installations.

In a circuit breaker, atmospheric air is used for both cooling and insulation. At high altitudes, the density of air is less resulting in poor cooling and poor insulation. Therefore, derating of circuit breakers is to be considered. The derating information is given in standards IEC 694 as well as in ANSI C37.04.

This derating information is available with manufacturers of switchgears, who offer it to users on request.

10.1.4 System Earthing Considerations

Both solidly grounded as well as ungrounded/non-effectively grounded systems are used for earthing.

A solidly grounded system produces ground fault currents of sufficient magnitude to operate the earth fault relay of the affected feeder. This leads to tripping of correct circuit breaker and isolating the faulted portion of the system without interruption of power to the unfaulted portion.

An ungrounded system has to provide continuity of service, in case of temporary ground fault. This system is employed where continuity of supply is of the utmost importance and, if interrupted, will cause a large financial loss. In such cases, the system will have to be continuously monitored by means of a ground detector. The system will continue to work with a fault which can be rectified during the next planned shutdown. Core balance current transformers (normally known as CBCTs) used in conjunction with sensitive earth leakage relays, or potential transformers with open delta secondary winding used in conjunction with neutral voltage displacement relay, help in the timely detection of these faults.

10.1.5 Seismic Considerations

Initially the concept of seismic withstand criteria was evolved for equipments used in nuclear power generating stations. But the human loss and the extensive property damage including heavy damages sustained by electrical equipment caused by major earthquakes in various parts of the world aroused widespread attention towards the need for the proper seismic design of equipment, which influenced the initiation of seismic requirement for non-nuclear related equipment also. The different locations are classified into five zones on the basis of the seismic intensity for each of these locations.

Verification is done by seismic testing which is performed on a shaker table, test bed or suitably constructed test fixture.

10.1.6 Over-voltage Considerations

Surge suppressors (also called surge arrestors) are used to protect electrical equipments such as transformers, motors, capacitor banks against the system over-voltages or over-voltages developed due to a particular type of switching medium employed. Commonly used surge suppressors are the capacitance-resistance combination type of surge suppressors and gapless zinc oxide surge suppressors. They are discussed in detail in a different chapter in this book.

10.2 APPLICATION OF SWITCHGEARS IN POWER SYSTEM

Switchgears are necessary at every switching point in the power system. Some of their applications are given below

10.2.1 Generator Circuit Breakers

Generator circuit breakers are installed between the generator and the transformer. These are indoor switchgears. These breakers are required to interrupt very high fault current of the generators and the operating speed for fault clearance has to be very high (within four cycles). This rapid clearance of the fault current helps to avoid expensive damage of power plant equipment and consequently long downtime for repair. The design of the circuit breaker becomes highly complicated due to this requirement.

Generator circuit breakers should incorporate a breaker, disconnect switches, starting disconnecting switches, current transformers, potential transformers, and surge capacitors/surge arrestors all in a common cubicle type assembly.

With the successful certification of 160 kA generator circuit breaker, generator circuit breakers are now available for generating units upto 1400 MW. Another new development has been the integration of all the associated items of the switchgear such as series disconnector, earthing switches, short-circuiting switches, current transformers, single-pole-insulated voltage transformers, protective capacitors and surge arrestors, within the generator circuit breaker enclosure as an option to separate installation.

The typical ratings of generator circuit breaker are given in Table 10.1.

Sl. No.	Voltage	Normal Current	Breaking Capacity	Generator Rating
1	27.5 kV	11000 A	100 kA	500 MW
2	25.3 kV	13000 A	100 kA to 120 kA	500 MW
3	21 kV	7700 A	63 kA	250 MW
4	30 kV	24000 A	160 kA	1400 MW

Table 10.1 Typical Ratings of a Generator Circuit Breaker

10.2.2 Switchgears for Power Plant Auxiliaries

Switchgears used for power stations will have high current ratings of the order of 2500 A to 3150 A and a fault level of the order of 44 kA to 50 kA. The rated voltages for the switchgears are 3.3 kV, 6.6 kV or 11 kV. These are indoor switchgears.

10.2.2.1 Station Switchgears

These switchgears are installed on the secondary side of the station transformers and are used to isolate the load side from the transformers in the event of a fault on the load side. During the initial stage of the setting up of power plants, power will be brought from the existing grid through station transformers and station switchgears These will have high current rating (of the order of 2500 A to 3150 A) and high fault level (of the order of 40/44 kA). The rating of the station switchgears depends on the total load of the power plant.

Station switchgears are indoor switchgears in board formation. Depending upon the application, the switchgears can be incoming feeders, tie feeders to and from other switchboards in the power plant, outgoing feeders, feeders controlling transformers, motors, etc. Various feeders differ in the current rating and protection requirement.

The inter-connection between various switchboards may be through bus ducts—top bus duct entry/bottom bus duct entry or top cable entry/bottom cable entry depending upon the layout designed by the power system engineers.

10.2.2.2 Unit Auxiliary Switchgears

The bulk of the power generated by the generating station is fed to the grid through grid transformers. However, the power required for the various auxiliaries in the power station is fed to the unit switchboards through unit auxiliary switchgears. There can be more than one unit auxiliary switchboard for one unit.

The various auxiliaries in a power station are the boiler feed pump (BFP) motors, mill motors, induced draught (ID) fan motors, forced draught (FD) fan motors, primary air (PA) fan motors, condensate extraction pump (CEP) motors etc.

10.2.2.3 Grid Transformer Switchgears

These switchgears are used for transferring the generated power to the grid.

10.2.2.4 Distribution Switchgears

The power is fed to various distribution purposes within the power plant through distribution switchgears, viz. the coal handling plant, ash handling plant, water treatment plant, etc.

10.2.3 Switchgears for Transmission Sub-stations

There are a number of sub-stations for transmission of power at various voltage levels and switchgears are employed at every sub-station. These switchgears are normally for voltages of 66 kV, 132 kV, 220 kV, 400 kV and 765 kV. These are outdoor structure mounted switchgears. The fault-clearing medium is SF6. Initially air blast, bulk oil and minimum oil circuit breakers were employed for voltages up to 220 kV but now these are becoming obsolete due to their inherent disadvantages of high maintenance cost and large space requirements.

The rating parameters generally used for different voltage levels are given in Table 10.2.

Normal Voltage	Highest System Voltage	Current Rating	Fault Current
66 kV	72.5 kV	2000 A	40
132 kV	145 kV	to	to
220 kV	245 kV	2500 A	50 kA
400 kV	420 kV		
765 kV	800 kV		

Table 10.2 Rating Parameters for Different Voltage Levels

The switchgears in transmission sub-stations can be indoor or outdoor depending upon the customer choice. Sometimes the switchgears and control and relay panels (CRPs) are located outdoors. Sometimes, however, only switchgears are located outdoors and the CRPs are located indoors.

10.2.4 Switchgears for Distribution Applications

These are vacuum circuit breakers and are normally rated for voltages of 11 kV and 33 kV. The current ratings are normally 400 A, 630 A, 800 A, 1250 A, 1600 A and 2000 A. The fault level ranges from 13.1 kA up to 25/26.3 kA. Until the late 1970s, bulk oil circuit breakers (BOCBs) or minimum oil circuit breakers (MOCBs) were specified for this application. But since the early 1980s, circuit breakers employing vacuum and sulphur hexaflouride (SF6) as switching mediums were developed. Ultimate utilities started specifying VCBs or SF6 switchgears and slowly oil type circuit breakers (BOCBs and MOCBs) are being phased out.

10.2.5 Switchgears for Industrial Applications

Switchgears for industrial application like steel plants, cement plants, aluminum plants, and furnace applications have to be heavy duty and suitable for a large number of switching operations.

The ratings of the switchgears are 3.3 kV to 6.6 kV, 630 A to 2000 A with fault levels of 25 kA to 31.5 kA.

10.2.6 Switchgears for Rural Applications: (Auto-reclosers and Sectionalisers)

The overhead distribution line network commonly used for rural electrification is vulnerable to climatic influences and other mechanical damages. Experience has shown that over 80 per cent of the faults are transient in nature. Protection by drop-out fuses cannot discriminate between transient and sustained faults. This leads to frequent blowing of fuses which causes inconvenience to users and also increasing high maintenance cost.

Auto-reclosers and sectionalisers are employed in the system for fast and automatic restoration of supply following transient faults in the system. The design should be such that the desired number (one, two or three) of fast auto-reclosing operations are followed by one or two delayed operations to be performed to discriminate between transient and sustained faults before lock-out.

The auto-reclosers are installed outdoors and are either pole-mounted or pedestal-mounted. These mainly consist of the following two main units:

1. The high voltage circuit breaker
2. The control unit.

These are normally used for 11 kV lines and the normal current rating is 400 A, with a breaking capacity of 6 kA and an STC rating of 6 kA for three seconds. A microprocessor-based control unit is used for the control and relaying functions. The control unit is mounted on the same pole as that of the auto-recloser at a convenient height. After the fault is cleared by the auto-recloser, the faulty sections are isolated by the sectionaliser, which is an isolating switch. It comprises of an automatic switch with its own control unit and a built-in control transformer. This is also pole-mounted like the auto-recloser. As a sectionaliser is not required to interrupt the fault current and always opens with a time delay after the line has been de-energised by the auto-recloser, it does not break the load current and acts as an isolating switch.

10.2.7 Circuit Breakers for Earthing Applications

We have seen that for most of the indoor applications, switchgears are arranged in the form of switchboards having a common busbar and individual feeders and tie switchgears to and from switchgears of other switchboards in the same station. When maintenance work is to be carried out, it is essential to provide a facility for the earthing of busbars, incoming feeders and all outgoing feeders of each switchboard. Earthing breakers are used for this earthing.

Earthing breakers are available in different designs. In one design, the earthing breaker is mounted on the moving portion like a conventional breaker with only one set of contacts engaging the fixed isolating contacts. (Top contacts engage with the bus side isolating contacts and bottom contacts are shorted and connected to earth for busbar earthing. Similarly bottom contacts engage with the cable side isolating contacts and top contacts are shorted and connected to earth for feeder earthing). The normal service breaker from the switchgear is withdrawn after the breaker is switched off and then the earthing breaker is inserted in its position and switched on for the purpose of earthing. In another design, the earthing device is integrally mounted inside the switchgear cubicle itself.

In both the cases, the design should have safety provisions before the earthing is done, like ensuring that all incoming supplies to the board are switched off before earthing busbars or tie/incoming feeders. These are achieved by means of mechanical interlocks and the provision of auxiliary switch contacts of upstream breakers in the closing/tripping circuits of the breaker to be earthed.

10.3 PROTECTION REQUIREMENTS

As we know, the basic function of a switchgear is to sense the fault through protective relays and to trip the faulty feeder. Generally they differ from one application to another in terms of rating parameters and protection requirements. In some applications like switchgears controlling motors and dry type transformers, the use of surge suppressors should be considered as during switching operations, the over-voltages generated may be more than the withstand capability of these equipments. Surge suppressors are employed to limit the voltage to a pre-determined level. In addition to being used in switchgears, these surge suppressors are also used at the terminals of electrical equipment.

The protection requirements for various equipment such as incoming/tie feeders, transformers, motors, capacitors, etc. is covered in Chapter 13—Protection Schemes for MV Switchgears.

Chapter 11

ENERGY METERING

S.G. Deshpande

Man has become highly dependent on electricity, gas, water, petrol and diesel fuels. The availability of these energy sources is taken for granted. It is, therefore, essential to ensure that control is exercised on the use of these energy sources and more and more economical use of the latter is encouraged.

11.1 NEED FOR METERING

Metering of electrical energy is essential to:
- Monitor its use;
- Decide about day-to-day and time of the day generation;
- Decide about loadshedding hours, if required;
- Plan for future addition of generating units (demand forecasting); and
- Charge the consumer as per the applicable tariff rates.

11.2 TYPES OF ELECTRICAL ENERGY METERS

'Energy meter' is a general term. These meters are also called integrating type meters since electric energy E is given by the relation:

$$E = \int_0^T v(t) \cdot i(t) \, dt \qquad 11.1$$

where $v(t)$ and $i(t)$ are the instantaneous values of voltage and current respectively as functions of real time and T is the period over which the energy is consumed. When the meter is designed to read kWh, kVArh and kVAh, it is called a 'trivector meter'. A simple kWh meter is called an active energy meter. Both electromagnetic and static energy meters are available in the market. The classical electromagnetic energy meters having relatively poor accuracy are normally considered adequate for the domestic consumer's energy metering or for indicative purposes. However, for the metering of bulk consumption of electricity, much better accuracy and other sophisticated features are essential, which are normally available in static type meters.

11.3 FEATURES OF MODERN METERS

Modern state-of-the-art static meters are built with many sophisticated features such as:
- Facility for active energy measurement (both export and import);
- Facility for reactive energy measurement (both export and import);
- RS 232/RS 485/optical communication ports for connecting to data acquisition systems;
- Period block energy reading for pre-defined periods (typically for 15/30 minute periods);
- Relay output for alarm/tripping based on power factor/maximum demand/overload, etc.;
- Load surveying for demand forecasting;
- Time of day tariff in various time zones for better availability of power (if higher rates are charged during peak load periods, the consumers will be forced to disconnect their non-essential loads during such periods); and
- Local and remote communication features.

Besides, in addition to indicating the active and reactive energies, they can also perform the following measurements and indicate them through LCD or LED displays on the meter window by taking only voltage and current inputs from the system, either through CTs and PTs or through transducers:
- Voltage
- Current
- Power factor
- Frequency
- Active power
- Reactive power
- Apparent power
- Maximum demand (in kW or kVA) (maximum demand during 15/30 minutes interval)

Meters can also be provided with unique alphanumeric identification codes permanently embedded in non-volatile memory. This feature is useful when a central controller wants to access a particular meter in the field through remote communication.

11.4 MECHANICAL REQUIREMENTS

The energy meter should be so designed and constructed as to avoid any danger in normal use. The following aspects need to be ensured:

- Safety of personnel against electric shock;
- Safety of personnel against effects of excessive temperature; and
- Protection against spread of fire.

11.5 AUXILIARY SUPPLY REQUIREMENTS

The global practice for all bulk supply points is to provide external auxiliary supplies (sometimes both ac and dc for redundancy). The provision of external auxiliary supply solves the practical problems related to meter reading, remote communication, local communication, etc., which may not function during voltage transformer outages, e.g., during PT fuse failure, etc.

11.6 ACCURACY OF METER AND LIMITS OF ERROR

The electromagnetic type of energy meters have an accuracy class ranging from 1.0 to 2.5 which is considered adequate for the domestic consumer's energy metering or for indicative purposes. However, for energy consumption metering of bulk consumers, much better accuracy (as good as 0.5 S or 0.2 S) is essential.

The limits of error corresponding to class 0.2 S and 0.5 S are stipulated in IEC 60687 for various values of input current and power factor of the system.

Many factors influence the error in reading of an energy meter, the details of which are given in Table 11 of IEC 60687. This table also stipulates the maximum allowable limits of errors under different conditions. Some of these major factors which introduce errors in meter readings are:

- Measuring circuits voltage variation;
- System frequency variations;
- Harmonics in the input voltage waveform;
- Auxiliary supply voltage variation; and
- External magnetic field coupling the metering circuit.

11.7 BALANCED/UNBALANCED LOAD METERS

Energy meters take voltage and current as input. Normally for medium and high voltage circuits CTs and PTs are employed and meters have built-in multiplying factors so as to directly indicate the power in terms of the primary quantities. In case of balanced three-phase loads, it is adequate to take current input from two phases only. Such meters are designated as three-phase, three-wire (two-element) type meters. However, for unbalanced three-phase loads, it is essential to take current inputs from all the three-phases. Such meters are designated as three-phase, four-wire type (three-element) type meters. Figures 11.1 and 11.2 show typical connection diagrams of a modern state-of-the-art static meter suitable for balanced/unbalanced loads respectively.

11.8 COMMUNICATION FEATURE

Modern solid state meters are equipped with optical ports for remote data communication to SCADA or central monitoring system via optical fibre cables. Telephone lines can also be used for data communication using appropriate modems. The communication speeds are typically 300 to 2400 bps and communication protocols such as IEC870 or SCTM are generally in use.

Meter reading instrument (MRI) is a device for downloading the data from a state-of-the-art static energy meter. This data is then uploaded on to a PC for analysis of the active and reactive energy consumption trend through a special software. A centrally located PC can be configured to access any meter in the network and download its data for analysis and records through special software. Pulse outputs with a specified number of voltage pulses per kWh or KVArh are also available for recording by the SCADA system.

A typical block diagram depicting meters and remote communication is shown in Fig. 11.3

11.9 ONLINE ENERGY MONITORING SYSTEM

The "online energy monitoring system" is an automated online data acquisition support and reporting system which shall gathers data from meters (as per technical specification attached for meters) on an Rs 485 serial communication system using MODBUS RTU protocol.

The system shall be user friendly, flexible, and easy to understand. It can be customized to suit the individual needs of the "USERs".

The system shall be capable to access the data simultaneously/parallelaly from networked chains of various meters, installed at different locations within a plant.

The online energy monitoring system shall help perform the functions of:

(a) Energy audit
(b) Energy accounting
(c) Energy conservation
(d) System management
(e) Data logging
(f) Trending
(g) Check the consumption/billing of tariff meter
(h) Reporting and analysis

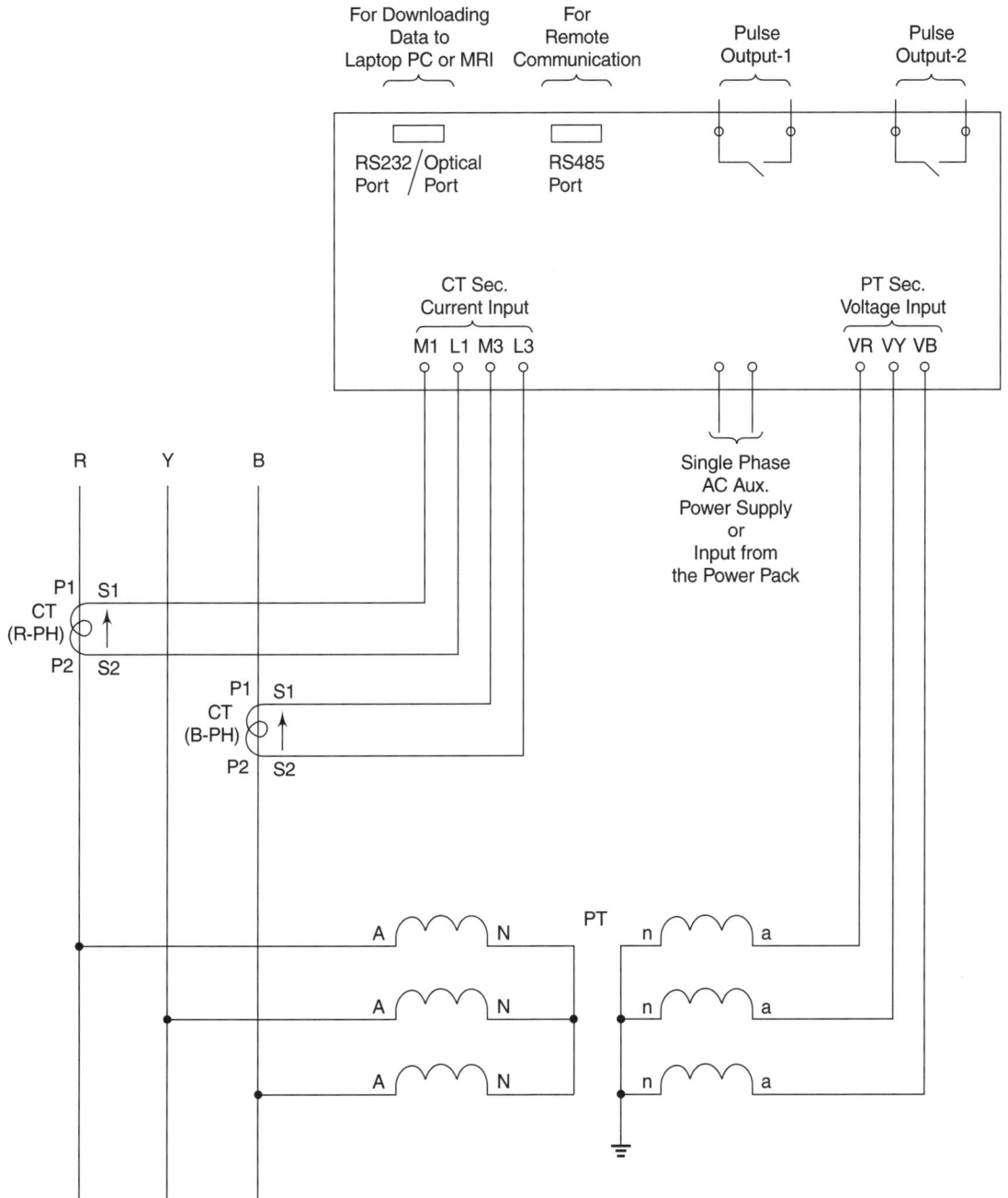

Fig. 11.1: Typical connection diagram for three-phase, three-wire meter

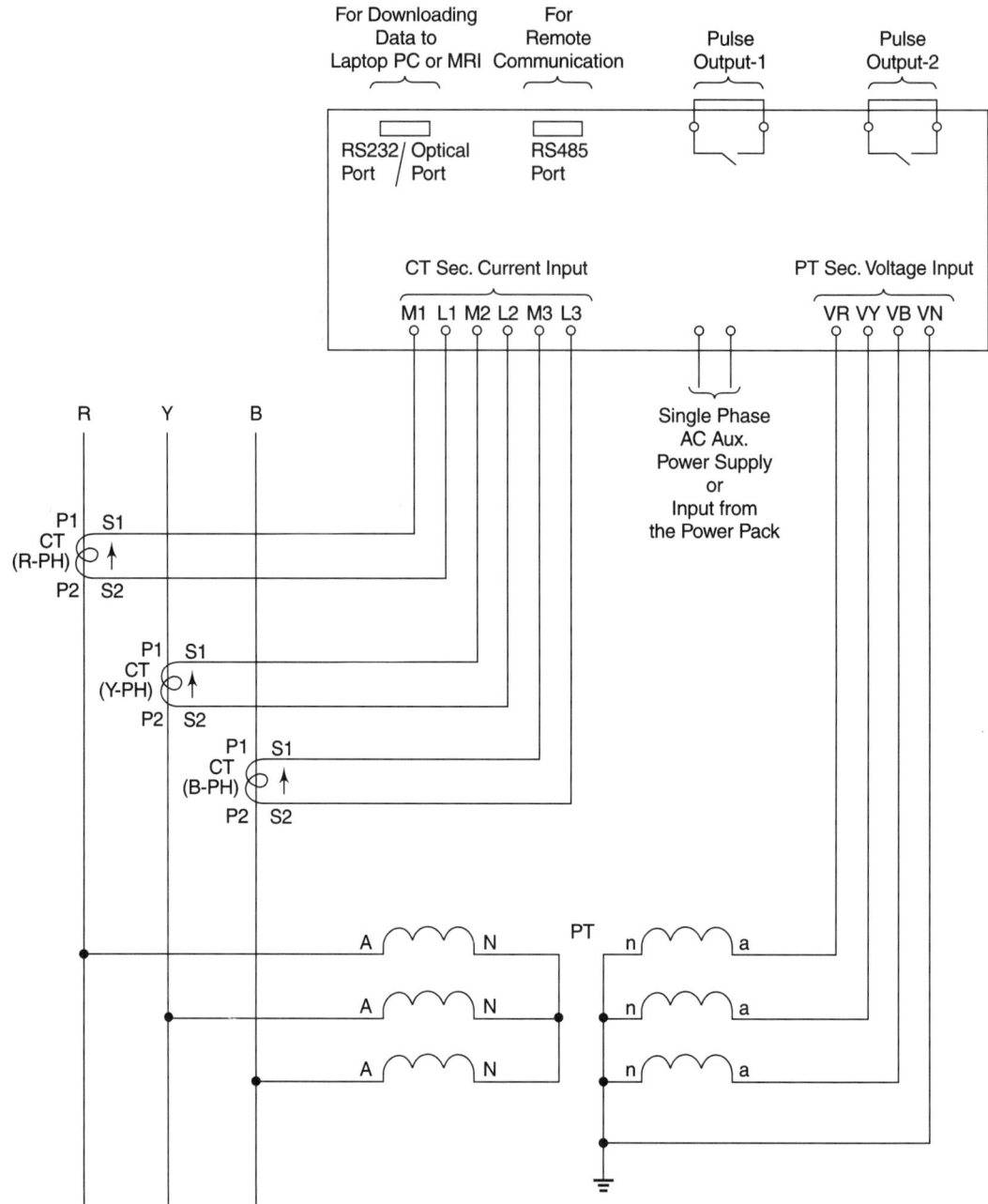

Fig. 11.2: Typical connection diagram for three-phase four-wire meter

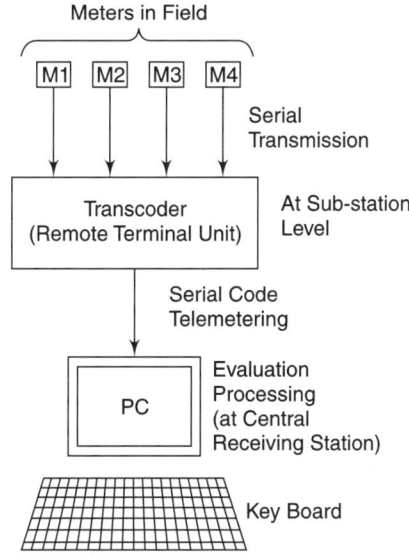

Fig. 11.3: Meters with remote communication

11.9.1 System Topology

The meters shall be connected through a multi-drop system using RS485 MODBUS protocol, to PC/server. The PC/server software, which shall run on Windows NT environment, shall gather information from individual meters ONLINE and also shall log the same into its memory.

The system shall also allow more than one client (workstation) to simultaneously access the data.

11.9.2 System Features

1. The data which shall be required to be viewed/logged can be configured by the USER, as per his need.

2. The ONLINE data shall be viewed from the various workstation (PCs) in TABULAR FORMAT (Spread Sheet format) or as TREND GRAPHS.

3. The software shall also allow the user to define upper and lower limit to various parameters in the individual meters. Any violation or deviation from the defined limits shall be promptly reported by the ALARM PROCESSING MODULE.

4. Mimic representation of the system, with data assigned to respective points in the same, shall also be reported.

5. The collected data shall also archive in an in-built data base for future access, queries and report generation.

6. The software shall also allow for the data related to the production, in the unit to be entered so as to relate it to the data logged by the system, while generating reports.

7. A library of report formats shall be available for the USER to choose from, while reporting so as to meet the function requirements of the various hierarchical levels of its management.
8. The software shall also provide for a USER overview, training, trouble shooting and help modules.
9. Database of the system shall be secure to support standard query language to generate different reports. No programming/macro shall be required/done by the user to get the reports.
10. Facility to export the data to other application software, e.g. ERP (enterprise resource planning) shall be available.
11. Facility to purge, store and back up of the desired data shall be provided.
12. Data will be collected in a fast, stable rugged and dynamic database for future access, queries and producing various reports.
13. Online software shall have a facility to online trends view up to 6 parameters on single screens.
14. It shall be possible to see up to six different parameters of a single meter on single screen.
15. It shall be possible to see main online parameters, i.e. (voltages, currents, power factors, frequency) in graphical view with indicators and all main energies shall be available in Text mode on single screen.
16. All online view screens shall have alarm LED; which shall blink to indicate when limits defined by user deviate.
17. All online screen shall have to take a print out for reference and record.
18. It shall be possible to select the particular meters for seeing the online gauge view.
19. Online monitoring software shall have a facility to generate a reports minimum, maximum and average of voltage, current and power factor.
20. Software shall have a feature to define the shift.
21. Software shall have a support of define the meters in groups and generates the report groupwise.
22. It shall have to generate report power consumption per unit based on user input (Production Quantity).
23. It shall be possible to see the historical data in trends as selectable user selectable time duration.
24. It shall be possible to generate an alarm report.

11.9.3 Remote Access

The online monitoring software shall have a feature to support the remote access the server (main system) as a client via PSTN/LAN network. Remote PC unit shall be worked as a client.

The monitoring software shall have a feature to access the data from the main server, i.e. online view and reports, etc. from remote location via PSTN/LAN network. This remote access PC shall be workstation.

11.10 STORAGE AND PROCESSING OF METER READINGS

The processor of a state-of-the-art solid state energy meter should be capable of processing the meter readings for billing. The required quantities for billing a bulk consumer are generally:

- Cumulative total kWh;
- Cumulative total kVArh;
- Cumulative total kVAh;
- Cumulative total kWh in different rates;
- Cumulative total kVArh in different rates; and
- Cumulative total kVAh in different rates.

11.11 MAIN AND CHECK ENERGY METERS

Sometimes, for better reliability, two meters (one main + one check) are employed in tandem. In such cases, the main meter's processor synchronises the check metering processor with a contact output closing at the end of a pre-defined integrating period.

11.12 GRID FREQUENCY MANAGEMENT

Excessive loading can result in lowering of the system frequency beyond acceptable limits. Hence, the present trend is to apply a tariff system, which discourages excessive loading. By linking the tariff rates with the prevailing grid frequency, the bulk consumers can be discouraged from overloading the system. With higher rates becoming applicable for drawing power under poor grid frequency conditions, the bulk consumers will be forced to disconnect their non-essential loads.

11.13 FUTURE TRENDS

In the developed countries, the supply of electrical energy through pre-paid meters is in vogue. This eliminates theft and non-payment of the electricity charges. The flow of energy stops automatically once the pre-paid amount has been consumed. The meters can be recharged through smart cards available at authorised stores and supermarkets. Efforts are on to adopt this system in India also.

Chapter 12

CONTROL AND INTERLOCKING SCHEMES FOR MEDIUM VOLTAGE SWITCHGEAR

S.K. Sahasrabudhe
A.K. Murgai
M.K. Srivastava
P.K. Upadhyay
Sanjay S. Dhavle
Manisha Singh

Medium voltage switchgears are used for switching medium voltage loads through associated control, measuring, protective and regulating equipment. This operation also involves considerable manual intervention which therefore necessitates the fulfilment of safety requirements laid down in Indian/International Standards.

A variety of control and interlocking schemes are used to achieve the above stated objectives. This chapter covers following schemes in detail:

(a) Safety schemes using position limit switches;

(b) Voltage selection schemes;

(c) Tripping schemes;

(d) Trip circuit supervision schemes;

(e) Alarm schemes;

(f) Synchronising schemes; and

(g) Automatic supply transfer schemes.

Schemes (a) to (d) are the frequently used schemes and are provided in most medium voltage switchgear applications, whereas schemes (e) to (g) are application-specific schemes.

Even though the functional requirements of schemes remain the same, there is always a variation from project to project depending upon the following factors:

(a) Distribution philosophy of ac, dc supply;

(b) System/bus configuration;

(c) Number of breaker trip coils;

(d) Grouping of alarm signals and trip/non-trip segregation; and

(e) Tripping logic.

The above aspects are generally finalised in agreement with the customer and based on experience gained over a period of time.

12.1 SAFETY SCHEMES USING POSITION LIMIT SWITCHES

The switchgear panel, irrespective of its 'make', is designed to perform the function of ensuring continuity of supply under normal conditions and disconnecting the supply when any abnormal/fault condition arises or it is switched off intentionally. Every switchgear panel is also designed to provide facilities for testing, maintenance and earthing.

However, the method of achieving these can vary from design to design. Hence, in order to comprehend the reasoning behind the provision of various interlocks and the method of achieving these in a particular design of switchgear panel, it is also necessary to understand the features that are unique to a particular design of switchgear. Various interlocking arrangements are provided in the designs of medium voltage indoor type switchgear panels. It may be noted that most of the interlocking requirements are generic in nature and are provided with all designs of switchgears.

As per the recommendations of IEC298/IS3427, the following requirements are mandatory for factory assembled metal-enclosed switchgears:

(a) Provision for earthing;

(b) Safety while withdrawing/inserting the removable/withdrawable parts; and

(c) Preventing operation of the switching device at certain positions/locations of the withdrawable parts.

These are achieved by providing interlocks in the switchgear panels. The method of achieving these, however, may vary depending upon the design, construction and application of that particular switchgear panel. Before we describe the purpose and methods of achieving various interlocking arrangements, let us familiarise ourselves with switchgear panels (see Fig. 12.1), which has the following distinct parts:

(a) House

(b) Breaker truck

(c) Instrument panel

The house consists of a chamber in which the breaker truck is inserted. At the rear top of the house, a busbar chamber is provided and at the rear bottom of the house, a cable chamber is provided. At the top of the breaker house, one instrument panel is provided which contains metering, protection and indicating devices as also the control switch for switching the breaker on and off. The connection between the instrument panel and the breaker truck is via a plug and

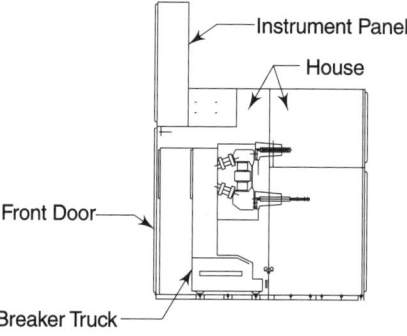

Fig. 12.1: Parts of a switchgear panel

socket arrangement. The plug and sockets are to be engaged when the truck is inserted in the house in the test position so that control supply is available to the closing and tripping coils of the breaker and to the spring charging motor. The auxiliary contacts provided on the breaker truck are also brought to the terminal blocks provided in the instrument panel for the purpose of interlocking. The breaker is provided with incoming and outgoing main contacts. The main contacts get engaged with the incoming and outgoing main contacts (provided in the busbar and cable chamber respectively) only when the breaker truck is inserted up to the service position. In the test position, only the control circuit of the breaker is 'through' making it possible to close and trip the breaker.

12.1.1 Test and Service Position Limit Switches

The test and service limit switches are provided to meet the requirement of IS:3427 which states that 'the operation of a circuit breaker shall be impossible unless it is in service or test position.' Although mechanical means are provided to achieve this, the test and service limit switches are provided to achieve this requirement electrically. The electrical closure of the circuit breaker is possible only in the two distinct positions, i.e. test or service.

The test position in the switchgear panel is known as the position where only the control circuit of the breaker is through but the main (power) contacts of the circuit breakers are not made. A limit switch is provided in the house, which gets pressed when the breaker truck reaches the test position.

The service position in a switchgear panel is known as the position where the control circuit of the breaker as well as the main (power) contacts of the circuit breakers are made. A limit switch is provided in the house, which gets pressed when the breaker truck reaches the service position. It is to be noted that either the test or service limit switch is pressed at a time depending upon whether the breaker truck is in the test or service position.

The various interlocks are described.

12.1.2 Test and Service Position Limit Switch Interlocks

These interlocks are used under the following conditions:

(a) When it is required to check/test only the control circuit operations of the breaker without connecting the power circuit, this interlock is used. The normally open contact of the test position limit switch is wired in the closing circuit of the breaker with the local position contact of the local/remote switch so that the breaker can be closed from the panel and testing can be done.

(b) The service position interlock is used to ensure that the breaker can be closed in the service position only. In the series of service position interlock, the remote position contact of the local/remote switch is also wired so that the breaker, when in the service position, can be closed from remote only.

12.1.3 Breaker Moving Portion Auxiliary Switch Interlocks

This breaker auxiliary switch is provided on the breaker truck and operates with the breaker main contacts irrespective of the breaker truck position, i.e. in test or in service. These contacts are normally used for the following interlocks:

(a) One NC contact is provided in the closing coil circuit of the breaker so that as soon as the breaker closes, the closing coil gets disconnected from the supply. This also ensures that the closing command to the breaker can be given only when it is in the off position.

(b) One each of NO contact is provided in the anti-pumping contactor circuit for anti-pumping and trip coil circuit for trip-free operation of the breaker.

(c) One NO contact is provided in the breaker on lamp circuit and one NC contact is provided in the breaker off lamp circuit.

12.1.4 Breaker Fixed Portion Auxiliary Switch Interlocks

This auxiliary switch is provided in the instrument panel and operates only in the service position with the breaker main contacts. These contacts are normally used for interlocks which are required to be activated only in the service position of the breaker.

12.1.5 Earthing Truck (Test to Service) Limit Switch

This limit switch is provided in the breaker house and operates as soon as the feeder type earthing truck reaches the test position and remains operated even when the earting truck is further pushed to the service position.

The contacts of this limit switch are wired in the closing and tripping circuits of the upstream breaker to ensure safety. This is to ensure that if the feeder earthing truck is inserted in the live panel by mistake, the incoming breaker to this panel trips and as long as the earthing truck remains inserted in this panel, the incoming breaker cannot be closed.

12.2 VOLTAGE SELECTION SCHEMES

Voltage signals to instruments and meters mounted on switchgear panels are derived from potential transformers (PTs). These PTs can be either bus-connected or feeder-connected. Feeder-connected PTs are normally provided on incomer feeder panels only. In normal service, the PT of each bus section feeds the voltage signal to feeders connected to the respective bus sections. However, in order to take care of a fault in the incomer breaker panel, the PT voltage signal to one section may have to be fed from the PT of another section. This has led to the need for voltage selection schemes.

For the bus sections separated by a bus coupler and having separate incomers on either sides, in the normal course, line PTs of either side located in the respective incomer, feed their own sections, with the bus coupler being open. In case of opening of any one of the two incomers, the PT on the other incomer takes over automatically and starts feeding to the complete board, with bus coupler being closed. This is achieved by using the breaker auxiliary as shown in Fig. 12.2.

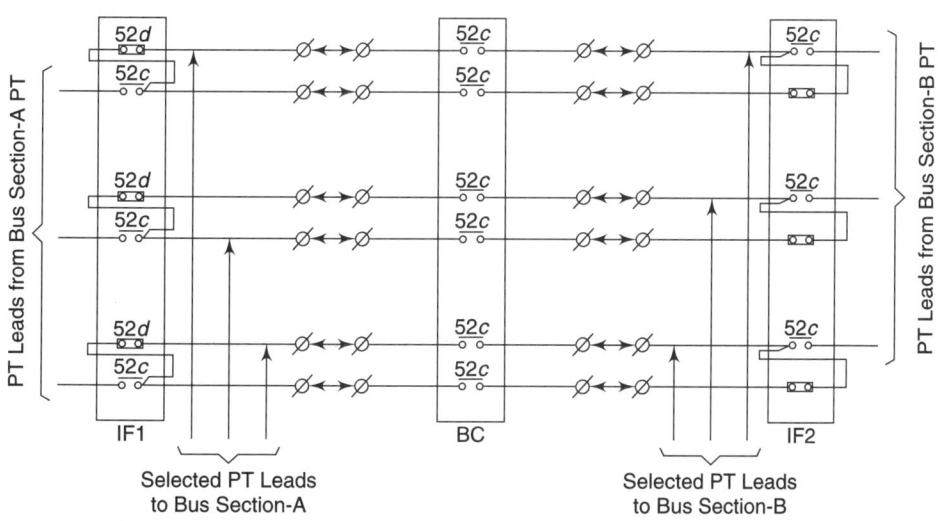

Fig. 12.2: Voltage selection using bus-coupler auxiliary swich

Owing to the requirement for large numbers of breaker auxiliary contacts and excessive inter-panel bus wiring involved in this scheme, an alternative and cost-effective scheme is sometimes employed, which uses only one PT to feed the entire switchboard. It uses under-voltage relays.

The relay is energised by using the PT supply itself and the PT supplies of both incomers are fed to the bus using contacts of the relays, one through all n.o.s and the other through all n.c.s. This scheme can be achieved using either single or twin relays as per the requirement. In a single relay scheme, normally that incomer's PT supply feeds to the bus which is given through NO contacts. When the relay is de-energised due to switching off of the PT input, the other PT, connected through NC contacts, takes over and starts feeding the bus. In a twin relay scheme, two separate relays are energised from both the PTs. To whichever incomer the supply comes first, it starts feeding the board and blocks the supply to other relays through its NC contact. The schemes for both the above selections are shown in Fig. 12.3(a) and 12.3(b).

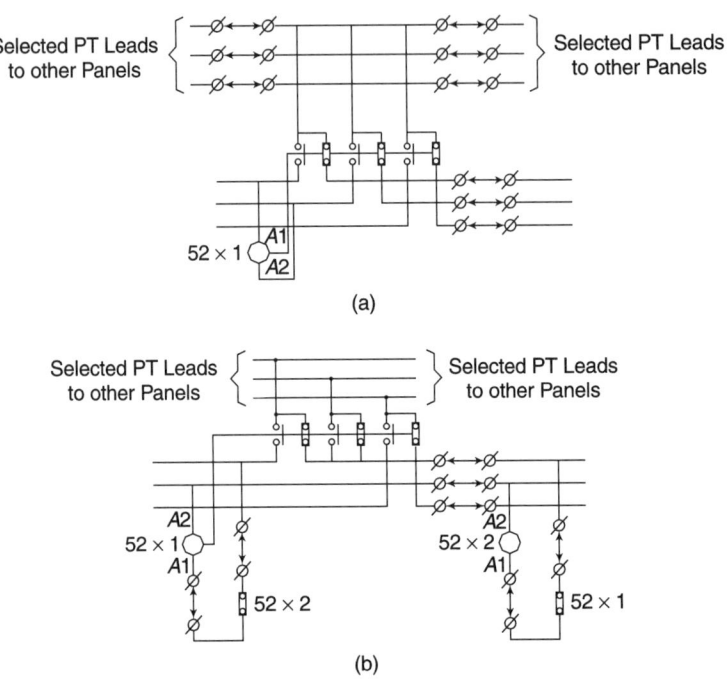

Fig. 12.3: (a) Voltage selection using one relay (b) Voltage selection using two relays

12.3 TRIPPING SCHEMES

A protection relay is usually required to trip a circuit breaker (CB). The power required by the trip coil of the CB may range from 50 W for a small distribution CB to 3000 W for a large EHV CB. Where such appreciable current-carrying capacity is required, interposing contactor type elements will normally be used. This interposing contactor type element is connected in any of the three arrangements discussed below.

12.3.1 Shunt Tripping Scheme

This is the most commonly used tripping scheme. The protective relay (PR) contact is arranged directly to trip the circuit breaker and it simultaneously energises an auxiliary unit X which then reinforces the contact that is energising the trip coil. The scheme is shown in Fig. 12.4.

All the above-mentioned tripping schemes envisage the use of separate dc supply for tripping. An alternative scheme using the fault current to trip the circuit breaker is called a series tripping scheme.

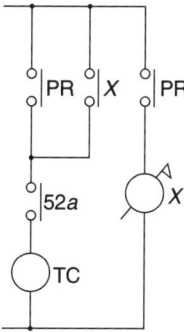

Fig. 12.4: Shunt tripping scheme

12.3.2 Series Tripping Schemes

These schemes operate by using different components as discussed below.

12.3.2.1 Using Relays

Here the series trip coil is normally kept shorted through the NC contacts of the series tripping relays. The trip coil comes into the circuit and trips the circuit breaker when the relay contact opens at fault. The circuit is shown in Fig. 12.5.

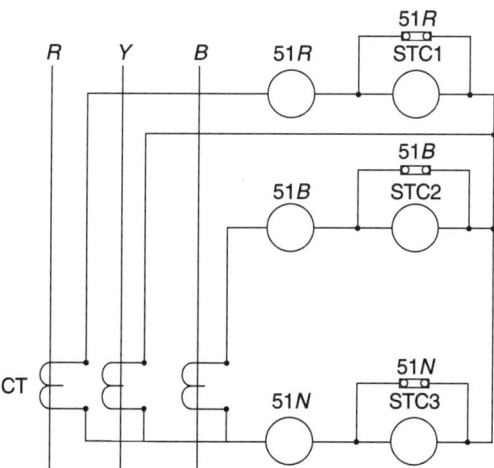

Fig. 12.5: Series tripping scheme

12.3.2.2 Using Summation CT

This scheme is similar to the one using relays but here a summation CT is used to summate the current of all the three phases, thereby minimising the requirement of three series trip coils to one. The summation CTs are LT CTs which sum up the secondary current output of HT CTs of three phases and provide a proportional single-phase output. The series trip coil is connected to the output of the summation CT and normally shorted through relay NC contacts. The scheme is shown in Fig. 12.6.

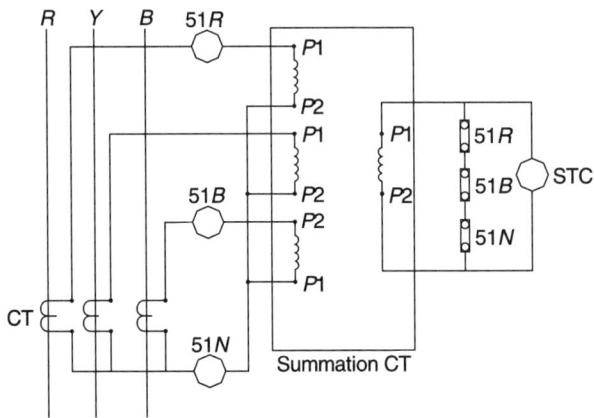

Fig. 12.6: Series tripping using summation CTs

12.3.2.3 Using Motor Protection Circuit Breaker (MPCB)

MPCBs are manual motor starters with thermal and electromagnetic trip features. Normally the trip coil is kept shorted through the MPCB. The MPCB blows off during a fault, and thereby pushing the fault current through the trip coil which, in turn, trips the circuit breaker. This process is shown in Fig. 12.7.

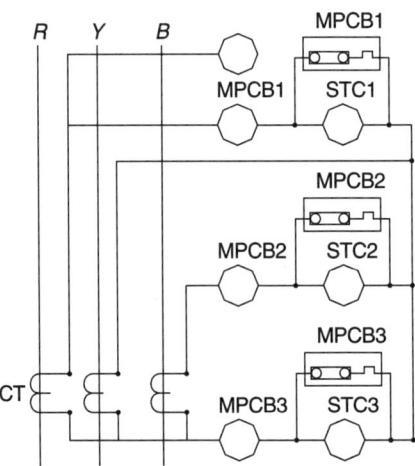

Fig. 12.7: Series tripping using MPCBs

12.3.2.4 Using Time Limit Fuses

This is similar to the MPCB scheme. Here, in the place of MPCBs, fuses with definite time characteristic are connected across the trip coil.

12.3.3 Capacitor Tripping Scheme

For installations where dc supply is not available or where it is uneconomical to provide battery/battery charger for dc supply or where the stations are unattended and battery maintenance cannot be guaranteed, a circuit using capacitor banks is employed to provide tripping energy to the breaker trip coil. It employs a three-phase full wave rectifier bridge with diodes for dc output and capacitive circuit for storing energy. The diode rectifier bridge is powered by the secondary of the HT PT. In normal service, tripping on faults is done through the diode bridge rectifier's dc supply. In abnormal conditions of failure of HT PT supply, the tripping energy is derived from energy stored in the charged capacitor banks. Normally the capacitors are rated to store energy for two trip and one close operation. The scheme is shown in Fig. 12.8.

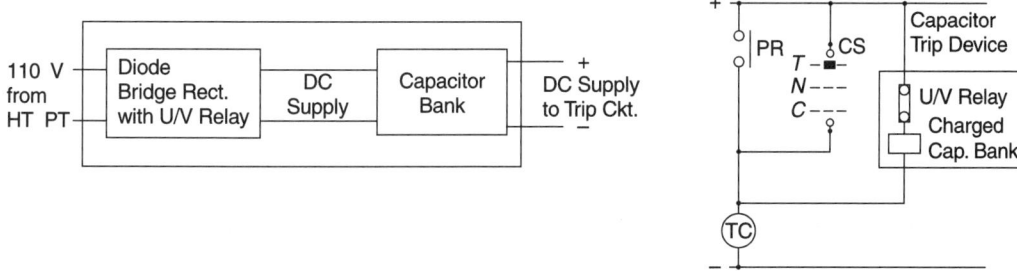

Fig. 12.8: Capacitor trip circuit

12.4 TRIP CIRCUIT SUPERVISION SCHEMES

The trip circuit extends beyond the relay enclosure and passes through more components such as fuses, links, relay contacts, auxiliary switch contacts and so on, and in some cases, through a considerable amount of circuit breaker wiring with intermediate terminal boards. These complications, coupled with the importance of the circuit, have directed attention to the need for its supervision.

The simplest arrangement contains a healthy trip lamp, as shown in Fig. 12.9. The resistance in series with the lamp prevents the breaker from being tripped by an internal short-circuit caused by failure of the lamp. This provides supervision while the circuit breaker is closed. Figure 12.10 shows how, by the addition of a normally closed auxiliary switch and a resistance unit, supervision can be undertaken while the breaker is both open and closed. In either case, the addition of a normally open push button contact in series with the lamp will make the supervision indication available only when required.

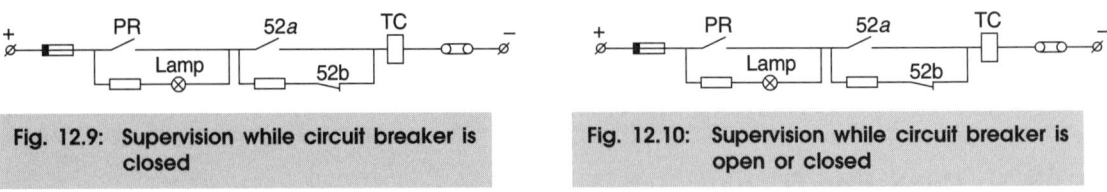

Fig. 12.9: Supervision while circuit breaker is closed

Fig. 12.10: Supervision while circuit breaker is open or closed

Schemes using a lamp to indicate continuity are suitable for locally controlled installations, but when control is exercised from a distance, it is necessary to use a relay system. Fig. 12.11 illustrates such a scheme, which is applicable wherever a remote signal is required.

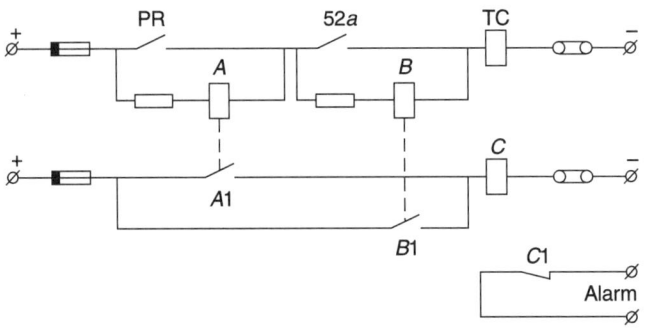

Fig. 12.11: Supervision while circuit breaker is open or closed with remote alarm

Under healthy conditions with the circuit breaker closed relays A and C are energised. If the trip circuit opens or the trip supply fails, relay A drops off and opens contact A1 to de-energise relay C. When the circuit breaker is open, relay B is also energised via the normally closed auxiliary switch of the circuit breaker and relay C is held in by contact B1. Relay B will detect trip circuit abnormalities with the circuit breaker open in a similar manner as relay A with the circuit breaker closed. Relay C is time delayed on drop-off by means of an RC circuit for a total time of 350 to 800 milli-seconds, to prevent a false alarm due to voltage dips caused by faults in other circuits or during a normal tripping operation, when relay A is momentarily short-circuited by the self re-set tripping relay contact. If the trip relay fails to re-set, possibly due to the failure of the circuit breaker tripping mechanism, the alarm is initiated. The alarm supply should be independent of the tripping supply so that an indication will be obtained in the event of the failure of the tripping battery.

12.5 ALARM SCHEMES

Prompt detection and rectification of any mal-operation is of vital importance for any continuous process or power plant. With increasing automation and complexity of process and power plants, the need for constant monitoring and indication of abnormal conditions has gained paramount importance to ensure the reliable and safe operation of power plants. Centralised annunciation systems fulfil this need and are increasingly finding large scale applications in power and process industries.

The following two types of alarm schemes are normally employed:

1. Alarm cancellation scheme; and
2. Alarm annunciation scheme.

12.5.1 Alarm Cancellation Scheme

A simple alarm cancellation scheme can be provided by using a bell and an auxiliary relay. Contacts of all protective relays are paralleled to initiate the auxiliary relay. The auxiliary relay, when energised, will initiate an audible bell and lamp for indication and will be latched by its own contact. The audible alarm can be cancelled by de-energising the auxiliary relay with the NC contact of the alarm cancellation push button. After removal of the fault which initiated the alarm scheme, the auxiliary relay should be hand re-set. The fault condition which initiated the alarm can be known by the flag provided on the protective relay. This scheme is shown in Fig. 12.12.

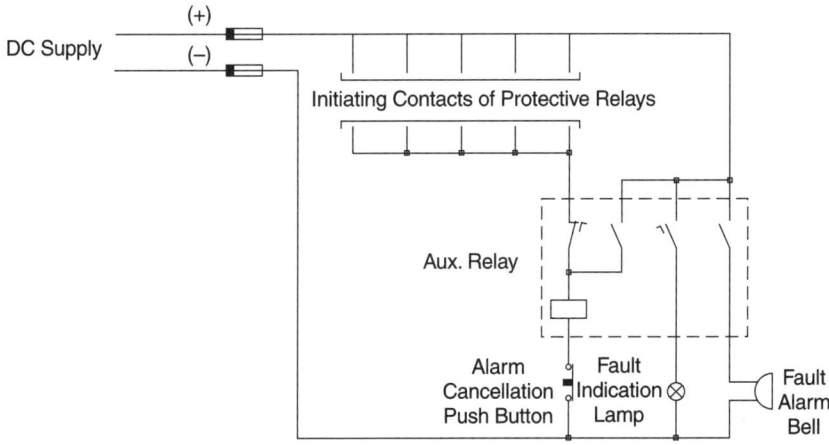

Fig. 12.12: Alarm cancellation scheme

12.5.2 Alarm Annunciation Scheme

An alarm can also be provided by continuously scanning inputs for fault conditions which may occur during any process operation. It can have any number of inputs for monitoring faults and a window is provided corresponding to each input. The change of state from normal to abnormal is displayed visually in the form of bright illuminated flashing windows and an audible device. The annunciator point is set for operation from potential-free normally open contacts which 'close on fault.' Also it is possible to change any alarm point from the 'close on fault' to 'open on fault' type of initiation. The fault initiating conditions, which also initiate tripping of the breaker are known as trip alarms and those which do not initiate tripping of the breaker are known as non-trip alarms. Separate potential-free contacts for initiating an external audible device are provided in the alarm annunciator for trip and non-trip alarms. Trip and non-trip alarms are segregated by red and amber coloured illumination of text in the window. Common silence/accept/re-set/test push button switches are provided as part of the alarm annunciator or they can be mounted externally on the control panel. The standard sequences of operation available in alarm annunciators are given in Tables 12.1, 12.2 and 12.3 as under.

Field Condition	Signal Contact	Facia Lamp	Audible
Normal	Open	OFF	OFF
Abnormal	Closed	Flashing	ON
Normal before Silence	Open	Flashing	ON
Silence	Closed/Open	Flashing	OFF
Accept	Closed/Open	Steady ON	OFF
Re-set before Normal			
(a) Sequence M	Closed	Steady ON	OFF
(b) Sequence N	Closed	Flashing	ON
Silence again (for sequence N only)	Closed	Flashing	OFF
Accept again (for sequence N only)	Closed	Steady ON	OFF
Return to Normal	Open	Steady ON	OFF
Re-set	Open	OFF	OFF
Lamp Test	Open	Steady ON	OFF

Table 12.1 Standard sequence M: manual re-set; Standard sequence N: M + repeat alarm

Field Condition	Signal Contact	Facia Lamp	Audible
Normal	Open	OFF	OFF
Abnormal	Closed	Flashing	ON
Silence	Closed	Flashing	OFF
Accept	Closed	Steady ON	OFF
Return to Normal	Open	OFF	OFF
For Momentary Alarms			
Normal before Silence	Open	Flashing	ON
Silence	Open	Flashing	OFF
Accept	Open	OFF	OFF
Lamp Test	Open	Steady ON	OFF

Table 12.2 Standard sequence S: self re-set

Field Condition	Signal Contact	Facia Lamp	Audible
Normal	Open	OFF	OFF
Abnormal:			
(a) First	Closed	Flashing	ON
(b) Subsequent	Closed	Steady ON	ON
Normal before Silence			
(a) First	Open	Flashing	ON
(b) Subsequent	Open	Steady ON	ON
Silence			
(a) First	Closed/Open	Flashing Steady	OFF
(b) Subsequent	Closed/Open	ON	OFF
Accept	Closed/Open	Steady ON	OFF
Re-set before Normal	Closed	Steady ON	OFF
Return to Normal after Accept	Open	Steady ON	OFF
Re-set	Open	OFF	OFF
Lamp Test	Open	Steady ON	OFF

Table 12.3 Standard sequence "F": first alarm

12.5.3 DC Fail Annunciation

Since the performance of the alarm annunciation scheme is wholly dependent on the availability of dc supply, the security of this dc supply is of paramount importance. A separate alarm scheme is normally provided to set off an alarm in case of failure of the dc supply. The failure of dc is displayed visually in the form of a bright illuminated steady window and an audible device. The visual and audible device is rated for 240 V AC supply and is initiated by potential-free contact available in the dc fail annunciator.

12.6 SYNCHRONISING SCHEMES

In order to ensure that two ac supplies are correctly paralleled, the following three conditions must be satisfied:

(a) The voltages of the two supplies must be within acceptable limits;

(b) The frequencies of the two supplies must be within acceptable limits; and

(c) The phase difference of the two supplies must be within acceptable limits.

In order to synchronise the two ac supplies, it is therefore essential to monitor the voltage, frequency and phase difference of the two supplies. The PT secondary voltage is used for monitoring purposes. The reference/bus voltage is termed as the 'running voltage' while the supply being connected to the bus supply is termed as the 'incoming voltage'. The incoming supply may be a generator source or some other section of a power system. The methods adopted for synchronisation are discussed below in order of their complexities.

12.6.1 Manual Synchronisation

In order to monitor the voltages, frequencies and the phase difference of incoming and running voltages, two ac voltmeters, two frequency meters and a synchroscope are used. The pointer of the synchroscope rotates in the clockwise or anti-clockwise direction depending upon whether the incoming supply frequency is lower or higher than the running supply frequency. When the two supply frequencies are equal and the voltages are in phase, the synchroscope pointer becomes steady and points at the 12 o'clock position. In actual practice, the two supply frequencies will never equal and the synchroscope needle will rotate at slow speed. At every null point crossing, the two phasors will be in synchronism for a short time and just before this instant, the closing command is given. The synchroscope is normally short time rated and should be switched on only when a synchronising operation is required.

An indicating lamp is connected across the incoming and running supply phasors. Under the condition when the two supplies are at the same potential, this indicating lamp will receive zero potential and hence will not glow. The lamp will glow the brightest when the two supplies are completely out of synchronism. This scheme is called 'dark at synchronism'. A dark at synchronism LED is provided on the synchroscope as well.

However, to distinguish between the 'dark at synchronism' condition and indicating lamp 'filament blown condition,' a 'bright at synchronism' scheme is commonly adopted. A phase reversing VT is used across one of the ac supplies to simulate a condition when the indicating lamp glows the brightest when the two supplies are in synchronism.

Thus, in the above scheme, before giving a closing command to the circuit breaker, the operator has to establish that the incoming and running voltages and frequencies are equal, that the synchroscope pointer is rotating at slow speed and that the 'bright at synchronism' lamp glows.

12.6.2 Check Synchronising Feature

A check synchronising relay is used to prevent inter-connection of two badly synchronised supplies. Its dual purpose is to:

(a) Safeguard manual synchronising; and

(b) Function in auto-reclosing sequence along with the auto-reclose relay.

The check synchronising relay contact is normally connected in series with the circuit breaker closing circuit. It ensures that the difference in the incoming and running voltages, frequencies and phases are within the pre-selected limits before switching on an incoming generator or interconnecting the two sections of a power system. For generator synchronising, the phase difference adopted is normally 20 degrees for small generators and 10 degrees for large generators.

The ac and dc circuits of a manual synchronising scheme with check feature are shown in Figs 12.13 and 12.14 respectively. A lamp is provided for 'synchronising in limits' operated by the check synchronising relay contact. In this scheme, the operator should establish the following conditions before giving a closing command to the breaker:

(a) The incoming and running voltages and frequencies are equal on the meters;

(b) The synchroscope pointer is steady at the 12 o'clock position;

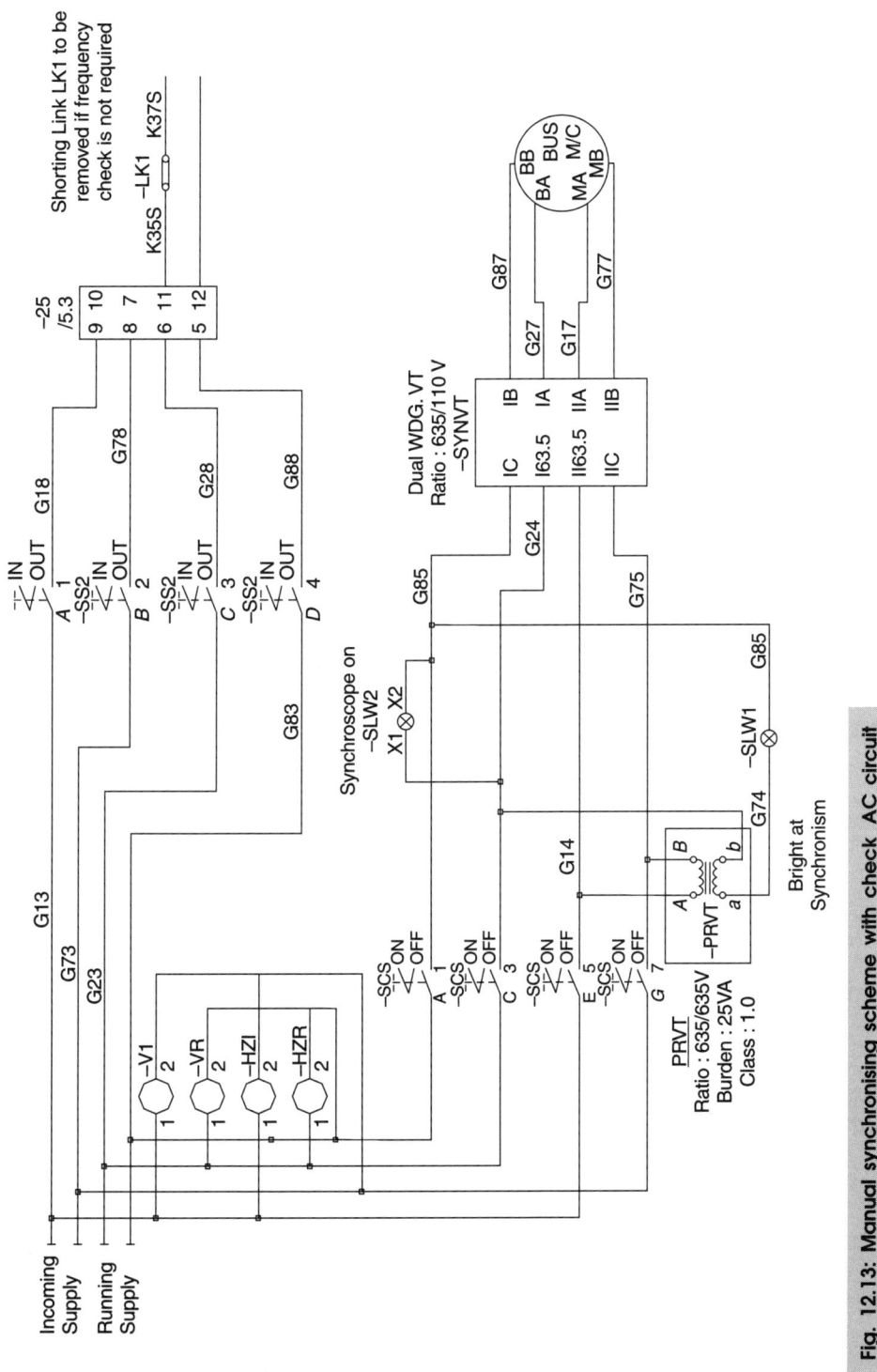

Fig. 12.13: Manual synchronising scheme with check AC circuit

12.16 Handbook of Switchgears

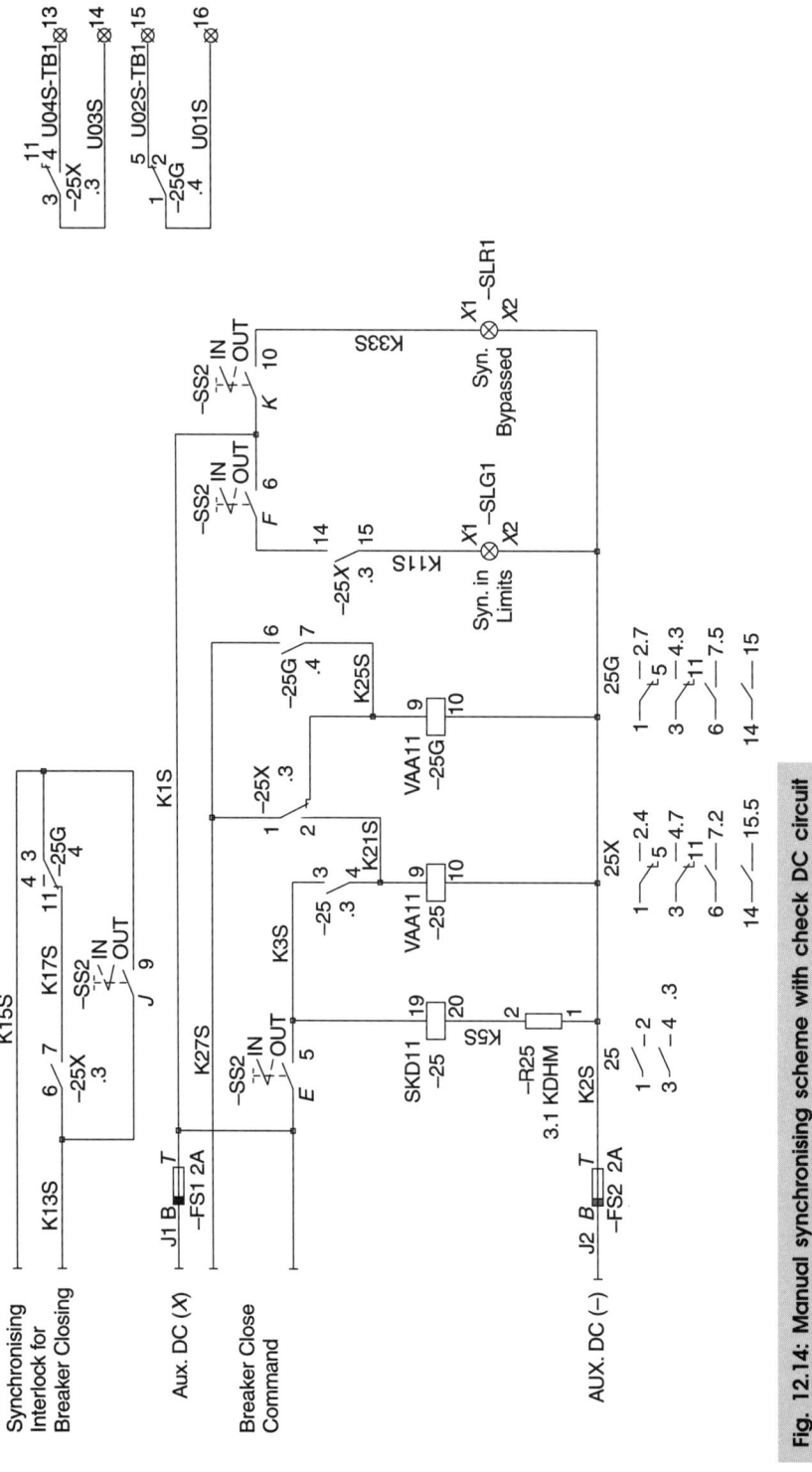

Fig. 12.14: Manual synchronising scheme with check DC circuit

(c) The 'bright at synchronism' lamp glows; and

(d) The 'synchronising in limits' lamp glows.

A guard relay is provided to prevent the operator from keeping the breaker close switch pressed under an asynchronous condition and allowing the breaker to close when the two supply phasors momentarily cross over. A contact of guard relay is wired in series with the check synchronising relay contact in the breaker closing circuit.

It may be necessary to bypass the check synchronising relay interlock in the breaker closing circuit under the following conditions:

(a) Connecting a dead line/other section of a power system to a live bus;

(b) Connecting a generator source to a dead bus; and

(c) Connecting a dead line/other section of a power system to a dead bus.

The above conditions are established by dead line/live bus, dead bus/live line or dead line/dead bus relays. These relays are nothing but voltage relays, which monitor the voltage levels of line/incomer and bus PT voltages. Under such conditions, the check synchronising relay interlock in the breaker closing circuit is bypassed by using a 'synchronising bypassed' selector switch.

The synchronising equipment is normally common for a sub-station as at any point of time, only one circuit is to be synchronised with the bus. The synchronising equipment is mounted on a swing type panel or a movable trolley so that the indicating instruments are visible from each circuit panel where the breaker control is provided. In case of a board with swing panel, a synchronising switch is provided on each breaker panel. All synchronising switches are equipped with removable handles or can be locked with a common key, with the key removable in the OFF position and trapped in the ON position. This is to ensure that only one circuit is synchronised at a time. The synchronising trolley is provided with a synchronising plug, which gets inter-connected to each breaker panel through a synchronising socket provided on each panel.

12.6.3 Auto-synchronising

This system is primarily intended to synchronise a generator with the power system grid. It automatically adjusts the prime mover speed and the generator voltage to match the busbar conditions and then give a closing signal to the circuit breaker when the voltage, frequency and phase errors are all within safe pre-set limits for smooth synchronising.

The frequency and voltage correction signals given by the auto-synchroniser are in the form of potential-free relay contact signals to give raise/lower signals to the prime mover speed governor motor and motorised potentiometer of the automatic voltage regulator.

The frequency control circuit of certain auto-synchronisers is designed to give one speed correction pulse to the prime mover speed governor for each cycle of the slip. The system provides continuous correction when the speed error is high, converting to pulses given at a slower rate as the synchronous speed is approached. The width of the pulses can be adjusted to optimise with the governor response to achieve the shortest possible synchronising time.

The voltage control circuit provides continuous raise/lower correction signal to the motorised potentiometer of the automatic voltage regulator whenever the generator voltage is outside the pre-set voltage matching limits.

When the frequency, voltage and phase difference of the incoming and running supplies are brought within permissible limits, a single pulse contact signal is given to close the breaker. The breaker closing pulse is of the order of 700 milli-seconds. A typical auto-synchronising scheme is shown in Fig. 12.15.

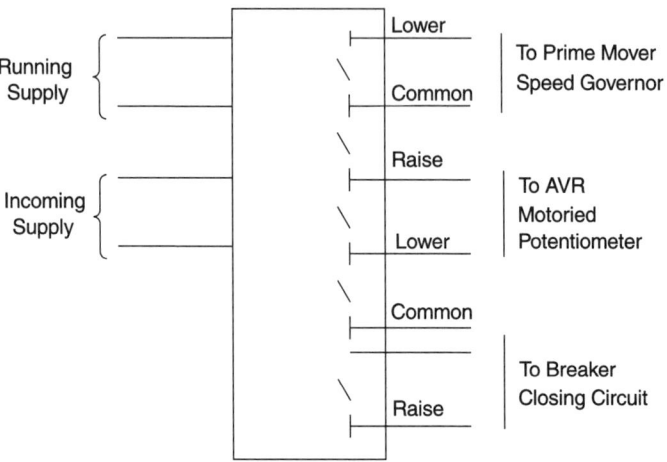

Fig. 12.15: A typical auto-synchronising scheme

All the operating functions of the auto-synchroniser are inhibited after the breaker closing command. The incoming and running voltage supplies must be disconnected to re-set the auto-synchroniser before another attempt can be made at synchronising the generator.

12.7 AUTOMATIC SUPPLY TRANSFER SCHEMES

In order to ensure uninterrupted supply to connected loads in thermal power stations or process industries, two types of automatic supply transfer schemes are normally used.

One is the automatic bus transfer scheme, used in medium voltage switchgear of thermal power stations, which is required for transferring load from the unit transformer bus to the station transformer bus and vice versa.

Another the second is the auto-changeover scheme, used in medium voltage switchgear of process industries/refineries, which is required for transferring load from one incomer bus to the other through a bus coupler breaker, in a two out of three breaker arrangements.

12.7.1 Automatic Bus Transfer Scheme

Automatic bus transfer schemes are normally required for switchgear employed in large thermal/nuclear power plants, where it is absolutely essential to provide uninterrupted power to the power plant auxiliaries, even during failure of their primary source of power, viz. the unit or station auxiliary transformer.

The requirements of a bus transfer scheme can be better understood by first understanding the arrangement and operation of medium voltage switchgear in large power plants.

Medium voltage switchgear employed in large power plants are mainly divided into two main categories, viz. unit switchgear and station switchgear.

12.7.1.1 Unit Switchgear

The unit switchgear receives power from the unit auxiliary transformer (UAT), which is directly connected to the output of generator of the same unit. Hence it is called a unit switchgear. It mainly supplies power to all the auxiliaries of the power plant that are essential for the running of that particular unit. Some examples of important auxiliaries are boiler feed pump, circulating water pump, ID fan, FD fan, PA fan, etc.

12.7.1.2 Station Switchgear

The station switchgear receives power from the station auxiliary transformer (SAT), which is fed from an outside source. This switchgear is mainly used for start-up of the power plant for power plant/township lighting and other supplies, and to feed auxiliary switchboards for coal handling, ash handling, etc., i.e. auxiliaries that are common to more than one unit.

An illustration of the typical arrangement of a unit and station switchgear, and unit to station bus transfer scheme, in a power station, is given in Fig. 12.16.

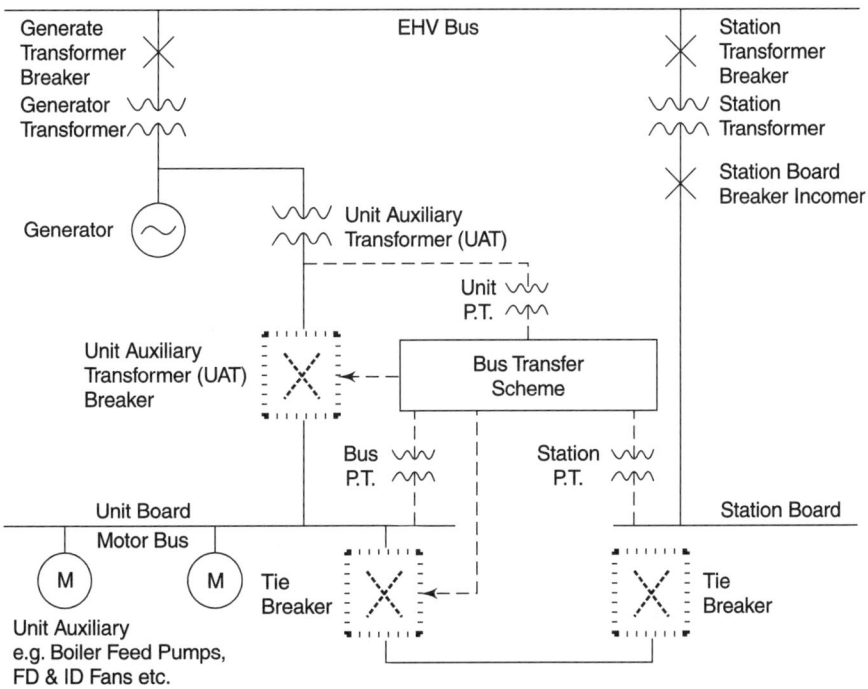

Fig. 12.16: A typical arrangement of unit and station switchboards, with unit to station bus transfer scheme

Before the power plant starts generating power, it is required to start-up the auxiliaries like the boiler feed pump, CW pump, etc. During start-up, as the unit is still not generating, the unit switchgear bus is fed from SAT through a tie connection to the station switchgear bus.

Once the unit starts generating power, the unit switchgear bus is disconnected from the station switchgear bus, and starts taking power from the unit itself, through UAT. Thus the power plant is run from the output of its own generating unit.

In case the unit is tripped due to some fault or planned shutdown, the incoming supply to the unit switchgear bus fails. In this condition also, it is essential to keep the power plant auxiliaries in a running condition. Hence the power supply to the unit switchgear bus is again taken from SAT, through the tie connection to the station switchgear bus. This is called a unit to station bus transfer.

Once the fault is cleared and the unit starts generating power again, the incoming supply to the unit switchgear bus is again shifted from SAT to UAT, by opening the tie breaker between the station and unit switchgear buses, and closing the incomer from UAT. This is called a station to unit bus transfer.

Sometimes, due to a fault in one station transformer, it may also be required to shift supply of the station switchgear from one SAT to another, through tie breakers provided between two station buses. This is called a station to station bus transfer.

This entire operation of shifting of the power from unit to station bus and vice versa or from one station bus to another, has to be very fast and accurate. Also, care has to be taken to ensure that such a bus transfer should not take place if there is any fault on unit or station bus, or if the station supply is not available. There are many other interlocking conditions, which govern the functioning of bus transfer. The transfer of bus can sometimes be manual and sometimes automatic, depending upon the operating conditions.

All these functions are carried out through the use of special bus transfer schemes, which are explained below.

The medium voltage unit and station switchboards are provided with bus transfer facility which could come into operation either manually (planned operation) and/or automatically. During the normal running of the unit, unit board is supplied from the unit transformer and the station board from the station transformer, as explained above. Both the switchboards are provided with bus tie breakers for inter-connection.

The bus transfer could be manual (pre-planned) in the event of the transfer of unit load from station to the unit transformer as is the case immediately after the unit has been started. The operation could also be vice versa. Similarly, in the event of one station transformer being taken out for some reason, the corresponding section of the station board is fed from the other station board. The transfer could also be automatic in the event of a fault characterised by the operation of protection relay. The manual and automatic bus transfer can be arranged in such a way that any of the following modes of operation is possible.

12.7.1.3 Modes of Bus Transfer

A. Manual Bus Transfer

   (i) Without Voltage Interruption

   This means allowing brief parallel operation of the two sources.

   (ii) With Voltage Interruption

   (a) Slow changeover; and

   (b) Fast changeover.

B. Automatic Bus Transfer (under fault condition) with Voltage Interruption

   (a) Slow changeover

   (b) Fast changeover

The requirements under various modes of bus transfer are detailed below.

12.7.1.4 Manual Changeover without Voltage Interruption (Involving Brief Parallel Operation)

This mode of operation is not used in the normal course because it involves the paralleling, however briefly, of the station and unit transformers or of the two station transformers. Such paralleling is not permissible because it will result in the fault level rising beyond the specified withstand and rupturing the capacity of medium voltage circuit breakers. The interlocking of the auxiliary system is so devised as not to permit, in the normal course, such a parallel operation. However, in case of an emergency, when the other modes of bus transfer with voltage interruption are not available, this mode can be adapted with due precautions.

For adopting this mode of transfer, there should be a facility bypassing the normal interlocking against parallel operation.

The changeover would involve check synchronising the two sources, whereafter the breaker for new sources of supply would be closed. Closure of the breaker will automatically give tripping command to the breaker for the existing supply automatically.

The bus transfer can be operable in either direction, i.e., transfer of unit load from the unit transformer to the station transformer and vice versa or transfer of the station load from one station transformer to the other station transformer.

12.7.1.5 Manual Changeover with Voltage Interruption

12.7.1.5.1 Slow Changeover This changeover consists in switching off the existing source of supply and allowing the residual voltage of motors connected to the bus to decay sufficiently, whereafter only impulse goes to the breaker of the new source of supply for closure. This signal for closing the new supply source breaker goes only after the residual voltage has reached the pre-set value, which is about 20 per cent of the rated voltage.

Necessary VTs and fast acting voltage sensing relays are used for sensing the appropriate voltage condition for bus transfer. This ensures that at the moment of changeover, even if the residual voltage of motor is in phase opposition to the voltage of the new source of supply, the motors will not be subjected to magnetic shocks which could prove dangerous. Reduction of residual motor voltage to 20 per cent of the rated voltage ensures that the closure of new supply even in phase opposition to the residual motor voltage will be harmless.

Since the residual motor voltage has dropped down to as low a level as 20 per cent, it presupposes that the motors have slowed down considerably. As such, on closure of the new sources of supply, there will be a danger of heavy inrush of starting current which will last till the motors are accelerated to the normal speed.

This starting current inrush, though somewhat lower than when starting from standstill, will yet be high enough to cause a heavy voltage drop in the transformer impedance of the new source which may be severe enough to cause the motors to pull out. In order to safeguard against such a contingency and to make bus transfer purposeful, it is therefore essential that simultaneously with the tripping of the existing source of supply, a trip command also goes to a pre-selected number of non-essential auxiliary motors on a pre-calculated basis so that inrush current at the time of closure of the new supply will be so restricted as to limit the voltage drop to not more than 15 per cent of the rated voltage. This mode of bus transfer can also be available for bus transfer in either direction, i.e. from the unit transformer to the station transformer and vice versa.

12.7.1.5.2 Fast Changeover This changeover ensures that the bus transfer takes place so rapidly that the residual motor voltage vector has fallen behind the incoming voltage vector only so little, that whereas on one hand, closure against phase opposition is ruled out, on the other hand, at the time of the closure, the connected motor speed has dropped down only marginally with the result that at the moment of transfer of connected motors to the new source of supply, there is only a small rise of current as compared to the normal running current (this marginal rise could be of the order of 50 per cent). The voltage drop during the bus transfer is therefore marginal and the bus transfer is satisfactory.

Since the closing time of the auxiliary breaker is considerably longer than that of the tripping time, in this mode of transfer, a fast time relay gives the closing command to the new supply breaker first and only after the pre-set time interval does a tripping command go to the existing supply breaker. The interval between the two commands is so arranged that the actual interruption time is limited to a few cycles (definitely less than seven cycles). This is supervised by a sensitive phase comparison relay. This bus transfer facility can also be available in both directions from the unit transformer to the station transformer and vice versa.

12.7.1.5.3 Phase Comparison Relay A sensitive phase comparison relay, which is the most vital element of the fast changeover scheme, should be used to permit the rapid changeover in the event of the two source voltage difference, phase angle difference and frequency difference is in pre-set limit before issuing closing command to the new source.

This relay otherwise blocks the closing command thereby preventing the changeover under unfavourable conditions of the new source and the residual voltage of the decaying bus. The phase comparison relay supervises the two sources continuously and is fast operating with the minimum operating time.

12.7.1.6 Automatic Changeover with Voltage Interruption, under Fault Condition (Fast Changeover followed by Slow Changeover)

As compared to the bus transfer described under Section 12.7.1.3 above, the changeover in this mode of transfer is initiated automatically on sensing a fault to the existing voltage supply. Thus with the help of appropriate sensing relays, the bus transfer takes place automatically when the existing voltage drops below a certain voltage level for a pre-set time (about 85 per cent of the rated voltages). Before the bus transfer is initiated in such a condition, appropriate relays are used to discriminate between the condition of fault in the bus voltages because of external fault and because of internal bus fault. In case of an internal bus fault (including uncleared feeder fault), the bus transfer system is locked out. Only in case of an external fault responsible for the voltage drop, such as a fault in the differential protection zone of the unit (generator transformer and unit transformer), is bus transfer initiated.

After having established that the conditions for bus transfer exist, the initiating command first goes automatically to fast bus transfer described under Section 12.7.1.3.2. Only if this circuit is found to not respond to the command, it automatically goes to slow transfer described under Section 12.7.1.3.1. The transfer is initiated after tripping the existing supply as also the pre-selected feeders to limit the inrush currents on the transformer. This mode of bus transfer can be available for bus transfer from the unit transformer to the station transformer.

12.7.1.7 General Requirements

The equipments used for the bus transfer scheme described above are normally wired in separate panels, which are placed in the control room. These are called bus transfer panels or BTS panels. The equipment includes selection switches, indicating lamps, transfer circuit, supervising relays, fast acting voltage sensing and phase comparison relays. Necessary voltage transformers of the required accuracy class are provided in appropriate medium voltage switchgear panels.

Testing facility is provided such that it enables the online testing of each of the operating circuits of each bus transfer scheme in working condition short of the actual bus transfer. However, in case the need arises for automatic changeover, while the testing of the scheme is in progress, the automatic changeover operation overrides the testing by re-setting the testing circuits.

Necessary selector switches are provided to select the mode of transfer, i.e. manual or automatic. In case of manual transfer, there is an additional selector switch for selecting manual without voltage interruption, manual fast and manual slow changeover with voltage interruption. In case of automatic selection, the transfer is normally initiated with fast mode. In case of this system being unhealthy and not responsive to the command, the selection automatically gets transferred to the slow mode. Mode selection and bus transfer initiation switches are located in the control panel.

Nowadays microprocessor-based fast bus transfer schemes are also available, where phase comparison and calculations, transfer mode selection, are done through software, and commands are issued for closing/tripping of breaker. No separate synchronising relay, auxiliary relay, mode selector switches, etc. are required in this scheme.

12.7.1.8 Blocking of Fast Changeover

In a condition when one station transformer out of two transformers is out of service, and one unit is being fed from the station supply, fast auto-changeover on the other unit is blocked. The manual changeover is not blocked in the above condition to give extra flexibility to the operator.

12.7.2 Auto-changeover Scheme

Auto-changeover schemes are normally required for switchgear employed in process industries, refineries, etc., where two incomers and one bus coupler are used in the two-out-of-three breakers configuration.

In such cases, the bus is divided into two parts, each fed by one incomer. The two buses are coupled through a bus-coupler breaker, which is normally kept open. This means that in normal condition, the two incomers are ON, feeding their respective bus sections, and the bus-coupler breaker is OFF. When incoming supply to any one of the bus section fails, and the incomer breaker of that bus section trips on under-voltage due to tripping of its upstream breaker, the supply to this bus is transferred to the other healthy incomer, by closing the bus-coupler breaker. This closing of the bus-coupler breaker is achieved automatically through the use of the auto-changeover scheme.

When the supply to the unhealthy incomer is restored to normal, this incoming supply is synchronised with the existing incoming supply, through synchronising relay, and the incomer is switched ON. The bus-coupler is then opened, after momentary paralleling of two incomers through the synchronising relay.

In some cases, it may be required to intentionally trip an incomer breaker for maintenance. In such cases also, the two incomers are momentarily paralleled by closing the bus-coupler after synchronisation, and then tripping the selected incomer breaker.

The auto-changeover scheme uses contacts of under-voltage relays, tripping relays, synchronising relays, and the auxiliary switches of the incomer and bus-coupler breakers.

12.7.2.1 Main Features of Auto-changeover Scheme

The main features required for the auto-changeover scheme are:

1. Auto-changeover between two incomers and one bus-coupler, such that, any two breakers are ON at a time (two-out-of-three arrangement);
2. Manual closing of bus-coupler with momentary paralleling of both incoming supplies to facilitate taking out of any one incomer breaker for maintenance; and
3. Blocking of auto-transfer if any incomer breaker trips due to fault or the incoming bus is not healthy.

In order to facilitate better understanding these features, the normal operating conditions, conditions for initiating or blocking of auto-changeover, and conditions for manual transfer are explained below.

12.7.2.1.1 Normal Operating Condtion The two incomer breakers are 'ON' and the bus-coupler breaker is 'OFF'. The two bus sections are running independently.

12.7.2.1.2 Conditions for Initiating/Effecting Auto-changeover These include the following situations:

(a) The incoming supply to any one incomer fails, and that incomer is tripped on under-voltage. The residual voltage on this bus section is below 40 per cent of the normal voltage.

[This is achieved through the use of an under-voltage relay, and a timer to introduce a suitable time delay. Time delay is introduced to avoid tripping on momentary voltage dips. Time delay range is normally 0.5 to 5 seconds.]

(b) The voltage on healthy bus section has been normal, or above 80 per cent of the normal voltage for a specified duration. The incomer breaker of a healthy bus is closed and is in the service position.

[This is achieved through the use of a healthy voltage check relay, and a timer to introduce a suitable time delay. The time delay range is normally 0.5 to 5 seconds.]

(c) The bus coupler is open and in service condition.

(d) The auto/manual switch is set for 'AUTO' operation.

If all the above conditions (a) to (d) are met, auto-transfer is effected, and the bus coupler breaker gets closed automatically.

12.7.2.1.3 Conditions for Specific Blocking of Auto-transfer The bus coupler should not close if any of the incomers trips due to a fault or the other bus section is not healthy.

(Transformer fault, i.e. 63TX, 64R or any upstream fault (before the breaker) should not block auto-transfer.

12.7.2.1.4 Conditions for Manual Transfer

(a) For maintenance: It should be possible to trip any one incomer after closing the bus coupler, to facilitate maintenance as required.

(b) Restoration of normal operating condition, after auto-changeover: When the supply is again available to the tripped incomer breaker, after effecting auto-changeover, it is possible to restore the system to the normal operating condition, by closing this incomer and tripping the bus coupler. Thus the two bus sections again run separately, and the bus-coupler breaker remains open.

Manual transfer permits momentary paralleling of two incomers. A synchronisation check relay is used to ensure safety of the system.

Chapter 13

PROTECTION SCHEMES FOR MEDIUM VOLTAGE SWITCHGEAR

S.K. Sahasrabudhe

The purpose of an electrical power system is to generate and supply electrical energy to consumers. Profitable operation requires this work to be done reliably, economically and with minimum interruptions. In this, power system protection plays a significant part for the efficient operation of the power supply system by improving the availability of power supply while at the same time reducing damages due to faults. In order to fulfil the requirement of protection with the requirement of profitable operation, any protection system is required to satisfy the 5-S principles listed below.

13.1 5-S PRINCIPLES

1. **Security:** Protective system should be reliable so that security of supply is ensured.
2. **Sensitivity:** Protective system should be able to sense minimum value of fault current, thereby reducing the consequent damage.
3. **Speed:** Protective system should be able to isolate fault in the shortest possible time.
4. **Selectivity:** Protective system should be able to select and trip only the nearest circuit breaker.
5. **Stability:** Protective system should not operate for external faults.

A protection system is required to limit the disturbances in a power system. These disturbances, if allowed to persist, may damage plant and interrupt the supply of electric energy. They are described as faults.

13.2 FAULTS

Power systems are subject to many kinds of faults. The principal types are three-phase with or without earth, phase-to-phase, single-phase-to-earth and two-phases-to-earth. In addition, generators transformers and motors are subject to short-circuits between turns of the same winding.

A fault is defined as any abnormal condition, which causes reduction in the basic insulation strength between current-carrying conductors, or between the current-carrying conductor and earth. In practice, this reduction is not regarded as a fault unless it results in an excess current or in a reduction of impedance to a value lower than the load impedance.

By determining the currents and voltages at the fault point, it is possible to define the fault. It is also essential to know the effect of the fault in branches of networks other than the faulted branch, so that protection can be applied correctly to isolate the section of the system directly involved in the fault. This approach of the fault studies on the basis of its effect on networks in a power system makes fault detection easier.

13.3 FAULT DETECTION

In order to facilitate rapid removal of a disturbance from a power system, the system is divided into protection zones. Relays monitor the system quantities (current, voltage) appearing in these zones; if a fault occurs inside a zone, the relays operate to isolate the zone from the remainder of the power system.

The operating characteristic of a relay depends upon the energising quantities fed to it such as current or voltage, or various combinations of these two quantities, and on the manner in which the relay is designed to respond to this information. Various methods are used for deriving these energising or relaying quantities.

13.4 DERIVATION OF RELAYING QUANTITIES

A majority of the system faults originate as faults between one phase and earth, developing later into faults involving one or both of the other phases. If a fault can be dealt with quickly, the damage and disturbance to the system can be limited in many cases. In order to apply protection relays, it is usually necessary to know the limiting values of current and voltage, and their relative phase displacement for various faults. There are cases where the system parameters are such that the current produced by fault conditions differs little in magnitude from the load current. In such cases, current magnitude discrimination becomes difficult or even impossible to apply. Discrimination between a fault and normal or overload condition can be possible if suitable circuitry is used through which a three-phase system can be resolved into their positive, negative and zero sequence components.

Zero-phase Sequence Systems

Faults involving earth give rise to zero-sequence currents, whereas in a healthy three-phase system with balance load currents, zero-sequence currents are negligible. This can be used as a basis for discrimination, which allows the protective system to ignore load current, thereby permitting the use of earth fault settings lower than the load current values.

Three common ways of applying zero-sequence discrimination are:

1. Putting a suitable current sensing device, e.g. CT between the power systems star (neutral) point and earth. However, this arrangement can only be used at points near an earthed neutral. The stator winding protection of generator usually employs this method of earth-fault detection. Various methods of generator neutral earthing and consequent forms of earth-fault protection are shown in Fig. 13.1.

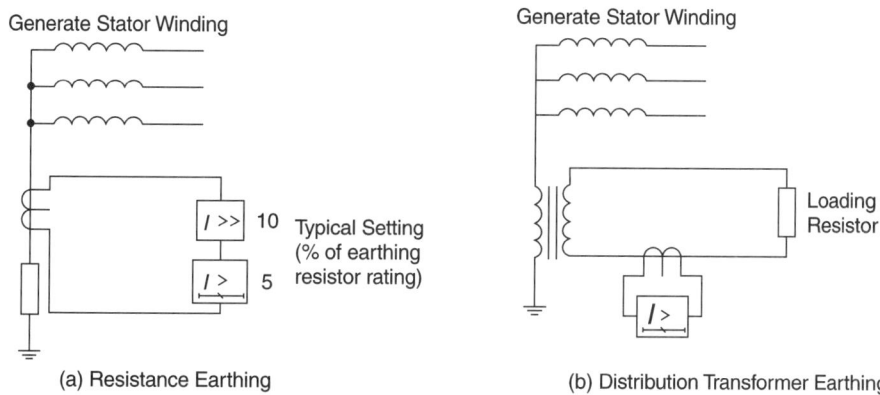

Fig. 13.1: Methods of generator earthing

2. Zero sequence currents can be detected by using core balance CT (CBCT) where zero-sequence currents will flow only in the event of unbalance in the phase currents. In this arrangement, it is important to ensure that no other conductor such as a cable sheath is passed through CBCT. This may cause incorrect operation, as the sheath will also be carrying some currents. However, this can be neutralised by deliberately bringing the sheath earthing conductor back through the CBCT opening. The positioning of core balance current transformers is shown in Fig. 13.2.

3. Zero-sequence currents can also be detected by using residual connection of three-phase current transformers. This is also known as a current balance system in which the secondary of current transformers is connected in residual connection, with relay in the fourth wire. Various forms of residual connections along with their setting limitations are shown in Fig. 13.3.

The other two systems, viz. negative phase sequence system, and positive phase sequence system, being of lesser use in medium voltage switchgears, are not being elaborated here.

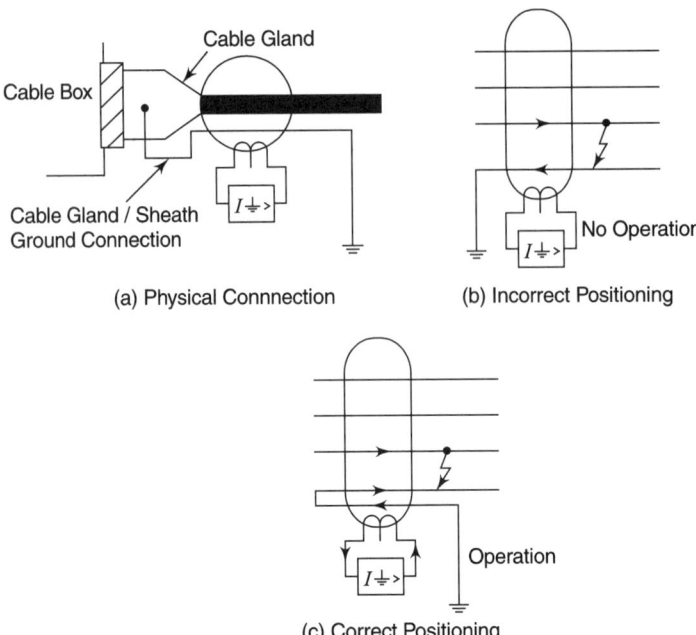

Fig. 13.2: Positioning of core balance current transformers

The four major and most commonly used protection schemes in the medium voltage switchgear area, namely motor protection, transformer protection, feeder protection and capacitor protection schemes are described below.

13.5 MOTOR PROTECTION

In the case of motor protection, the function of protection is the early detection of fault condition so as to limit to an absolute minimum the damage caused to the affected equipment. In the case of motor protection, since motors are designed on a continuous maximum rating (CMR) basis, the discrimination of fault even though important, the need for rapid operation is more important than discrimination.

13.5.1 Types of Faults

In general, it is necessary to protect a motor against abnormal running and fault conditions arising from:

(a) Prolonged overload as a result of excessive mechanical load;

(b) Single phasing caused by rupturing of a fuse or by open-circuiting of one phase;

(c) Short-circuits between phases or between phase and earth in the motor winding or its connections;

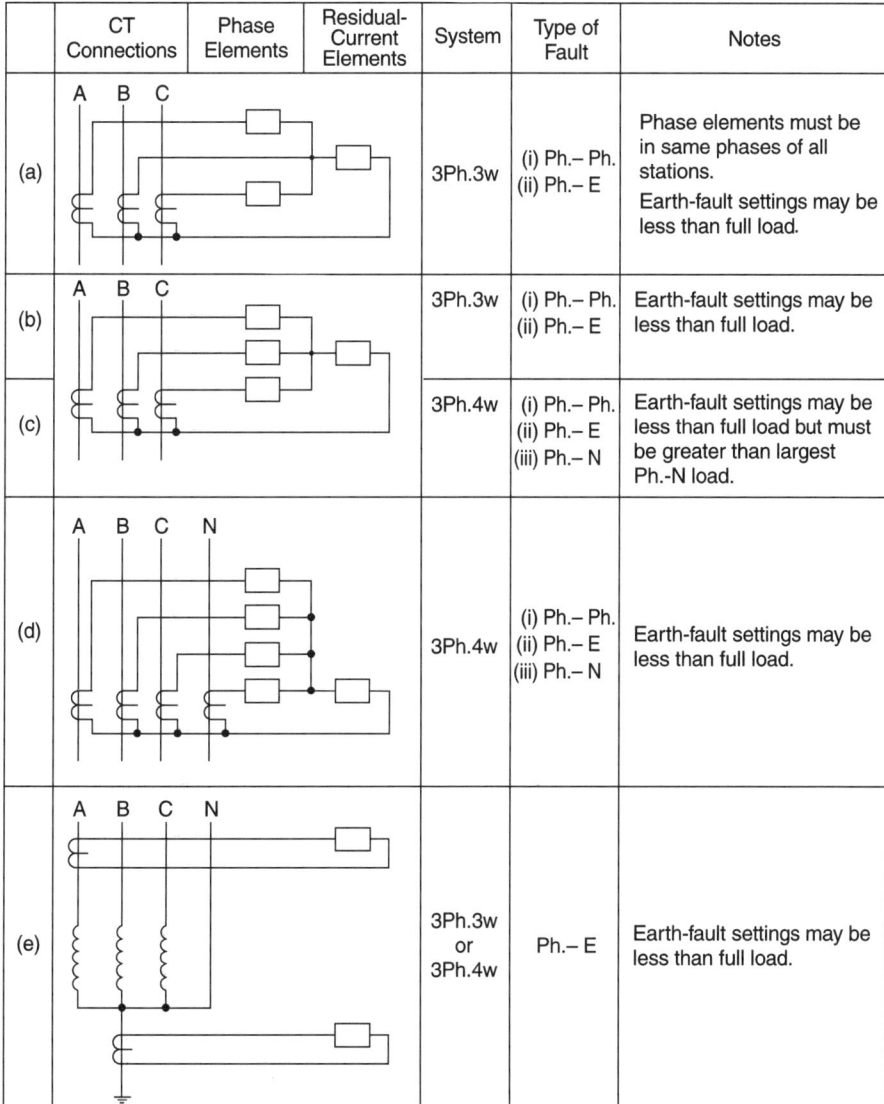

Fig. 13.3: Types of residual connections and their setting limitations

(d) Partial or complete collapse of voltage; and

(e) Locked rotor.

These are described below:

13.5.1.1 Overloading

Motors draw a starting current from the supply which is several times their normal full load current, and it is essential for the protection to be unresponsive to this starting surge provided that the motor current returns to its running value within the time determined by the design of the motor. On the other hand, the protection must not be given a setting greatly in excess of the full load or it will be unable to safeguard the motor against over-current conditions.

The above two opposing requirements are met in a thermal relay. The essence of the thermal relay is the specially designed element which closely simulates the changing thermal conditions in the motor, thus allowing the motor to be retained in service up to the point beyond which damage would probably be caused.

13.5.1.2 Single-phasing

The ability of a three-phase motor to continue to run with an open-circuit in one phase causes problems in providing appropriate protection. Single-phasing condition causes the motor current in the remaining two phases to increase to 1.73 times the normal motor full load current and ultimately leads to overloading. The protection against single-phasing is based on the measurement of negative sequence component and motors are tripped on single phasing/unbalance depending upon the negative and positive sequence components and their consequent heating.

13.5.1.3 Short-circuit Protection

Protection against short-circuits in the motor winding, or the connections to it, is provided by high-set instantaneous over-current elements. The typical settings are 4 to 16 times the full load for over-current elements and 1 to 4 times the full load for earth fault elements.

13.5.1.4 Under-voltage Protection

The main reason for employing under-voltage protection is to ensure that the motor circuit breaker is tripped on a complete loss of supply, so that when the supply is restored, it is not overloaded by the simultaneous starting of all the motors. It is also necessary to avoid danger to the operating staff when the supply is restored without their prior knowledge. Under-voltage protection is normally provided through an under-voltage relay with a time lag. In order to reduce the risk of unwanted tripping of entire groups of motors, it is necessary to ensure that the relay will not operate when momentary voltage dips are experienced. A sufficiently low voltage setting with an adequate time setting is therefore applied so that the protection remains inoperative during transient disturbances.

13.5.1.5 Stalling of Motors (Locked Rotor)

Should a motor stall when running or be unable to start because of excessive load, it will draw a current from the supply equivalent to the locked rotor current. It is obviously desirable to avoid damage by disconnecting the machine as quickly as possible. It is not possible to distinguish this condition from a healthy starting condition purely on current magnitude basis. The only way to overcome this problem is to ensure that the protective relays disconnect the motor if the current continues for longer than the normal starting time. In the case of motors driving high inertia loads, the starting time may be nearly equal or even more than the safe stall time, making the problem of discrimination much more difficult. In such cases, speed switches to distinguish between the starting and running conditions are employed. A typical interlocking scheme is shown in Fig. 13.4.

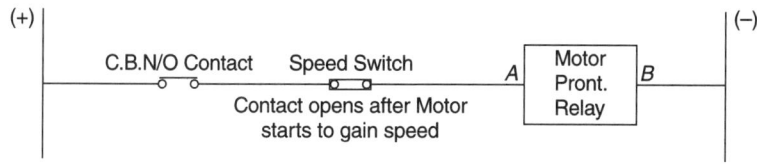

Fig. 13.4: Locked rotor protection inhibit scheme

Under healthy starting, speed switch contact opens before stall withstand time is lapsed and no tripping takes place. Under locked rotor condition, the speed switch contact always remains closed, hence after the safe stall withstand time, the motor trips through command on terminal A.

13.6 TRANSFORMER PROTECTION

The protective system in a transformer is aimed at reducing the downtime to a minimum by the rapid clearance of all faults, while at the same time minimising the damage caused to the unit, which indirectly means low cost of repair and downtime. In designing a protective system emphasis is laid on such criteria which provide early warnings and do not result in an immediate shutdown. A typical example can be the Buchholz alarm whose warning can indicate a suitable corrective action to prevent shutdown as the result of a major fault.

13.6.1 Philosophy of Protection

A transformer can experience broadly two types of faults, i.e. internal and external. External faults are cleared by feeder protection schemes whereas in the case of internal faults, isolation is ensured by the unit protection scheme. These are described below.

13.6.1.1 External Faults

13.6.1.1.1 Faults on Transformer Terminals These constitute system fault and flashover. They do not pose the threat of extensive damage but may constitute a serious hazard to the stability of the power system.

13.6.1.1.2 Fault External to the Transformer These are normal system phase and earth faults and are cleared by appropriate external protection system. The only precaution to be taken is that the 'unit protection' given for faults within the transformer tank must remain immune to these external faults.

13.6.1.1.3 Incipient Faults These faults are initially minor in nature but could grow into major faults, if left undetected. They include core faults, failure of cooling arrangements and blockage of insulating oil.

13.6.1.2 Internal Faults

These can be faults in windings, tap changer faults, core over-heating, insulator or bushing failure. In case of all these faults, the risk of fire enhances the possibility of damage. Since the faults within the transformer tank are more common, they are described in detail below.

13.6.1.2.1 Over-current and Earth Fault Protection This relay protects the transformer against excess overload as well as short-circuits within the transformer winding. IDMT relays are used for protection against excess overload whereas high set instantaneous relays are used for protection against short-circuits.

13.6.1.2.2 Differential Protection The protection provided for a power transformer depends, to some extent, upon its size and rating. The possibility of damage is high for internal faults, e.g. a short-circuit of a few turns of the winding will produce a heavy current in the faulted loop but a very small terminal current. Similarly faults near the neutral end can cause dangerously high fault currents. A differential protection scheme as shown in Fig. 13.5 is employed which compares HV and LV currents.

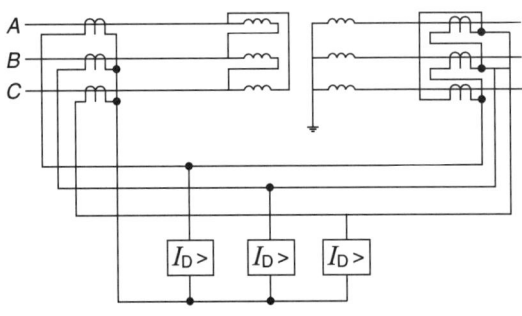

Fig. 13.5: Differential protection for two-winding delta star transformer

Since HV and LV currents are in a known relationship under a healthy condition, they are used for detecting inter-turn short-circuits since this changes the effective overall transformation ratio. The transformation ratio also gets affected due to: (a) transformer connection, (b) tap changing, and (c) magnetising inrush current during switching on. In order to prevent mal-operation, differential protection is provided with a proportional bias, which, in effect, exceeds the maximum deviation caused because of (a) and (b) above, whereas for (c) the second and fifth harmonic restraint feature is provided. The time of operation and setting current can be made independent of other protection systems as differential protection is a unit protection scheme, thus allowing low setting and fast operating time.

One of the most important considerations in any differential protection scheme is its stability under through-faults, i.e. faults outside the protected transformer. When a through-fault current traverses a zone protected by a differential system, the transient flux produced in the CTs is not detrimental as long as it remains within the linear range of the magnetising characteristics. The magnetising characteristic of a CT has a knee point beyond which for a 10 per cent increase in secondary EMF, a 50 per cent increase in exciting current is required. This extreme high increase of exciting current causes mismatches between two groups of current transformers and this spill current may cause false tripping of the differential relay.

13.6.1.2.3 Restricted Earth Fault Protection A single over-current and earth fault system is not able to provide good protection cover for a star-connected primary winding, particularly if the neutral is earthed through impedance. The degree of protection is greatly improved by the provision of restricted earth fault protection. A high impedance relay is connected across the summated output of three-phase CT balanced against the output of a CT in the neutral conductor as shown in Fig. 13.6.

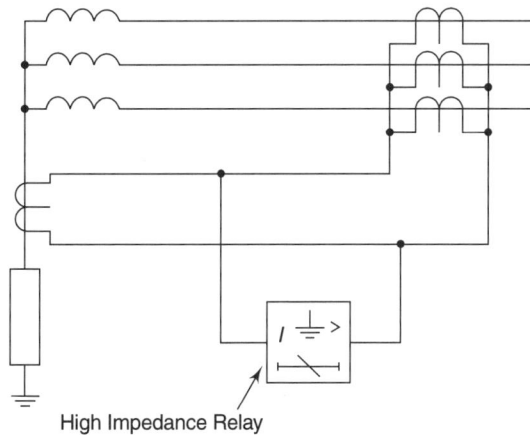

Fig. 13.6: Restricted earth-fault protection for a star winding

This caters to faults in the zone between the CTs only and hence use of the term 'restricted'. Both windings of a transformer can be protected separately with REF protection, thereby providing high-speed protection to the entire power transformer. The time of operation and setting current can be made independent of other protection systems as it is also a unit protection scheme thus allowing low setting and fast operating time.

13.6.2 Inter-tripping

Inter-tripping is an essential requirement for transformer feeders to ensure complete isolation from HV and LV sides. Some of these faults where inter-tripping is desirable are:

(a) Buchholz relay operation; and
(b) Earth faults on star winding of transformer.

13.7 FEEDER PROTECTION

When the first small power system was set up, the need to add automatic protection was soon realised. Equipment responsive to excess current was the obvious solution. Selective action was soon needed and the graded over-current protection has evolved to give discriminative fault protection. Over-current protection is different from overload protection in the sense that whereas overload protection is related to the thermal capability of the plant, over-current protection is for the correct clearance of faults. Different over-current protections for feeder cases are detailed below.

13.7.1 Over-current and Earth Fault Protection

Among the various possible methods used to achieve correct relay co-ordination are those using either time or over-current or a combination of both. The common aim of all three methods is to provide correct discrimination. Relays designed to provide an inverse characteristic provide more scope for precise grading. This leads to the evolution of normal inverse, very inverse and extremely inverse time over-current relays.

The time interval between the operations of two adjacent relays depends upon a number of factors. These are:

1. The fault current interrupting time of circuit breaker;
2. The overshoot time of the relay;
3. Errors in relay operation; and
4. Margin to be kept to take care of the above errors.

In the case of long transmission lines or transformer feeders, where the source impedance is small in comparison with the protected circuit impedance, a combined inverse and high set instantaneous over-current relay makes possible a reduction in tripping time at high fault levels and improvement in the overall system grading by allowing the inverse curve. Figure 13.7 shows typical inverse curves employed in various relays.

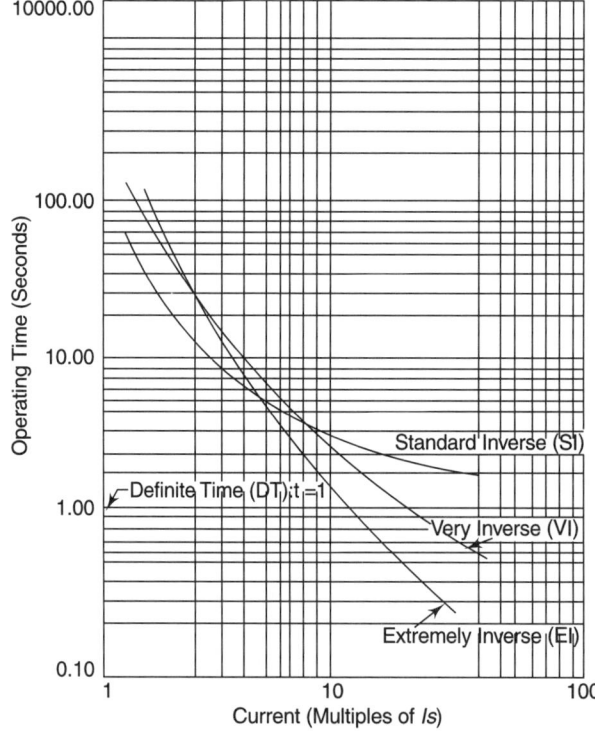

Fig. 13.7: Standard inverse characteristics as per IEC 60255

The inverse time relays described have their characteristic curves with X-axis being the plug setting multiplier and Y-axis being the operating time. The term plug setting multiplier (PSM), which is the ratio of primary current divided by primary setting current, is used to simplify the curves, otherwise relays having different settings would require as many time current curves.

Another term, viz. 'time multiplier setting' (TMS) is used for curves that are displaced with the basic curve downwards maintaining the original curve shapes. In electro-mechanical type relays, this is achieved by restricting the movement of the operating disc, by forwarding the backstop proportionately to reduce the operating time.

Finer time grading can be obtained for any fault through these two mechanisms.

13.7.2 Sensitive Earth Fault Protection

Normal earth fault protection is based on the assumption that fault currents which flow to earth will be determined by the earth fault impedance of the line and associated plant. In some localities, however, the nature of the ground may be such that it is difficult to make an effective earth connection. In such a case, a system fault to earth may result in a current, which is too small to overcome inherent inertia of any rotating disc or attracted armature of a protective relay. Also circumstances arise where an overhead conductor breaks and falls on ground of high resistivity or across a hedge or haystack. This has led to the use of extremely low burden sensitive earth fault protection relays which operate either in conjunction with suitable core balance CTS or with residually connected CTs.

13.7.3 Auto-reclosing

Since 80 per cent of the faults occurring on the rural feeder are of a transient nature and cause no permanent damage, it will therefore be apparent that an improvement in the continuity of supply to consumers and a reduction in operating costs will result if the supply can be restored automatically. Factors affecting the successful reclosing are the speed of operation of the circuit breaker and the re-setting of its mechanism. The time interval between tripping and subsequent reclosure is usually referred to as the 'dead time'. Figures 13.8 and 13.9 show timing diagrams of an auto-reclosing scheme responding to a transient fault and a permanent fault respectively.

The choice of suitable dead time depends upon the type of load fed, e.g. industrial consumers using induction motors can sustain a higher dead time whereas synchronous motors require a very minimum dead time. Hence a dead time of 300 milli-seconds is considered adequate for such consumers. Street lighting loads can be restored with a 1–2 second dead time. Loads fed to domestic consumers using TV sets require higher dead time, as here the supply restoration should be delayed by 2 to 3 minutes if it cannot be restored in 10 seconds. This gives a time limit of 10 seconds dead time.

Another important timing in a re-closing scheme is known as 'reclaim time', i.e. the time following a successful re-closing operation after which the scheme will provide a full operating sequence in the event of subsequent faults. This time is important under repetitive fault conditions like lighting storms or conductors clashing in high winds. Reclaim time in excess of the time interval between the type of successive fault conditions referred above may cause unnecessary lock-out. Reclaim time of the order of 5 seconds is sufficient under such conditions.

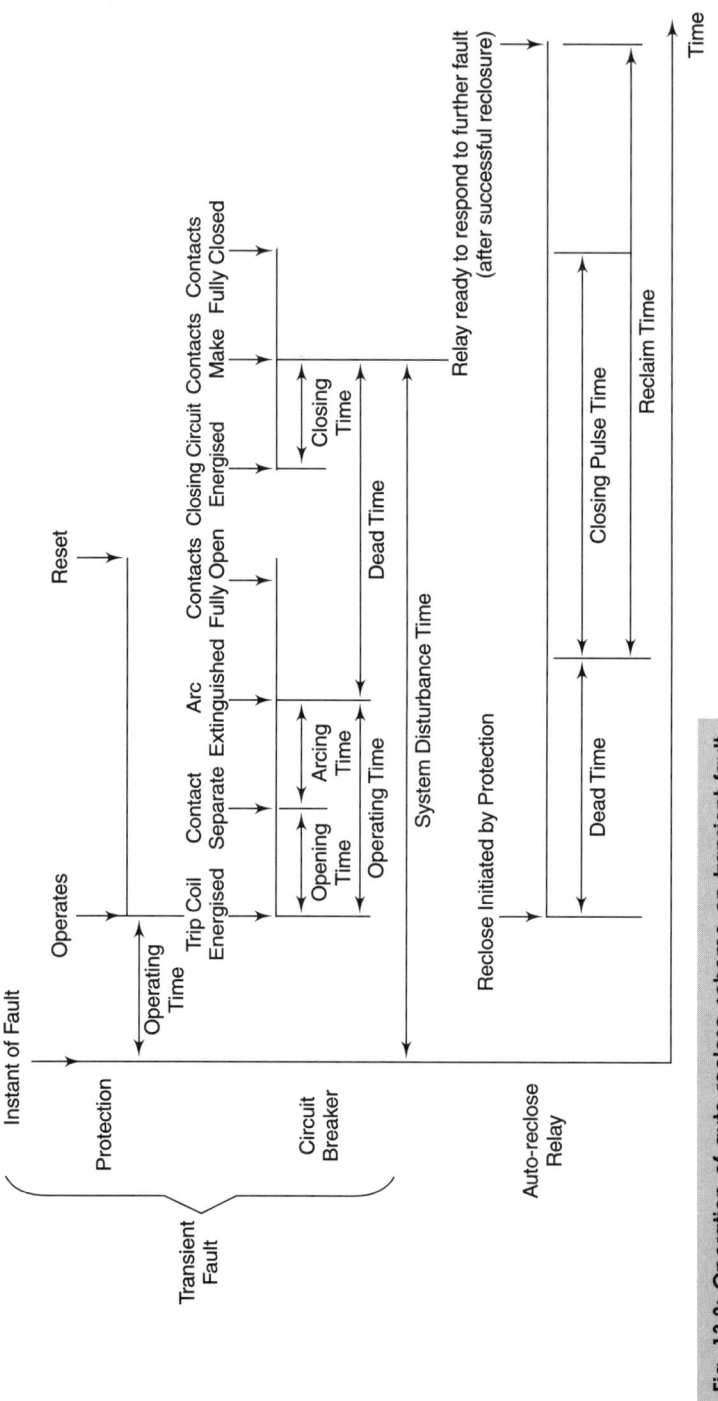

Fig. 13.8: Operation of auto-reclose scheme on transient fault

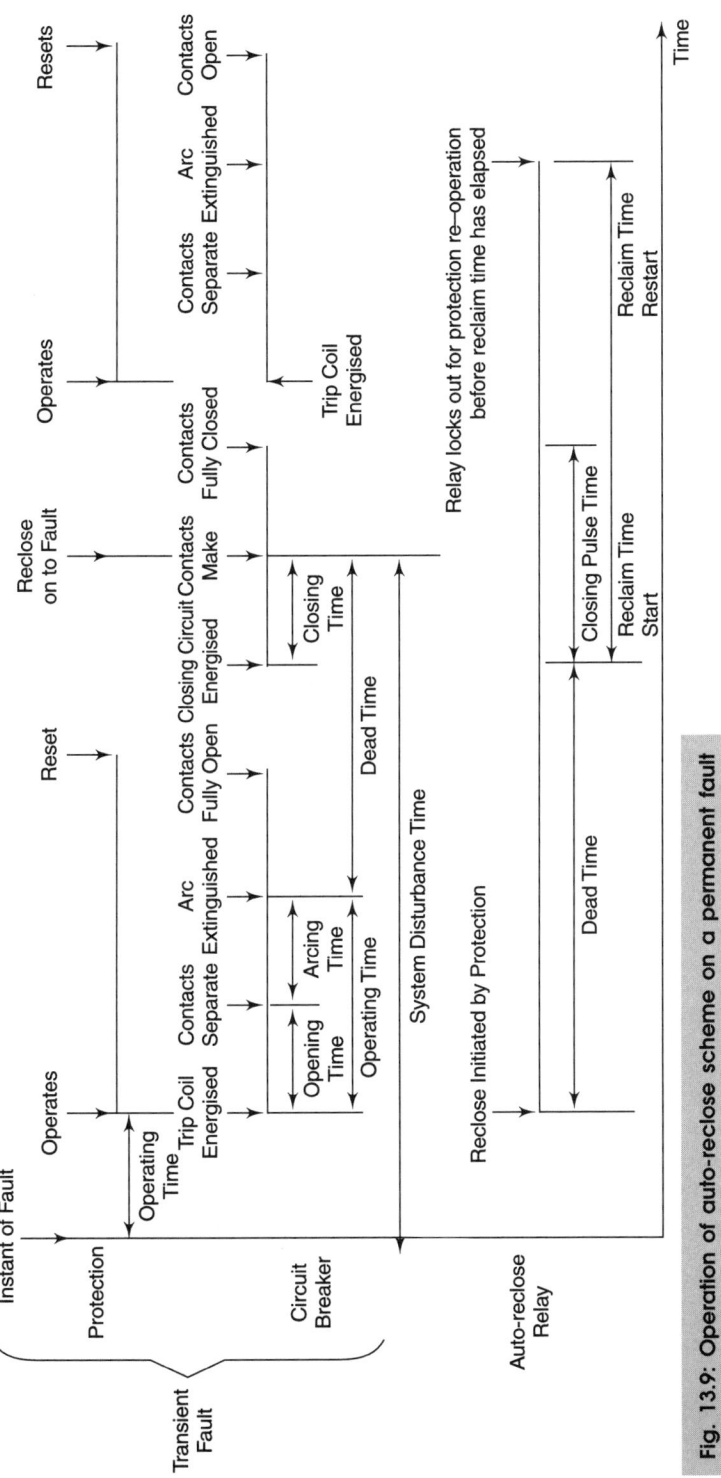

Fig. 13.9: Operation of auto-reclose scheme on a permanent fault

13.7.4 Protection of Parallel Feeders

Directional over-current relays at the receiving ends and non-directional over-current relays at the sending ends are normally used for protecting parallel feeders as shown in Fig. 13.10.

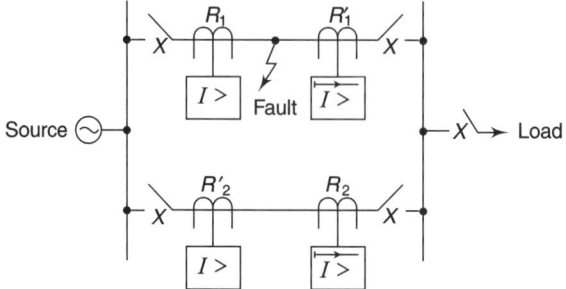

Fig. 13.10: Use of over-current relays and directional over-current relays for protection

The directional relays at the sending ends are normally given low settings, typically a 50 per cent current setting and 0.1 TMS. In an ac power system, the direction of flow of current is not an absolute quantity; it can only be measured relative to some reference. Normally this reference is the system voltage. In case of faults, the system voltage falls to zero at the point of short-circuit and therefore the relays develop a much lower torque. In order to correct this situation, healthy phase voltages are chosen to polarise each relay element so as to retain satisfactory phase relation to the current under all probable conditions. Combinations of relay connections are used to produce maximum torque for various feeder protection applications.

One of the major issues of protection which, though not forming part of any protection scheme, needs consideration.

13.7.5 Back-up Protection

There are two reasons for applying back-up protection to a power system. One is to ensure that in the event of failure of main protection, the fault will be cleared with complete discrimination or at least with minimum interruption of supply. The second is to cover those parts of the protective circuit, which are not getting covered by main protection due to the location of sensing element, namely current or voltage transformers. As far as possible, back-up protection should be independent of the main protection with as few common components as possible. The term 'back-up protection' is different from 'standby protection' in the sense that standby protection is normally out of service and is made operational when the main protection has been taken out for maintenance or investigation.

13.8 CAPACITOR PROTECTION

The major portion of loads in a power system consists of induction motors. They draw significant reactive power from the supply system, and a poor overall power factor may result. The flow of reactive power increases the voltage drops. In order to offset the losses and to assist with voltage regulation, shunt capacitors are often used. Normally large banks of capacitors connected in star or double star configuration are employed.

A capacitor bank usually consists of a group of individual unit capacitors, which are connected, in a series-parallel combination so as to give the required kVAr output. Normally each capacitor unit is provided with a primary protection fuse. However, besides this protection, it is essential to provide the following protections for a capacitor bank as shown in Fig. 13.11.

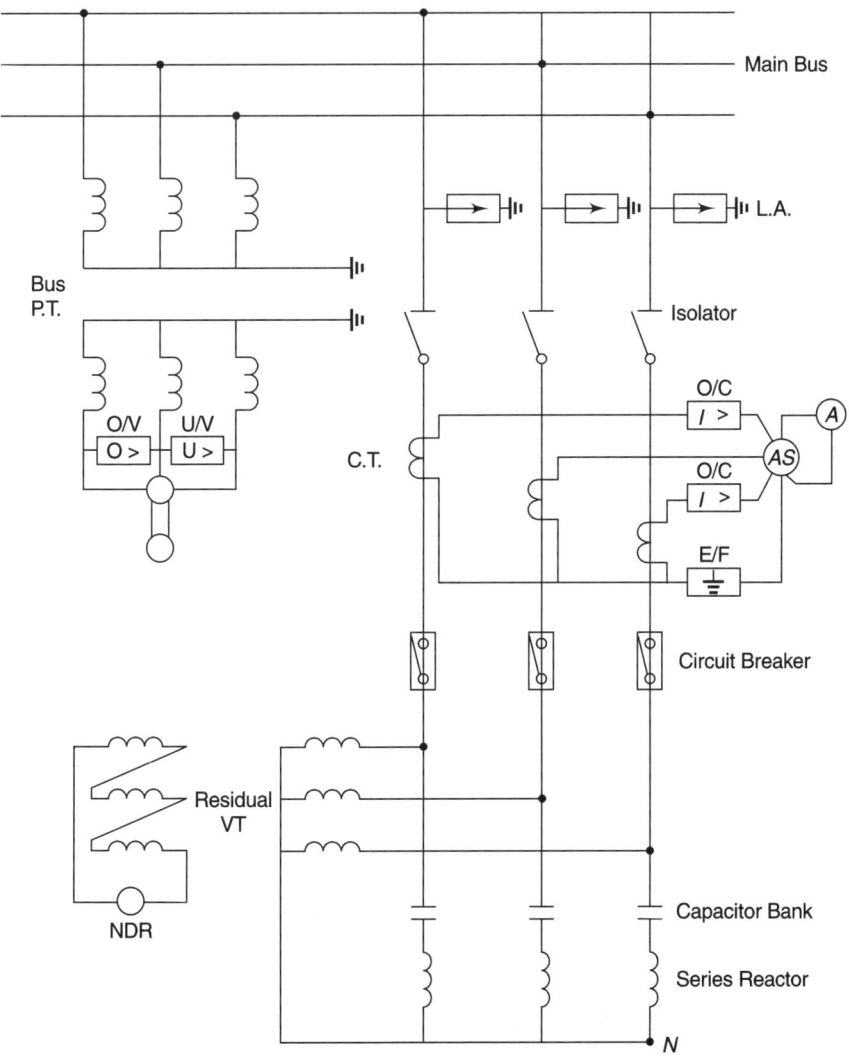

Fig. 13.11: Schematic diagram of capacitor bank protection

13.8.1 Over-current Protection

In order to guard against faults on capacitor leads between the circuit breaker and the units, it is normal to provide over-current and earth fault protection using IDMTL relays. Since the reactance of a capacitor decreases as the frequency increases, relatively small harmonic voltages produce high harmonic currents. These harmonics are generally produced by other equipment in the power system such as generators, transformers, arc furnaces, etc. Capacitors are designed to carry 130 per cent of its kVAr rating continuously. Hence over-current protection set to operate at 130 per cent of the full load current giving adequate overload protection to capacitors.

13.8.2 Over-voltage Protection

Capacitors are designed to withstand 110 per cent of the rated r.m.s. voltage continuously. The failure rate of the capacitor unit increases rapidly if it is subjected to over-voltages. In order to guard against this condition, it is normal to provide an over-voltage relay with a setting of 110 per cent and a slight time delay is provided to override the transients or alternatively an over-voltage relay with inverse time characteristics is used.

13.8.3 Unbalance Protection

As indicated above, capacitor units are connected in series-parallel combination, and the removal of one or more units as a result of individual fuse operation causes an increase in the impedance of that section. This causes higher voltage to appear across the remaining units in that group. This may cause a cascading effect if allowed to persist for some time leading to failure of a large number of units if the maximum limit of 110 per cent over-voltage is reached. To guard against this condition, it is normal to provide either an unbalance voltage detection relay connected to open delta winding of residual voltage transformer connected across the capacitor bank or by connecting a current relay operated off a CT in the floating neutral connection of a *Wye-Wye* capacitor bank as shown in Fig. 13.12.

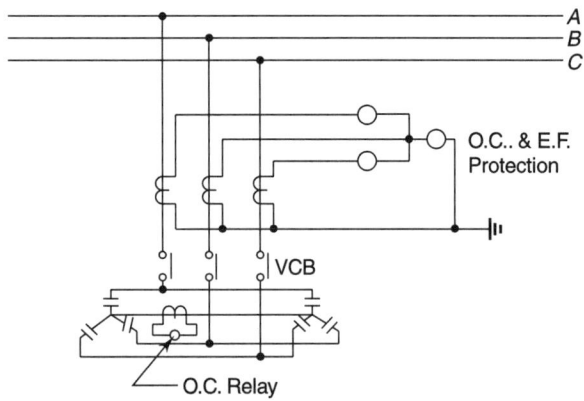

Fig. 13.12: Neutral unbalance protection for double star connected bank with floating neutral

13.8.4 Under-voltage Protection

This protection is provided to disconnect the capacitor bank under low voltage condition as also to provide interlock in the closing circuit. It is important to ensure that the relay provided for this application is connected to the bus VTs.

Chapter 14

GENERATOR PROTECTION

P.K. Upadhyay

The heart of an electrical power system is the generator. The prime mover which drives the generator provides the necessary input mechanical energy for conversion into electrical energy by the generator. The prime mover can be based on steam, gas, water power and diesel engines. The generator's ratings may vary from a few hundred kVA to as much as 500 MVA or more. The present largest installed single generator unit in India has a capacity as large as 500 MW. Future projects with unit capacities as large as 660 MW are being envisaged in the country. While the smaller generating sets may be directly connected to the distribution system, larger units are usually associated with generator transformers for connection to EHV transmission systems. The power for the auxiliaries is also drawn from the large generating units via auxiliary transformers. The composition of generator, generator transformer and unit auxiliary transformer is considered as a unit and the protection associated with it is referred to as unit protection. Figure 14.1 represents a typical SLD representation of the unit.

The failure rate in generators is low due to modern design practices and improved technology. However, the rare occurrences of failure may result in severe damage and long shutdowns for repairs. Hence the abnormal conditions must be recognised promptly to quickly isolate the system. In order to achieve maximum utilisation of installed generator capacity, the unit protection should be designed in such a way that the damage caused by faults is kept to a minimum. Some of the faults do not require the unit to be tripped automatically since in a properly attended station, they can be corrected while the machine remains in service. These conditions are signalled by alarms. However, certain faults require prompt isolation of the machine. These are:

— Faults in stator winding;
— Faults in rotor winding;

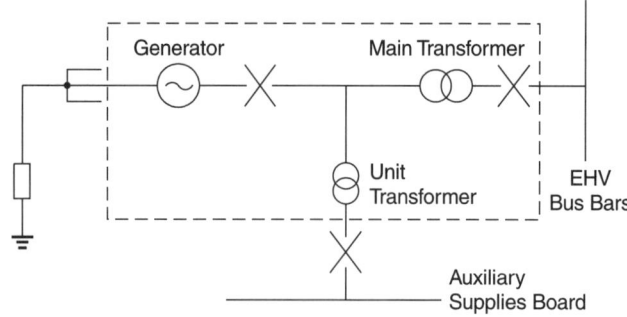

Fig. 14.1: Typical SLD representation of a generating unit

— Faults in transformer winding;
— Overload;
— Over-heating of windings/bearings;
— Over-speed;
— Loss of excitation;
— Motoring;
— Inadvertent energisation;
— Single-phase or unbalanced current operation; and
— Out of step.

14.1 CLASSIFICATION OF TRIPPING

The classification of tripping is categorised on the basis of the location and type of faults. The classes of tripping are different from one another in respect to the action to be taken. The objective is to trip only the absolutely essential equipment so that the damage is minimum and over-speeding of the machine due to sudden load throw-off is avoided. The idea is that the impact of tripping of a large generating set on the grid is minimum and if possible, the auxiliaries are kept energised so that the time to re-start the unit is normal. There are normally three types of classes of tripping, namely Class A, Class B and Class C.

14.1.1 Class 'A' Tripping

In this tripping mode, the generator and turbine will be tripped simultaneously and instantly. Generally, electrical faults of the generator, generator transformer and unit auxiliary transformer will lead to Class A tripping. This will cause tripping of the generator transformer HV side circuit breaker, de-excitation of the generator, tripping of LV side incomer breaker of unit auxiliary transformer, auto-changeover from unit to station supply for unit auxiliaries and tripping of turbine. For large generators (500 MW and above), Class A tripping is further sub-divided into Class A1 and Class A2 with class A2 excluding the generator transformer HVCB tripping and unit transformer LVCB tripping. However, Class A2 tripping takes place only when there is a generator circuit breaker.

14.1.2 Class 'B' Tripping

Class B tripping is followed for all turbine faults which are mechanical in nature and for some electrical faults of generator, generator transformer and unit auxiliary transformer for which it is safe to first trip the turbine and then the generator through low forward power interlock. This ensures that the unit does not over-speed due to trapped steam in the turbine during the shut-down and that the loss of power to the grid is not sudden.

14.1.3 Class 'C' Tripping

This class of tripping is followed for all faults beyond the generator system which can be cleared by tripping of the generator transformer HV side circuit breaker alone In case of a thermal unit, the TG set runs with an HP/LP bypass system in operation and the generator continues to feed the unit auxiliary transformer.

14.2 GROUPING OF PROTECTIONS IN CLASS A1, A2, B AND C TRIPPING

For a typical 500 MW generator, the recommended grouping of protections in various tripping classes is discussed in detail below.

14.2.1 Class 'A1' Tripping

These inlude:
— Generator back-up impedance protection;
— Generator dead machine accidental energisation protection;
— Generator earth fault protection;
— Generator transformer pressure relief valve protection;
— Generator transformer sudden pressure valve protection;
— Generator transformer Buchholz relay trip protection;
— Fire protection of the generator transformer;
— Generator transformer standby earth fault protection;
— Differential protection for generator transformer HV side to overhead transmission line link;
— Generator transformer differential protection;
— Differential protection covering the generator transformer and overhead connections to the unit auxiliary transformer;
— Generator breaker fail protection;
— Generator transformer over-fluxing protection;
— Unit transformer Buchholz protection;

- Unit transformer oil surge protection;
- Unit transformer pressure high protection;
- Unit transformer HV over-current protection;
- Unit transformer LV earth fault protection;
- Unit transformer differential protection;
- Unit transformer restricted earth fault protection;
- Unit transformer fire protection;
- Tee differential protection;
- Generator transformer HV circuit breaker pole discrepancy protection; and
- Generator transformer HV circuit breaker fail protection.

14.2.2 Class 'A2' Tripping

These include:
- Generator field failure protection (delayed tripping);
- Generator field failure in conjunction with generator under-voltage protection (instantaneous tripping);
- Generator negative phase sequence protection;
- Generator over-voltage protection stage 2;
- Generator 90 per cent stator earth fault protection;
- Generator inter-turn fault protection;
- Generator 100 per cent stator earth fault protection;
- Generator rotor earth fault protection;
- Generator differential protection;
- Generator low forward power protection (tripping after a sufficient time delay, allowing for Class B tripping first);
- Generator out-of-step (pole slipping) protection in conjunction with static offset mho relay operated condition; and
- Generator over-fluxing protection.

14.2.3 Class 'B' Tripping

These include:
- Generator low forward power protection after a short time delay;
- Turbine faults;
- Generator winding temperature high;
- Generator bearing temperature high; and
- Turbine bearing temperature high.

14.2.4 Class 'C' Tripping

This include:

— Generator under frequency protection stage 2 (stage 1 is for alarming).

In case of hydro units, the following faults are also covered in this Class C tripping (also called Class Z tripping):

— Reverse power fault;
— Over-fluxing fault; and
— Negative phase sequence fault.

14.3 PROTECTION PHILOSOPHY

For increased reliability, some vital protections are duplicated. However, since many protective relays either back-up or overlap, it is not necessary to duplicate all the protections. Higher reliability can be achieved by dividing the protective relays into two groups, with each group catering to the basic minimum protection of the unit. The typical groupings of protections for a large unit (500 MW) are given under Section 14.3.1.

In order to increase the reliability, it is also necessary to have independent cabling, CTs/PTs, dc auxiliary supplies, tripping relays and circuit breaker trip coils. This helps to build a fairly high degree of redundancy into the protection system. During normal operation, both the groups of protection will be in service. However, during maintenance, etc., the unit can still be in service with only one of the protective groups in operation.

14.3.1 Typical Groupings of Protections for a Large Unit (500 MW)

These groupings are delineated in Table 14.1.

Protection	*Group 1*	*Group 2*
Generator back-up impedance	✓	—
Generator dead machine	✓	—
Generator transformer internal pressure high	✓	—
Generator transformer sudden increase in internal pressure	—	✓
Generator transformer Buchholz relay trip	—	✓
Fire protection of generator transformer	✓	—
Standby earth fault protection of Generator transformer	—	✓
Differential protection of generator transformer to HV side OH line link	✓	—
Differential protection of generator transformer	—	✓
Differential protection covering generator transformer and unit auxiliary transformer	✓	—
Generator breaker fail protection	✓	—

(Contd.)

(Contd.)

Protection	Group 1	Group 2
Generator transformer over-fluxing protection	—	✓
Unit transformer Buchholz protection	✓	—
Unit transformer internal pressure high protection	—	✓
Unit transformer HV over-current protection	—	✓
Unit transformer differential protection	✓	—
Unit transformer restricted earth fault protection	—	✓
Unit transformer fire protection	—	✓
Tee differential protection	✓	✓
Generator transformer HV circuit breaker pole discrepancy protection	✓	—
Generator transformer HV circuit breaker fail protection	—	✓
Generator field failure protection	✓	✓
Generator negative phase sequence protection	—	✓
Generator over-voltage protection	✓	✓
Generator stator E/F protection (90%)	✓	—
Generator inter-turn fault protection	✓	—
Generator rotor earth fault protection	—	✓
Generator 100 per cent stator earth fault protection	—	✓
Generator differential protection	✓	✓
Generator low forward power protection	✓	✓
Generator pole slipping protection	✓	—
Generator over-fluxing protection	✓	—
Generator under frequency stage 2	✓	✓

Table 14.1 Typical Groupings of Protections for a Large Unit

14.4 LESS SEVERE FAULT CONDITIONS

As mentioned earlier, there are some faults, however, which do not require isolation of the unit and are to be signalled by alarms so that necessary corrective action may be taken with the unit in running condition. A typical list of such faults for a large unit may include the following (These faults are signalled as alarms on the digital distributed control and monitoring information system):

— Generator negative phase sequence stage 1;
— Generator over-voltage stage 1;
— Generator voltage transformer fuse failed;
— Generator integrated protection relay faulty;
— Generator protection group 1 or protection group 2 dc supply failed;
— Generator transformer oil level low;
— Generator transformer oil temperature moderately high;
— Generator transformer winding temperature moderately high;

— Generator transformer normal cooler failure;
— Generator transformer normal cooler oil flow low condition;
— Generator transformer Buchholz relay alarm (stage 1) condition;
— Generator transformer over-fluxing condition
— Disturbance recorder operated condition;
— Unit transformer on load tap changer (OLTC) control supply failure;
— Unit transformer bank fan failure;
— Unit transformer OLTC mechanism stuck-up;
— Unit transformer OLTC drive motor auto-trip;
— Generator CB trip circuit supervision relay operated; and
— Tripping relays' coil circuit interrupted (indicated by the coil supervision relay).

14.5 MODERN TRENDS IN GENERATOR PROTECTION

In latest power stations, the state-of-the-art integrated control and protection system is installed. The relay is multi-functional with a communication facility and is integrated with a digital distributed control and monitoring information system. However, for better reliability in larger units (210 MW and above), it is felt that two such integrated multi-functional generator protection relays should be used. The compact design and functional integration minimises panel space. It also maximises protection availability through extensive self-monitoring and system supervision.

References

1. Recent Trends in Electric Energy Systems, edited by Dr. J. Nanda and Dr. D.P. Kothari, Prentice-Hall of India, 1988.
2. IEEE Tutorial on the Protection of Synchronous Generators, IEEE Catalog No. 95 TP 102.

Chapter 15

EHV TRANSMISSION SYSTEM: CONTROL AND PROTECTION

A.K. Murgai

The demand for power is growing rapidly due to increased industrial and agricultural activities in the country. Generating sets of large ratings are being set up to meet this requirement. The generator, which is the source of power in networks, is a major component in electrical installations. Equally important is the transmission system, which is used for distribution and the proper utilisation of power generated by power plants. A fault or breakdown in the power plant or transmission network may have far-reaching consequences. It is therefore of paramount importance that the generators and transmission equipment are optimally used and efficiently protected.

Control and protection panels have a functionally important role. They perform the following functions:

(a) They provide facility for centralised control.
(b) They provide a point for centralised supervision at which all vital information relating to controlled equipment is received and assimilated.
(c) They provide for necessary protection and isolation facility of all power circuits like generators, feeders, transformers, bus-coupler, reactors, etc. The control and protection panels provide alarm and trip commands under abnormal conditions and hence function like a watchdog for the system.

15.1 TYPICAL CIRCUITS IN AN EHV TRANSMISSION SYSTEM

The circuits in a typical EHV system are:
 (a) Transmission lines;
 (b) Generator transformers;
 (c) Inter-connecting transformers;
 (d) Feeder transformers;
 (e) Reactors;
 (f) Tie/transfer bus-coupler;
 (g) Bus-coupler; and
 (h) Busbars.

15.2 TYPES OF EHV SYSTEMS

Two types of EHV transmission systems are generally employed in our country as also elsewhere. These are detailed below.

15.2.1 One-and-a-half Breaker System

It is a two-bus system. An arrangement of two circuits with the associated tie breaker is called a diameter. The diameter can be a line-line or line-transformer or transformer-transformer circuit diameter. Each diameter has three breakers for two circuits, hence the name one-and-a-half breaker system. Normally all the three breakers are closed and power is fed to both the circuits from two buses which operate in parallel. The middle breaker or the tie breaker acts as a bus-coupler for the two circuits.

The main advantage of this configuration is that a fault on any one bus is cleared instantaneously and yet all circuits continue to be fed from the other bus without any supply interruption. The same is true in case of a breaker stuck condition, but with a time delay. The system is otherwise more expensive as it employs a higher number of breakers.

In case of failure of the breaker of any one circuit, the power is fed through the breaker of the second circuit and the tie breaker. Each breaker, therefore, has to have a rating suitable for feeding both the circuits. A typical single line diagram of one-and-a-half breaker system is shown in Fig. 15.1.

15.2.2 Two Main and a Transfer Bus System

This system has two main buses and one transfer bus. In case of maintenance on any one breaker, the particular circuit is connected to the system through a transfer bus-coupler or transfer breaker. Complete control and protection is transferred in this case to the transfer breaker through the trip transfer selector switch. A typical single line diagram or the two main and a transfer bus system is shown in Fig. 15.2.

EHV Transmission System: Control and Protection

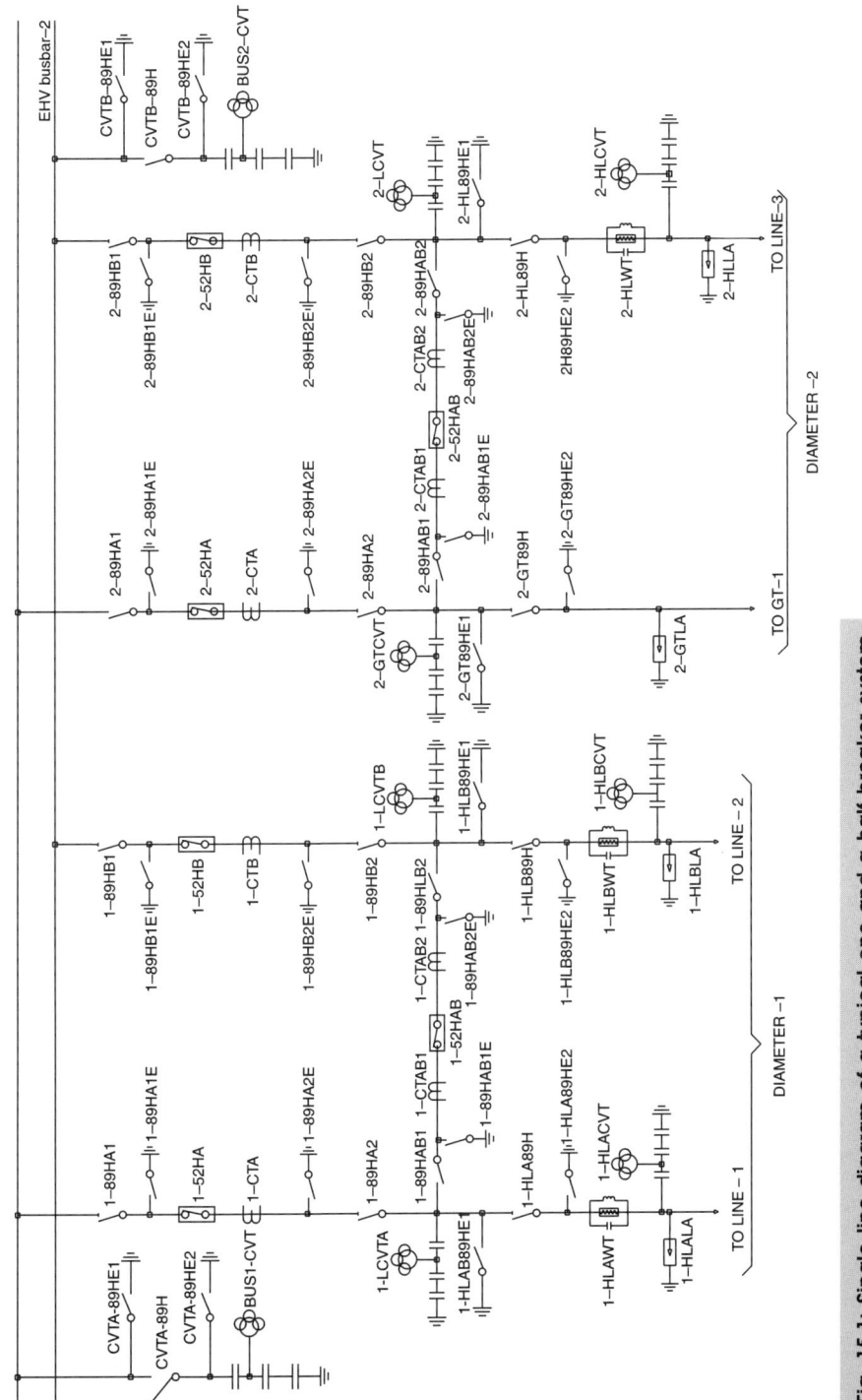

Fig. 15.1: Single line diagram of a typical one-and-a half breaker system

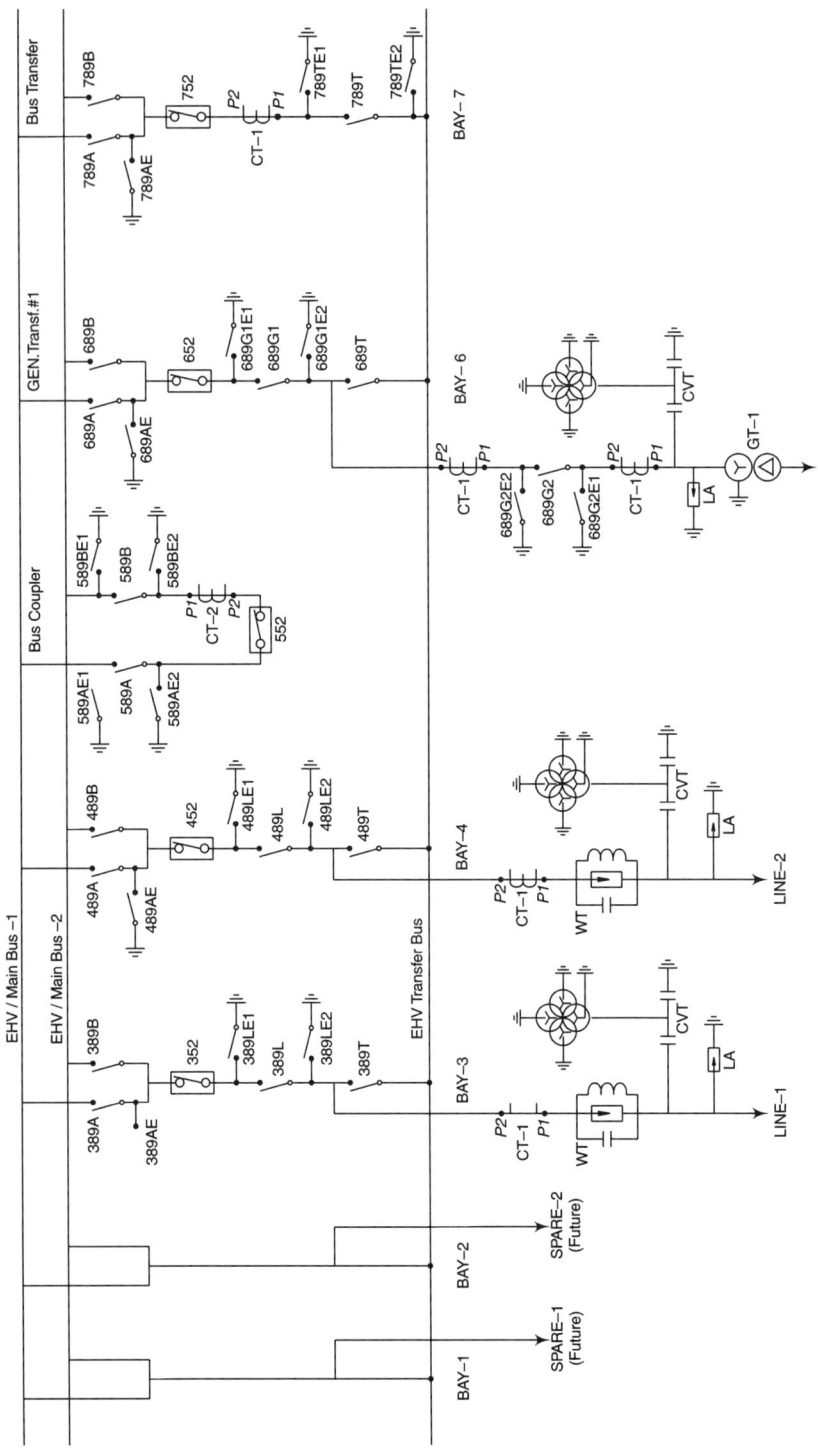

Fig. 15.2: Single line diagram of a typical two main and transfer bus system

15.3 CONTROL AND PROTECTION FUNCTIONS

These include:

(a) Control: This covers the ON/OFF functions using control/discrepancy switches for breakers and motor operated isolators. Suitable electrical interlocks are hardwired in the closing circuits of the breaker and isolators. The control is provided both locally and at remote as per the requirement.

(b) Indication: ON/OFF indication is provided for breakers, isolators and earth switches using lamps or semaphore indicators. Auto-trip of circuit breakers and panel dc fail indications are also desired.

(c) Protection: This is as described below against each type of circuit.

(d) Alarm Annunciation: Audio and visual alarms are provided to attract the attention of the operator in case of any fault. There are two types of fault alarms, viz. the trip alarms, which are due to serious faults causing tripping of the circuit breaker and the non-trip alarms due to faults, which are not so serious in nature and are annunciated to alert the operator to take corrective action before the trip condition arises. The alarm condition is inscribed on the facia window, which illuminates when the alarm occurs. Different coloured facias (red and amber) and/or different tones of audible devices are used to differentiate between the trip and non-trip alarms.

(e) Metering: Electrical parameters like current, voltage, active power, reactive power, frequency, etc. are monitored for the convenience of the operator. For transmission lines, energy is monitored at high accuracy of the order of class 0.2 for the purpose of tariff metering. Load profile data is also recorded for the purpose of analysis of energy flowing through the line.

(f) Event Logging: All fault conditions are time tagged with a real time clock and recorded as events on a printer for the purpose of fault analysis. Event recording can be provided as a built-in part of the numerical protection relays or by using a stand-alone event logger.

(g) Disturbance Recording: It is a very important tool for the system analyst. On the occurrence of a fault in the system, the pre-fault, during fault and post-fault data are captured and recorded. The data includes current, voltage, selected internal logic signals, digital inputs and outputs. The disturbance recording is triggered whenever any electrical parameter exceeds the threshold value or rate of change of frequency/voltage exceeds the pre-set value or any digital input and output (for which it is programmed) changes its state. The transient records are captured in the form of waveforms on a real time scale. These waveforms can be viewed on a personal computer using the disturbance analysis software. Disturbance recording can be provided as abuilt-in part of the numerical protection relays or as a stand-alone disturbance recorder.

(h) Time Synchronisation: Clocks of all devices like the event recorder, disturbance recorders, energy meters and numerical relays in a system should be synchronised with a common global position satellite (GPS) clock. The GPS clock supplies the system with an accurate date and time, common to all locations wherever implemented.

(i) Fault Location: This allows the operator to know the exact location of the fault on a transmission line. The distance to fault is indicated in kilometers or miles. Algorithm is written in the numerical distance protection relay or disturbance recorder software to calculate the distance to fault. It is based on the reactance measurement principle.

(j) Communication: Communication ports are provided for connection to a local PC and also for transmitting data to a remote integration system or supervisory control and data acquisition system (SCADA). A network interface device may be required between the numerical relay and SCADA. The following functions are possible while communicating with the numerical relay:

  (i) Retrieve settings from the relay;

  (ii) Do necessary setting changes and upload them to the relay;

  (iii) Retrieve/change/send programmable scheme logic to the relay;

  (iv) Monitor electrical parameters and alarms;

  (v) Retrieve events, fault records;

  (vi) Retrieve disturbance records from the relay for analysis; and

  (vii) Control breakers and other electrical devices.

(k) Synchronising: A circuit can be connected to grid only after proper synchronising. The circuit voltage, frequency and phase angle parameters are matched with the corresponding bus parameters before connecting to the grid. Synchronising is normally done using synchronising voltmeters, frequency meters, synchroscope and check synchronising relay. The synchronising function is bypassed during dead circuit or dead bus condition.

(l) Auto-reclosing: At times, the fault in a transmission line is of a transient nature. In view of this, when the breaker trips, it is normally closed immediately using the auto-recloser. In case of a transient fault, the circuit is restored after reclosing. However, in case of a permanent fault, the auto-recloser is blocked after the first shot while the circuit trips again. For single-phase faults, single pole tripping and high speed reclosing take place without check synchronising as the system remains in synchronism. In case of three-phase faults, reclosing is achieved through check synchronising.

15.4 AUXILIARY SUPPLY DISTRIBUTION PHILOSOPHY

Auxiliary supply for control and protection of any power system application is normally taken from the station battery. The dc auxiliary supply is considered to be much more reliable than the station ac auxiliary supply. The dc supply is used for control and protection so that it remains uninterrupted as far as possible.

In an EHV transmission system, there are normally two groups of protection for each circuit. Each group is designed to provide adequate protection to prevent the equipment from any damage. For this purpose, the normal practice is to bring two separately fused feeders from the dc distribution board for each circuit, one dc supply for each group protection. Thus, even in the case of failure of one dc supply or one group protection, the equipment still remains adequately protected. Each group dc supply is further segregated into sub-groups for protection function, breaker tripping, breaker closing, breaker/isolator/earth switch control and interlocking. Separate fuses are used for each function. Separately fused circuits also help in providing isolation during maintenance.

There are, however, certain functions that are common for the whole system. These include busbar protection, breaker stuck protection, alarm annunciation, event logging, etc. For such functions, the selection of two dc supply feeders is normally done using a switch and the selected dc supply is used. Separate fuses are used for each function. In case of failure of one feeder, the operator is expected to first identify and clear the fault and then select the second supply. The disadvantage of this method is that during some manual changeover period, the supply is not available. Another school of thought is to use the auto-changeover scheme or parallel the two dc supplies. The disadvantage in this case is that the cause of failure of the first dc supply can blow off the second dc supply also.

Each important sub-group dc supply is monitored by using voltage monitoring or trip circuit supervision relay. In case of failure of any supply or blowing off of fuses, an alarm is annunciated and the event is recorded.

The EHV breaker normally has two sets of trip coils, designated as trip coil 1 and trip coil 2. Each group of trip coil has three single-phase coils. The two trip coil groups operate on independent dc supplies. Separate sets of contacts of each protection are used for operating the two sets of trip coils. In order to save burden on the station dc battery, the station ac auxiliary supply is used for less important functions like panel interior illumination, space heater, etc.

15.5 BREAKER RELATED PROTECTION

For every breaker, normally one separate panel is provided to cater for breaker related faults. The panel mainly comprises the following relays:

(a) Local breaker back-up (LBB) or breaker fail relay, which is an important protection and comes into picture when the trip command is given and the breaker fails to open due to some mechanical problem. The protection is time delayed and initiated by all trip conditions. On LBB operation, busbar protection is initiated, as all the breakers connected to that particular bus to which the circuit is connected have to be tripped. It also trips the tie breaker and sends a direct trip command to the other end breaker. In the one-and-a-half breaker scheme, in case the tie breaker gets stuck, its LBB relay trips the other two breakers in its diameter.

(b) Auxiliary flag relays are provided for breaker incipient faults. The main protections are in the breaker. These relays are provided mainly for alarm and event logging. Considering the SF6 EHV circuit breaker, these faults can be due to the following reasons:

   (i) Loss of nitrogen, pole A/pole B/pole C;
   (ii) Loss of SF6, pole A/pole B/pole C;
   (iii) Nitrogen lock-out;
   (iv) Overload;
   (v) Breaker dc fail;
   (vi) MCB trip;
   (vii) General lock-out; or
   (viii) Pole discrepancy.

(c) Group A and group B trip relays.

(d) Busbar protection trip relay.

(e) CVT fuse failure relay for the metering core.

(f) Breaker auxiliary contact multiplier relays as required by the scheme.

(g) Breaker control switch contact multiplier relays as required by the scheme.

(h) dc supply supervision relays.

(i) Breaker trip coil supervision relays.

(j) Trip relay coil supervision relays.

15.6 TRANSMISSION LINE PROTECTION

Two groups of main protections are provided for the transmission line. Each group protection operates on an independent auxiliary dc supply. It is preferable to keep the two group protections physically separate in two different panels. The protections that are normally provided for transmission lines are detailed below.

15.6.1 Distance Protection

The main protection provided is the distance protection for each of the group A and group B trip relays. Normally the employed characteristics are mho or polygonal. For short lines with high resistive reach component, the polygonal characteristic is preferred, as the resistive reach component can be independently set. Mutual compensation is necessary in case of parallel lines, especially for fault location, to achieve better accuracy. In this case, the neutral current of the other parallel line is also fed as an analogue input to the relay.

The distance protection scheme is provided at each end of the line to effectively isolate the line on a fault. When the line breaker at one end trips due to a fault, the breaker at the other end of the line is also tripped.

15.6.2 Inter-tripping Schemes

An inter-tripping scheme is used to facilitate fast tripping of breaker at the other end of the line. The inter-tripping commands are transmitted by using power line carrier communication (PLCC) equipment.

Duplicated PLCC channels operating on two different auxiliary supplies are used for the purpose of high reliability. Each channel has three pairs of 'signal receive' and 'signal send' commands. Command 1 of each channel is normally used for direct trip due to operation of the LBB or busbar protection or any other three-pole trip function. Two direct trip commands are used for the purpose of redundancy. Command 2 of PLCC channel 1 is used for main 1 distance protection. Command 2 of PLCC channel 2 is used for main 2 distance protection. Command 3 of PLCC channel 1 is used for main 1 directional e/f protection. Command 3 of PLCC channel 2 is used for main 2 directional e/f protection. The health of the carrier is also monitored and in case of its failure, the carrier fail signal is used for alarm, event logging and other interlocking purposes.

One of the following inter-tripping schemes is selected on the distance relay:

(i) Zone 1 extension scheme;
(ii) Permissive under-reach (PUR) scheme;
(iii) Permissive over-reach (POR) scheme; or
(iv) Blocking scheme.

15.6.2.1 Zone 1 Extension Scheme

This scheme is normally switched ON under the PLCC carrier fail or carrier out conditions. As the name suggests, the zone 1 setting covers more than 100 per cent of the line length (see Fig. 15.3).

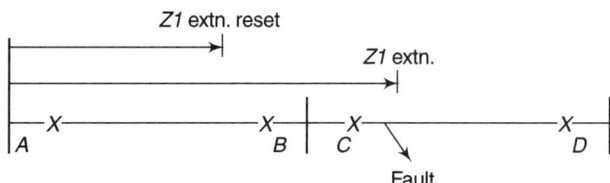

Fig. 15.3: Zone 1 extension scheme

Consider a fault close to the end C. The relay at end A will operate in zone 1 and trip the breaker at A instantaneously. The relay at end C will also cause tripping of the breaker at C instantaneously as it also sees the fault in zone 1. Relays at ends B and D will trip the through carrier. Auto-reclosing of breakers at ends A, B, C and D will take place. Immediately after auto-reclosing, a command is given to the distance scheme for zone 1 extension re-set. This is to ensure that if there is a permanent fault at end C, the breaker at end A does not trip again after the first recloser. Zone 1 extension re-set is also done when the auto-reclose scheme is selected for a non-auto condition. In this case, there is no option but to trip the breaker with a zone 2 time delay when a fault occurs in 80 to 100 per cent of the line length (basic scheme).

The PLCC scheme is sometimes switched over to zone 1 extension scheme on carrier fail/out condition.

15.6.2.2 Permissive Under-reach Scheme Logic (Fig. 15.4)

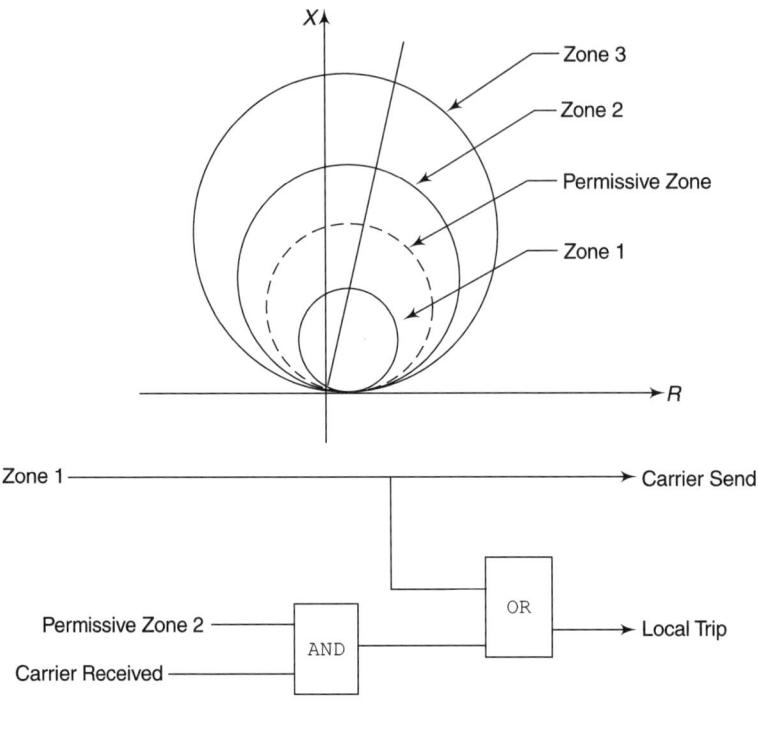

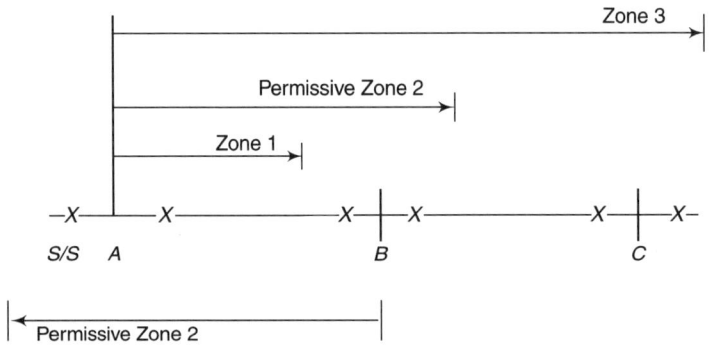

Fig. 15.4: Permissive under-reach scheme logic

15.6.2.3 Permissive Over-reach Scheme Logic (Fig. 15.5)

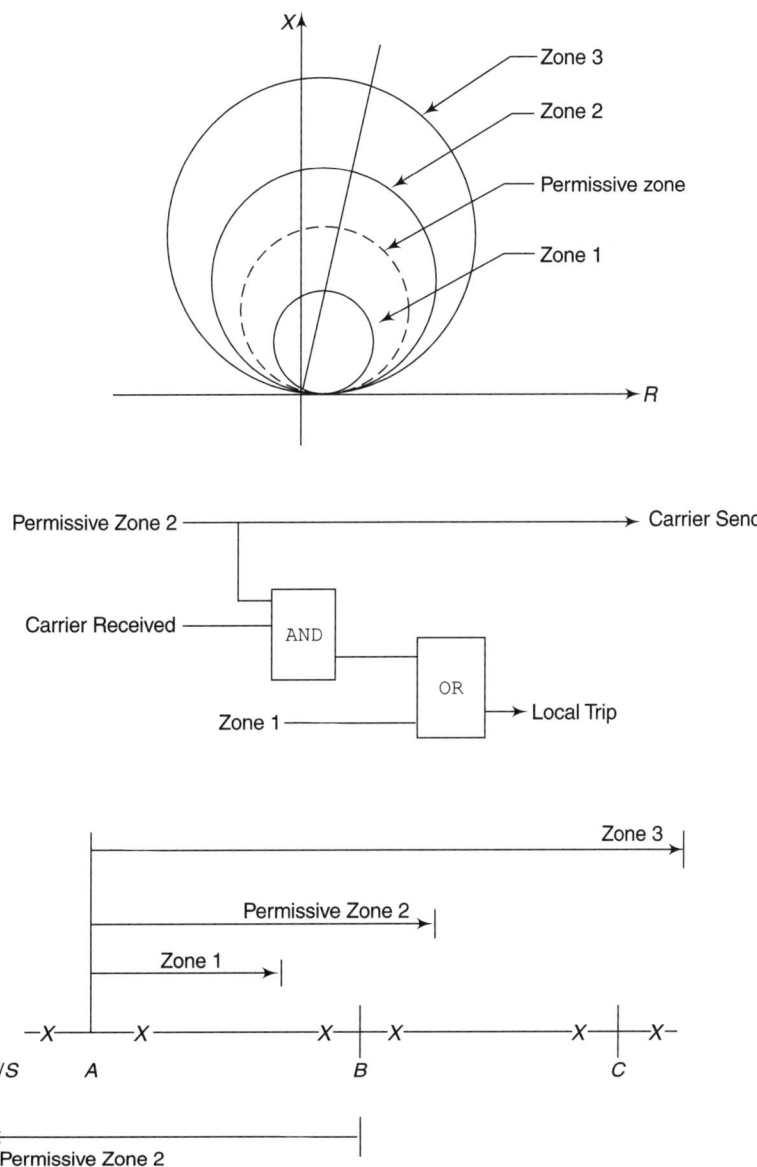

Fig. 15.5: Permissive over-reach scheme logic

15.6.2.4 Blocking Scheme Logic (Fig. 15.6)

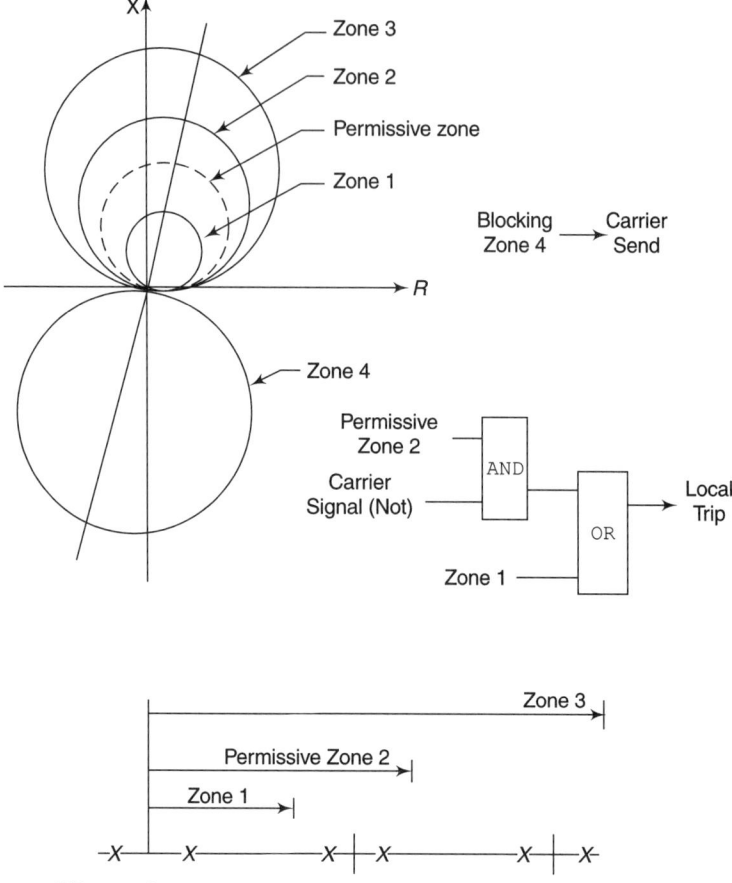

Fig. 15.6: Blocking scheme logic

The two distance schemes provided for line protection are normally programmed for two different carrier inter-tripping schemes, i.e. if main 1 is programmed for the PUR/POR scheme, main 2 is programmed for the blocking scheme.

15.6.2.5 Special Features

The more advanced distance schemes employ certain important additional features. This are detailed below.

15.6.2.5.1 Echo Feature to Detect Remote End Open Consider local end A and remote end B. If the breaker at the remote end B is open and the fault occurs at 80 to 100 per cent of the line, zone 2 of the relay at local end A will pick up. The relay at end B will not operate, as the breaker is open and there is no current in the forward direction. As such, no carrier signal will be sent by the remote end relay. The relay at local end A will therefore operate with a zone 2 time delay. In order to avoid this, carrier is sent by the relay at end B under the additional condition that the carrier is received from end A and the reverse looking zone of the relay at B does not operate. This sending back of the carrier signal is called 'echo feature'. Therefore, the carrier scheme logic is as seen in Fig. 15.8.

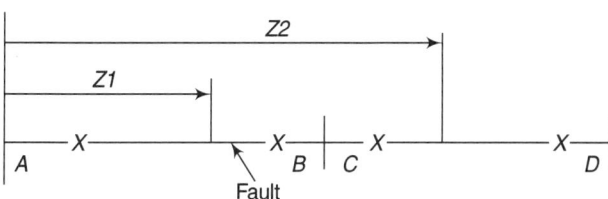

Fig. 15.7: Echo feature to detect remote end open

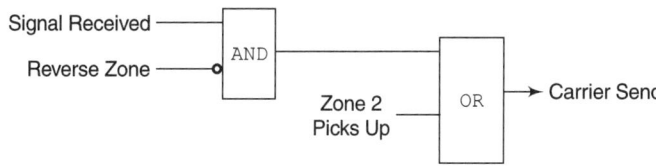

Fig. 15.8: Carrier scheme logic

15.6.2.5.2 Weak Infeed Logic Feature (Both End Breakers Closed) Referring to Section 15.6.2.5.1, suppose the source at end B is weak and the current seen by the relay at end B is not large enough for its zone 2 element to pick up. Under this condition, the carrier scheme adopted is as seen in Fig. 15.9.

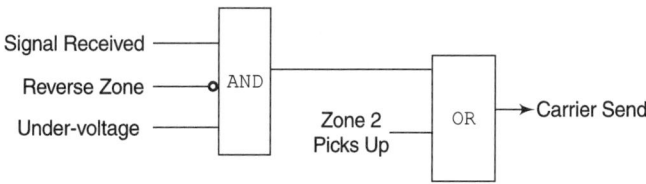

Fig. 15.9: Carrier scheme when zone 2 element does not pick up

Certain relays do not make use of the under-voltage interlock. However, the advantage of sensing UV is to prevent unwanted tripping due to mal-operation of the PLCC equipment. In such a case, under-voltage is the only positive interlock. A dip in voltage is bound to take place when there is a fault. The carrier send condition is also the trip condition for the relay at end B.

15.6.2.5.3 Current Reversal Logic This feature is important in case of short lines running in parallel, as shown in Fig. 15.10.

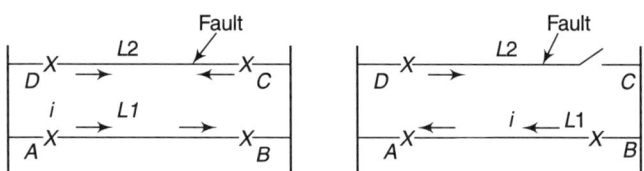

Fig. 15.10: Short time running in parallel

Note how after circuit breaker C on line L2 opens, the direction of the current flow in line L1 reverses. Consider a fault near the relay at end C. Zone 2 of relay at end A will pick up and send a carrier signal to the relay at end B. Zone 1 of the relay at end C will pick up and trip the breaker at end C instantaneously and send a carrier to end D. When the breaker at end C is open, the current will flow from B to A to D to the fault till the time the breaker at end D trips through the carrier received command.

In the absence of current reversal logic, the race between operation and re-setting of the distance elements can cause the permissive over-reach and blocking schemes to trip the healthy line (the setting of zone 2 exceeds 150 per cent of line impedance at each terminal in case of short lines). In case of the permissive over-reach scheme, zone 2 of the relay at end B will pick up on current reversal and with the availability of the carrier signal from relay A, trip the breaker at B instantaneously. If the blocking scheme is used, the reverse looking zone at end B will de-energise on current reversal and no signal will be received at end A. In this case, the relay at end A may operate and trip the breaker A. In order to prevent this inadvertent tripping, current reversal logic is built-in in the relay. In the permissive over-reach scheme, the current reversal guard logic is initiated when the reverse looking zone element operates on a healthy line. Once the reverse looking zone has operated, the relay's permissive trip logic and signal send logic are inhibited. In case of the blocking scheme, the current reversal logic is initiated when the blocking signal is received to inhibit the channel-aided trip. When the current reversal takes place and the reverse looking zone re-sets, the blocking signal is maintained by a timer.

15.6.3 Programmable Scheme Logic (PSL)

Logic functions like AND, OR, NOT, programmable gate, timer, etc. are available for designing user-based non-standard scheme logic. These are also used for configuring the opto inputs, contact outputs and LEDs on the relay.

15.6.4 Other Line Protection Functions

Other important protection functions, which are either achieved through discrete relays or as an integrated part of a distance relay, are:

(a) Over-voltage;
(b) Broken conductor;
(c) Negative phase sequence;
(d) Switch on to fault (SOTF);
(e) Directional earth fault (simple or channel-aided);
(f) Directional/Non-directional phase over-current;
(g) Power swing blocking; and
(h) CVT fuse fail.

15.6.5 Pilot Wire Differential Protection

In case of very short lines that are less than a kilometer long, distance protection may not give a very reliable protection on account of the limitation of settings. In such a case, the alternative to distance protection, viz. pilot wire differential protection is used. Differential relays are provided at each end of the line and they are connected using pilot wires. Pilot wire supervision relay is provided for monitoring open-circuit or short-circuit or any fault in the pilot wires. The insulation level of the pilot wires should be 15 kV with respect to the ground as high voltage of this order may appear at the time of the fault. Isolation transformers are used at each end to achieve this insulation level.

The voltage induced in the pilot is given by the following formula:

$$E = 0.232 I_n \log_{10}(D/S) \qquad 15.1$$

where

E = induced voltage per mile;

I_n = single-phase earth fault current;

D = equivalent depth of earth returned; and

S = distance of pilot from the phase conductor.

This formula is more applicable for overhead conductors. For underground cables, a thumb rule is used. For power cable applications, for instance, where lower resistance cable sheath material is used (e.g. aluminium at 275 kV or 400 kV), the level of pilot induced voltage is of the order of 0.03 V/A/mile. If lead is used as the cable sheath, the value turns out to be close to 0.2 V/A/mile. In the formula for induced voltage in the pilot wire, D implies the equivalent depth of earth returned in feet, which is equal to 2160 * (e/f), where 'e' is the soil resistivity in ohms/m^3 and 'f' is the frequency (50/60 Hz).

The specification of pilot cable used is as follows:

2.5 sq mm, 2 core;

Loop resistance < 1000 ohms;

Shunt capacitance < 5 micro-farads;

250 V grade inter-core insulation;

15 kV core earth insulation; and

Twisted pair cable is preferred.

15.6.6 Tee Differential Protection

In the one-and-a-half breaker system, duplicated differential protection is provided for protection of the circuit between the main breaker and the tie breaker. The common practice is to use a high impedance differential relay as main 1 and a biased differential relay as main 2 protection.

15.6.7 Auto-reclosing Requirements in One-and-a-half Breaker System

Auto-reclosing, check synchronising and dead line or dead bus charging functions can be an integral part of the distance relay for the main breaker. For the tie breaker, these functions are achieved by using discrete relays.

In the one-and-a-half breaker system, it is necessary to trip the main breaker and the tie breaker to clear any fault in the circuit. Auto-reclosing of breakers is associated only with line circuits. It is not applied in case of transformer circuits. A memory circuit using a timer is employed to check the breaker status prior to reclosing. In case the breaker is open, auto-recloser is blocked and the reclosing impulse is not issued to the breaker.

While reclosing the two circuit breakers in the one-and-a-half breaker scheme, a priority is selected. Usually the main breaker is allowed to reclose first and only after its successful recloser is the tie breaker allowed to reclose. If the fault persists after reclosing of the main breaker, reclosing of the tie breaker is blocked.

Auto-reclosing is blocked under the following conditions:

(a) Breaker fail;

(b) Broken conductor, over-voltage, etc.;

(c) Back-up over-current or earth fault;

(d) Tee differential protection;

(e) Direct trip signal received; and

(f) Bus bar protection.

It is initiated only under the distance trip condition.

15.7 TRANSFORMER PROTECTION

The following protections are commonly provided for the transformer:

(a) Biased differential protection;

(b) Restricted earth fault protection for HV and LV side;

(c) Over-current protection with instantaneous high-set for HV and LV side;

(d) Over-fluxing protection (on demand);

(e) Standby earth fault protection; and

(f) Transformer incipient fault contact multipliers.

All the above protections can be achieved by using discrete conventional relays or multi-functional numerical relays. Contacts of the transformer incipient faults like Buchholz, winding temperature, oil temperature, OLTC surge, etc. can be wired as digital inputs to the numerical relay for event recording or tripping.

In case of a conventional scheme, auxiliary CTs are required in the differential circuit for CT ratio error and vector group correction of the transformer. However, in case of the numerical transformer protection relay, these corrections are achieved through programming. For transformer feeder protection, distance protection is also recommended.

15.8 REACTOR PROTECTION

The following protections are commonly provided for the reactor:

(a) High impedance differential protection;

(b) Restricted earth fault protection;

(c) Reactor incipient fault contact multipliers; and

(d) Back-up impedance relay.

15.9 BUS BAR PROTECTION

Busbar protection is a very critical protection and has to be very reliable and highly secured as all circuits connected to the bus are tripped on the occurrence of a bus fault. Any mal-operation cannot be tolerated. Since the fault level of a bus fault is very high, the busbar protection scheme should be very fast acting. The operating time is of the order of 15 to 25 milli-seconds. Busbar protection is also initiated in the event of any breaker getting mechanically stuck and failing to trip on the occurrence of a fault in that circuit.

The following two types of busbar protections are normally employed:

(a) High impedance differential protection scheme; and

(b) Low impedance differential protection schemes.

Normally two busbar protection schemes operating on two different principles are employed. Thus one scheme can be a high impedance scheme and the second a low impedance scheme. The two schemes operate on two different sets of CTs and two different dc auxiliary supplies. For increased security, two out of two schemes are adopted, i.e. only if both the schemes operate is, tripping through busbar protection initiated.

In case of two-main and transfer system, CT switching is required depending upon which bus the particular circuit is connected to. Electrically re-set type relays are used for CT switching. Separate busbar differential protection relays are used on each busbar zone. The CT circuits of each zone are supervised by a definite time delayed CT supervision relay. In case of any CT circuit getting open-circuited, the CT supervision relay operates on unbalance voltage and the CT bus is shorted using a CT shorting relay. Operation of the CT supervision relay blocks the bus bar protection and issues an alarm.

The two dc supplies are also monitored independently. A busbar protection IN/OUT switch with indications is provided for each busbar protection relay. A dedicated busbar protection trip relay is used to facilitate the tripping of each circuit. In case of line circuit, auto-reclosing is blocked and a direct trip command to the other end is sent through PLCC on operation of the busbar protection trip relay.

An important aspect of the high impedance busbar protection scheme is that the knee-point voltage of the CT is high. Also, the CTs of all circuits should be identical otherwise there will be a spill current in the differential circuit and the busbar protection may mal-operate. In case of the low impedance scheme, auxiliary CTs are used in case of any CT mismatch.

15.10 ISLANDING SCHEME

In any transmission system, it is very important to ensure that the system frequency is maintained within tolerance limits. The system frequency is load-dependent and dips in case of an increase in load and vice versa.

Under-frequency relays are employed to monitor the dip in bus frequency. All loads connected to the sub-station are prioritised, i.e. more important circuits are given higher priority and less important circuits, lower priority. When the frequency falls, sequential tripping of the circuits is carried out, i.e. low priority circuits are tripped first and in case the frequency fails to restore, higher priority circuits are tripped.

A four stage frequency relay with the rate of change of frequency (df/dt) feature is used, one for each bus. The loads can thus be prioritised into three or four categories depending upon the requirement. For increased security, two-out-of-two schemes is adopted, i.e. load-shedding is done when under-frequency relays connected on both the buses operate. For higher reliability, duplicated frequency relays can be used. A typical islanding scheme logic is shown in Fig. 15.11.

15.11 INTEGRATED CONTROL AND PROTECTION

The need of the day is for a modern sub-station and for installing intelligent electronic devices like numerical relays, programmable logic controllers, etc. at the bay level. These are called 'bay

EHV Transmission System: Control and Protection **15.19**

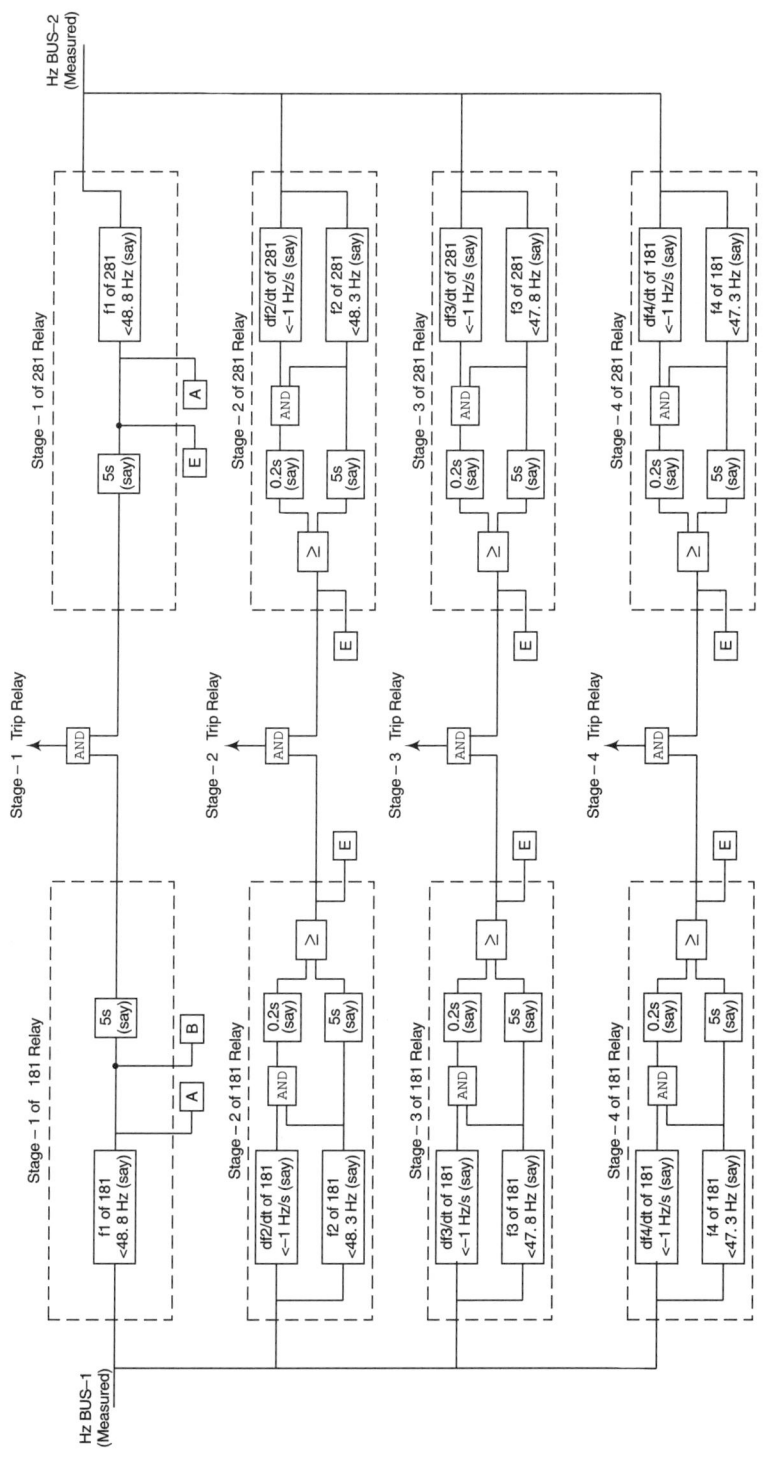

Fig. 15.11: Typical islanding scheme logic

controllers'. An integration system combines the hardware and software components needed to integrate the various bay controllers, providing data collection, control and monitoring from a central work station.

The equipment consists of network interface devices, and host computer and printer. The network interface devices provide interface between the bay controller specific protocol and the system's local area network (LAN). The system can thus integrate bay controllers of different manufacturers, with different protocols, to a LAN with a common protocol. In order to provide communication and control, the LAN information can be accessed by a local laptop computer via a local port or a remote computer via a modem.

The integration system software is windows-based and operates on standard PC hardware. The computer provides a user-friendly customised interface to the system operator. The software provides the graphical representation of the status of primary equipment, alarms, event recording and historical information. The following screens/pages are normally available:

(a) Login window;

(b) Main menu;

(c) Single line diagram;

(d) Annunciator panel;

(e) Alarm history;

(f) System architecture;

(g) Bay controller status (one screen for each bay controller);

(h) System status; and

(i) On-line system documentation.

Numerical protection, along with an integration system and a suitable communication medium, thus provides a total solution for transmission system control and protection. The integrated control and protection system is the latest trend in the world and the need of the day.

The efficiency and availability of electric power is greatly improved with an integrated system. The system analyst is provided with reliable and precise information for planning and operation at a centralised place and the job of decision making is highly simplified.

A typical configuration of integrated control and protection is shown in Fig. 15.12.

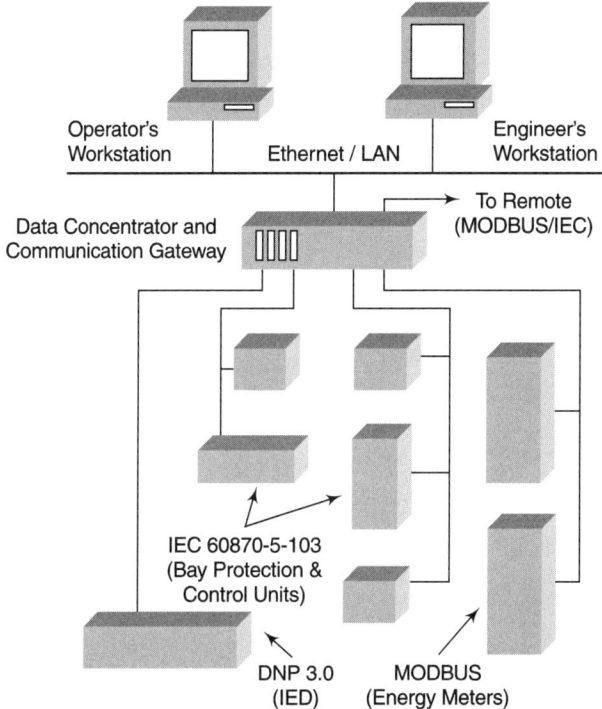

Fig. 15.12: Integrated control and protection (a typical configuration)

Chapter 16

INTEGRATED PROTECTION AND CONTROL SYSTEMS FOR SUB-STATION

Dr. M.P. Soni

Electrical power utilities have been using discrete electronics and electro-mechanical devices for power system protection, metering and supervisory control. Each device independently acquires and processes the power system data from current and potential transformers, circuit breakers, isolators, tap changers, etc., and performs the assigned function. Such a system suffers from two disadvantages. The first one is the cost associated with each device acquiring power system signals independently. The second and foremost disadvantage is that each device has only the information that it acquires directly for performing its function.

With advancements in computer technology, it has now become possible to introduce microcomputers, digital signal processors, and analogue to digital converters, optical transducer and fibre-optic communication systems to acquire and process electrical power system information in an effective manner and use it in the development of the integrated protection and control system for a sub-station. Such a system not only provides a cost-effective solution to the problems earlier faced by power utilities while using conventional protection and control equipments but adds the good features of MMI, disturbance records and event recording helpful for the post-fault analysis.

The integrated protection and control system provides the following features:

(a) Power system protection;
(b) Supervisory control and data acquisition;
(c) Statistical and revenue metering;
(d) Local control;
(e) Voltage regulator;

(f) Station battery monitoring; and

(g) Digital fault recording.

Figure 16.1 shows the system architecture for an integrated protection and control system for the sub-station. The above listed functions could be achieved by adopting three-level architecture which has been used to provide information at three distinct rates. The first level is assigned to the protection and metering system. Real time data processing is carried out at the highest rate for the protection and metering functions. For example, a sampling frequency of 800 Hz is required to discretise each current/voltage signal for computing the RMS values of that signal if the processing algorithm so selected is based on 16 samples per cycle. The control and monitoring of sub-station equipment is carried out at a relatively lower speed as compared to the protection system. Similarly the data analysis and archiving function is processed either after the critical event has passed or at a fixed interval of the order of hours. Thus the data is available at the highest rate in the first level (protection and metering), at a lower rate in the second level (control and sub-station monitoring) and at the lowest rate in the third level (data analysis and archiving). Each level incorporates processor(s) to perform the functions that are appropriate for the speed at which data is available at that level. The system is described below.

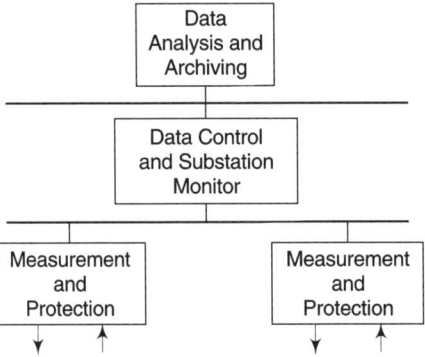

Fig. 16.1: Connections to primary equipments

16.1 PROTECTION AND METERING SYSTEM

A modern digital relay known as 'numerical relay' is most suited to perform the function of the protection and metering of the considered electrical equipment. Figure 16.2 shows the general block diagram of the digital relay. The primary voltage and current signals of the power system range up to hundreds of kilovolts and kiloamperes. The nominal voltage signals are usually reduced to 110 V and the current signals are reduced to 5 A or 1 A range using voltage and current transformers respectively. The signals from the secondary of the VTs and CTs are connected to the analogue input sub-system. The analogue input sub-system isolates and further scales down the signals using low VA burden instrument type voltage and current transformers. These transformers are placed in the relay cabinet and their function is to make voltage and current signals compatible with the input signal range of the analogue to the digital converter (ADC). The scaled signals are low-pass filtered to minimise aliasing problems.

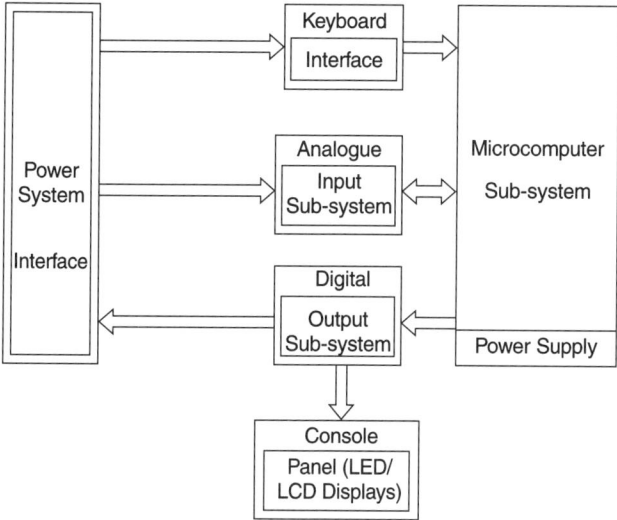

Fig. 16.2: Digital relay—General block diagram

The filtered signals are multiplexed and re-scaled, if needed, before sampling and analogue-to-digital conversion. The ADC converts the sampled analogue signals to digital values and transfers them to the microcomputer sub-system. The digital input sub-system provides the relay with the status of circuit breakers, nearby relays, voltage transformer, fuses and keyboard interface. Isolation and protection circuits are also provided at the digital input sub-system. The digital output sub-system is responsible for sending trip and alarm output contacts and for controlling the target LEDs and the LCD display for man-machine interface (MMI). The relay output contacts provide trip signals to circuit breakers, front panel trip targets, and alarm and annunciation signals to system operators.

The microcomputer sub-system is the heart of the digital relay. It is responsible for processing the input signals and making protective relaying decisions. The relay software executes a variety of signal processing algorithms and calculates several system parameters. The relay logic software compares these parameters against relay settings to detect fault and abnormal conditions. Communication ports provide remote communication capability to the outside world. The digital relay also requires a power supply to provide various dc voltages required for its normal operation.

Figure 16.3 shows the hardware block diagram of a typical stand-alone digital relay. The hardware configurations for digital relays range from a simple, single-processor configuration to a complex multiprocessor system. The various sub-systems of a typical relay are:

(a) Analogue input unit;
(b) Central processing unit;
(c) Digital input/output unit;
(d) Power supply unit; and
(e) Communication unit.

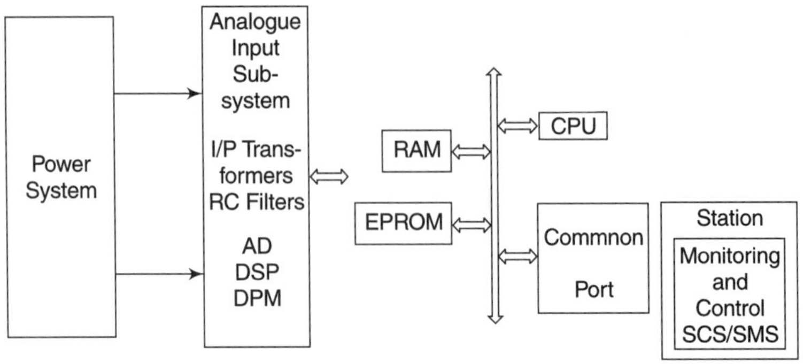

Fig. 16.3: Hardware details of the digital relay

The analogue input unit comprises the transformer which provides the electrical and static isolation between the analogue input variables and the internal electronic circuits, and adjusts the signals to a suitable level for processing. The input transformer unit can accommodate three-phase voltage and current signals of the transmission lines.

Every analogue variable is passed through a first order R/C low-pass filter on the main CPU unit to suppress high frequency interferences and to eliminate the aliasing effect. The output signal is then passed through the ADC and sample and hold circuit to convert the analogue signal to discrete values.

The digital signal processor (DSP) is used for digital filtering of the acquired data. This data is stored in the memory and processed by the main processor as per the protection algorithms.

The microcomputer system essentially comprises the main microprocessor for the protection algorithms and dual-ported memories (DPMs) for communication between the AD converters and the main processor. The main processor performs the function of protection and metering. It can also control the local MMI and interfaces to the station control system. Binary signals from the main processor are relayed to the corresponding inputs of the IO unit and thus control the auxiliary output relays and the light emitting diode (LED) signals. The main processor unit is equipped with an RS-232 serial interface for communicating with the outside world. Remote settings of the relay are possible through this port. Connectivity with the PC is also provided through this port. The fault data recordings stored in the memory are transferred to this PC. As shown in Fig. 16.3, there are two PCC slots and one RS-232 interface. These serial interfaces provide remote communication to the station monitoring system (SMS) and station control system (SCS) as well as to the remote input and output systems.

Figure 16.4 shows the signal data flow. As shown in the data flow block diagram, the analogue signals pass through the sequence input transformers, low-pass filter, i.e. anti-aliasing filter, multiplexer and AD converter and DSP. In their digital form, the filtering function is performed to compute the sin and cos components of the signals. Subsequently RMS values of the signal are computed which can be used in the protection system trip decision as per the characteristics. The binary signals from the opto-coupler inputs go straight to the main processor. The actual processing of the signals in relation to the protection algorithms and logic then takes place. The metering function computes the single phase rms values of the voltage, current, frequency, and real and reactive power.

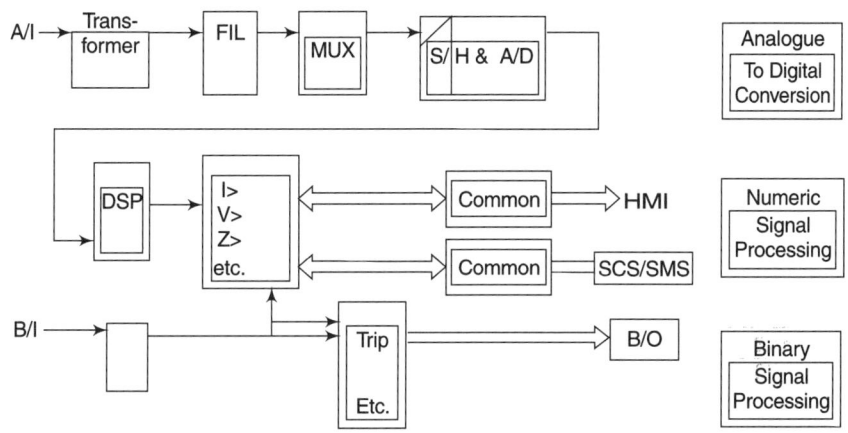

Fig. 16.4: Signal/data flow in digital relay

16.2 CONTROL AND SUB-STATION MONITORING

The control and sub-station monitoring functions are implemented in the second level of architecture status and analogue values are passed to the control and sub-station monitoring general purpose microprocessor on an exception basis, from the digital signal processor in the metering and protection level. The control and sub-station monitoring functions are implemented in this level because they do not require power system information as rapidly as the protection and metering functions. Rules and smart automation tasks, running in an expert system shell, implement the control and sub-station monitoring functions.

The control functions to be incorporated are:

(a) Opening and closing breakers;

(b) Automatic and manual tap changer control;

(c) Tap changer auto/manual selection;

(d) Enabling and disabling enclosure; and

(e) Acknowledge alarms.

In addition to the above, a screen and keyboard (terminal) provide the interface for local control. A modem, connected to a leased communication facility, provides communications to the supervisory master in the control centre.

The sub-station monitoring functions are:

(a) Trip and sustained alarm logging;

(b) Sequence of event recording; and

(c) Digital fault recording.

A local area network (LAN) interface provides a 10 M bit/sec Ethernet LAN connection to communicate between the control and sub-station monitoring level, and the data analysis and archiving level. The sub-station LAN is also used to communicate with a redundant IPACS system as shown in Fig. 16.3. This redundant architecture is used to provide a duplicated protection function.

16.3 DATA ANALYSIS AND ARCHIVING LEVEL

The data analysis and archiving functions are implemented in the third level. Power system data is stored in a relational database that is updated each time there is a power system event. The voltage and current waveforms are also archived each time one of the protection functions operates. Power system events include the operation of the circuit breakers and tap changes, power system faults, voltage sags and swells, etc. This information is used for sequence of events recording, circuit breaker predictive maintenance, digital fault recording and power quality monitoring. Power system data is also added to the database at pre-determined periodic intervals for bus voltage tending.

References

1. "Integration of Protection, Control and Monitoring Functions for Transmission and Distribution Sub-stations", G.R. Allen and R. Cheung, IEEE Trans. Power Delivery, Vol. 13, No. 1, Jan. 1998, pp 96–101.
2. "A Microprocessor-based System for Integrated Protection and Control", T.S. Sidhu, *et al.*, Power System Research Publication, University of Saskatchewan, May 1991.
3. "The Integration of Protection, Controls and Monitoring in a High Voltage Sub-station", John Burger, *et al.*, CEA (Canadian Electrical Association) Paper, March 1993.
4. "Integrated Protection, Control and Data Acquisition", J.C. Burger and J.C. McGough, Presentation to AEG Modicon 1994 Sub-station Focus Conference, June 1994.
5. "Integrated Sub-station Protection and Control", N.W. Eastopn, *et al.*, CEA Report ST-366, October 1993.
6. IEC Publication 255-3 (1989-05), Single Input Energising Quantity Measuring Relays with Dependent or Independent Time, Second edition.
7. The Application of an Adaptive Technology to Power System Protection and Control, Cigre Paper 34-03, August 1988.

Chapter 17

ERECTION AND COMMISSIONING OF SWITCHGEARS

H.R. Patel

Often the switchgears cannot be shipped in completely assembled condition due to limitations of transportation and packing. The erection includes placing the equipments on its foundation, completion of assembly, i.e. assembly of items/sub-assemblies supplied loose, their mechanical and electrical inter-connection, fitting of any extra instruments/devices, and completion of control and power circuits.

Once the equipment is erected, it needs to be commissioned through generally pre-defined checks and tests before finally putting into service. Testing is necessary to prove that the installation meets the required specifications and safety. The work content for the installation of outdoor switchgear is different from that of indoor switchgear, but the principles followed are the same.

The satisfactory performance of any installed equipment depends upon the quality of erection and commissioning. Hence erection and commissioning are very important activities that are carried out at site. The major requirements and precautions for erection and commissioning are covered in this chapter.

The availability of following documents should be ensured during erection and commissioning:

— Dispatch/delivery documents, i.e. Packing list;
— Instruction manual for operation and maintenance;
— Outline general arrangement, layout and floor plan drawing, Foundation drawing;
— Schematic and wiring diagrams; and
— Instruction manuals for relays, meters and other devices like switches, transducers, etc. fitted into the equipment.

17.1 DISPATCH

After manufacturing and completion of all routine tests and other quality and inspection checks, the switchgear is dismantled for packing and is made ready for dispatch. As many a time complete assembly cannot be dispatched due to transport limitations, these are made ready for dispatch after dismantling. Following procedure and precaution should be followed while dispatching switchgears.

17.1.1 Outdoor Switchgears

Outdoor switchgears are generally designed and dispatched so as to involve minimum amount of erection work. It is generally dismantled into four parts as follows:

1. Pole unit with inter-phase mechanism (Fig. 17.2);
2. Mechanism box with local control unit (Fig. 17.3);
3. Structure for circuit breaker, and
4. CTs and PTs.

(An assembled outdoor VCB is shown in Fig. 17.1.)

Fig. 17.1: Assembled outdoor VCB

Fig. 17.2: Pole unit with inter-phase mechanism

The inter-connecting drive rod with eyebolts (Fig. 17.5) is de-linked and kept tied inside the inter-phase mechanism. The connecting hoods are kept inside the mechanism house. Structure parts are dismantled and tied together. These parts are marked suitably to facilitate easy assembly at site. Openings on the top of the mechanism housing and bottom of the inter-phase house are covered with blanking plates in order to avoid ingress of dust, rain water and foreign bodies inside the units during transport and storage at site. These blanking plates are to be removed just before assembly.

Fig. 17.3: Mechanism box with local control unit

Foundation bolts, loose hardware, terminal clamps (if specified) and any other small items are packed in a small box which is fixed to the main packing box. The spring charging handle and slow closing handle are generally kept inside the mechanism house of each unit.

The structure assembly is packed separately with clamps for easy transportation. The pole unit with inter-phase assembly is packed in a wooden crate, while the mechanism house is packed in a separate wooden box.

In case of a breaker with outdoor CTs/PTs and associated mounting structure, the three single-phase CT/PT are packed in a separate packing case. Suitable pictorial marks are made on the packages to ensure that the handling and transport are done in an upright position. While lifting by a crane or hoist, wire ropes should be attached to the bottom of the crate.

17.1.2 Indoor Switchgears

Indoor metal-clad switchgears up to 33 kV class are dispatched in the following manner:

1. For easy transport and handling, the switchboard is divided into dispatchable units, and generally two or three (Fig. 17.4) panels are sent as one dispatchable unit depending upon the dimension of the panel, considering transport and shipping limitations.

2. As far as possible, the bus-coupler panel with trunking panel and incomer panel with separate feeder PT panels are sent as one unit respectively so that minimum connections are required to be done at site.

3. Panels are fully covered by polythene sheets to protect them from dust and moisture, and are packed in a wooden crate. Suitable felts are provided to protect the painted surface when it comes into contact with crates.

Fig. 17.4: Front view of set of two panels

4. (Before packing, the withdrawable VCB truck is kept in service position inside the panel and the circuit breaker contacts are kept in open condition.) The closing spring of the operating mechanism is kept in 'free' (discharged) condition.
5. Suitable pictorial marks are made on the packages to ensure that the handling and transport are done in upright position. When lifting by a crane or hoist, wire ropes should be attached to the bottom of crate.
6. For each consignment, a dispatch advice note and packing list are sent so that the units and corresponding accessories can be unpacked and identified. Each crate is marked with the Serial No. of the units kept within it.

17.2 INSPECTION ON RECEIPT AT SITE

Switchgear is often dismantled before being dispatched with the parts packed separately. All items of equipment should therefore be carefully inspected as soon as possible upon arrival at site to ascertain whether any items are missing or damaged. Hence, after unpacking, the contents should be checked against dispatch notes and the receipt of all items verified. The recipient should check for damages, if any, which might have occurred during transport or handling. In particular, he should carefully check for damages in the vacuum interrupter, inter-phase barriers, bushings, relay and epoxy insulators. If any damages are noticed, communication of the same may be sent to the transportation company, insurance company and a copy to the manufacturer.

17.3 STORAGE AND HANDLING

Delivery of the switchgear should be made to the site at a time convenient for erection to proceed immediately thus obviating the need for any storage facility. The storage of switchgear is very important if it does not have to be erected immediately. It is always recommended that all the parts of the circuit breakers are stored indoors.

During handling, one must ensure that the slings are of adequate strength and secured properly, that no parts are subjected to undue strain, and that the surfaces are neither scratched nor damaged. For circuit breakers, wire ropes should never be attached on conductors and insulating parts. It is recommended that the switchgear panel should be put immediately at its permanent location. If this is not possible and storage of the panel is intended for long durations, the certain precautions must be taken to ensure the proper storage of the units. These are detailed below.

17.3.1 Precautions to be Followed for Indoor Switchgears

The following precautions must be followed during storage of indoor switchgears:

- If the equipment is to be stored for more than three months or the weather is humid, switch on the anti-condensation heaters provided in the circuit breaker ensuring proper voltage supply to the appropriate terminal.
- The crates should be unpacked carefully after selecting a dust-free area. The dust sticking to the crates should be removed and the polythene cover taken off.
- All the hardware should be stored essentially in polythene bags and kept indoor.
- Circuit breakers should be stored in clean, dry airy place indoors and the site should be cool and free from dust, humidity and corrosive gases.
- When any work is done in the vicinity of the storage place, the panels should be protected with suitable covers.
- The packing case should be raised clear off the floor on bottoms so that air may circulate freely beneath it.
- Rats and other vermin are a source of considerable damage and periodic inspection is necessary to minimise the danger they pose.

17.3.2 Precautions to be Followed for Outdoor Switchgears

The following precautions need to be followed while storing outdoor switchgears:

- In case of high voltage circuit breakers beyond 33 kV, all the insulators should be stacked properly and stored in a safe place to prevent damages to petty coats. The surface of sealing flanges should be protected adequately.
- Rubber parts like oil seals, gaskets, washers, etc. should be stored in such a way as to avoid mechanical damages and should be smeared with chalk powder before being stored. They should be stored away from heat and moisture to avoid condensation.
- All the sealing surfaces and surfaces of terminal plates should be protected from damages during handling and storage.
- Oil circuit breakers must be filled with oil.
- Packing material such as wood should be removed as it may absorb moisture.
- The packing case should be removed if it is not designed for outdoor storage.
- During handling, it must be ensured that the slings are of adequate strength and secured properly and that no parts are subjected to undue strains.

- In the case of circuit breakers, wire ropes should never be attached on conductors and insulating parts.

17.4 INSTALLATION

One must check that the unit serial number packed in all the cases received for installation of each breaker is the same. This will ensure that factory settings are achieved automatically and no elaborate checking/setting of various parameters such as travel, spring compression, etc. would be required, which otherwise may be necessary.

17.4.1 Outdoor Switchgears

For outdoor switchgear the following installation checks should be done:
  (a) Before unpacking the packing cases for installation, ensure that the civil work of the foundation blocks with pockets as per the foundation drawings is completed.
  (b) The foundation bolt should project a minimum height as per the manufacturer's foundation drawing above concrete level.
  (c) Grout the foundation bolts and allow the cement to cure.

17.4.2 Indoor Switchgears

For indoor switchgear the following installation checks should be done:
  (a) Refer Outline and General Arrangement drawing, floor plan layout and foundation layout, to ensure the completion/correctness of the civil work of the foundation.
  (b) Check the serial number of panels and ascertain the position to be occupied by them.
  (c) Before installation of the equipment, it is necessary for the civil work of the building to be completed.
  (d) Do not hook or sling on support insulators for lifting the truck.
  (e) Draw out the trucks from panels and store them separately in a dry and dust-free room. Cover them with polythene covers.
  (f) Shift the panels to their specified location carefully so that no damage is caused to instruments and other parts.
  (g) Ensure that all cable trenches and floor openings for main and auxiliary cables are present as per the foundation drawings. Switchgear panels fitted with foundation channels are to be laid as per the dimensions of layout drawings.
  (h) Ensure that the foundation channel is levelled properly and that the surface to be covered by the panel base is at the main floor level. This is to ensure the free movement of the truck.
  (i) On level floors, panels can be moved on rollers or pipes of adequate length kept underneath. The truck can be moved on its own wheels.

17.5 ERECTION

17.5.1 Outdoor Switchgears

The following steps should be followed for erection of a typical outdoor circuit breaker:

1. Prepare foundation and place the support structure assembly with the mechanism house over the foundation blocks. Ensure that the vertical columns of the structure are perpendicular to the ground and use shims under the base angle wherever necessary. Tighten the nuts fully.
2. Open the covers of the mechanism and Pole Unit base frame.
3. Place the drive rod in position (Fig. 17.5) and connect to the operating shaft by pin and circlip. Do not alter the length of the drive rod, as any change will render the breaker inoperative. Ensure that the serial numbers of the structure, inter-phase/three-phase assembly, mechanism-house and drive rod are same. Place the pole assembly and base frame over the support structure, aligning the fixing holes of both assemblies. Ensure that the gasket over the mechanism house is in position. Do not hook or sling insulators while lifting the pole assembly for the sling position.
4. Insert all the fixing hardwares. Now gradually tighten the top fixing hardware to compress the sealing gasket between the base frame and the mechanism housings. Finally tighten the bottom sheet fixing hardware. Use plain washers and spring washers with all the hardware. Also check the tightness of all hardware on poles, structures, etc.

17.5.2 Indoor Switchgears

For indoor switchgears, refer the floor plan layout drawing, erection and commissioning instructions furnished in the instruction manual for erection to ensure the completion/correctness of erection. Spirit level and plumb line should be used to ensure that the equipment is erected vertically.

Move the panel to the appropriate position and locate on foundation channels. Ensure that the correct sequence of panels is maintained as per the layout drawings. Remove the back and top covers of the bus bar chamber of every panel. Connect the busbars between panels and align them. After the busbars have been aligned and tightened, weld the foundations on embedments and lighten the inter-panel bolts to complete the formation of the switchboard. Lay HT, LT cables and seal all cable entry holes using cable sealing plate and cable glands. Ensure that the fronts of all panels are in one line and that the front face is at a right angle to the floor. Earthing bars mounted on the rear of the breaker chamber should be inter-connected to the earth bar of the adjacent panel using the connecting strips provided and this earth bar should, in turn, be connected with the station earth terminal. While connecting the main busbar, ensure that the disc washers are on the nut side and after tightening fully to flatten the disc washer, turn back the nut by a half-turn or use the torque spanner for tightening the busbar with the torque as suggested by the manufacturer. After busbar connections have been made, assemble busbar shrouds. Smear petroleum jelly after cleaning on the mating surfaces of conductors during the jointing process. On completion of the erection work of panels, trials should be carried out with the respective trucks by inserting and withdrawal inside the panel to ensure that smooth movement of the breaker truck is achieved.

17.8 Handbook of Switchgears

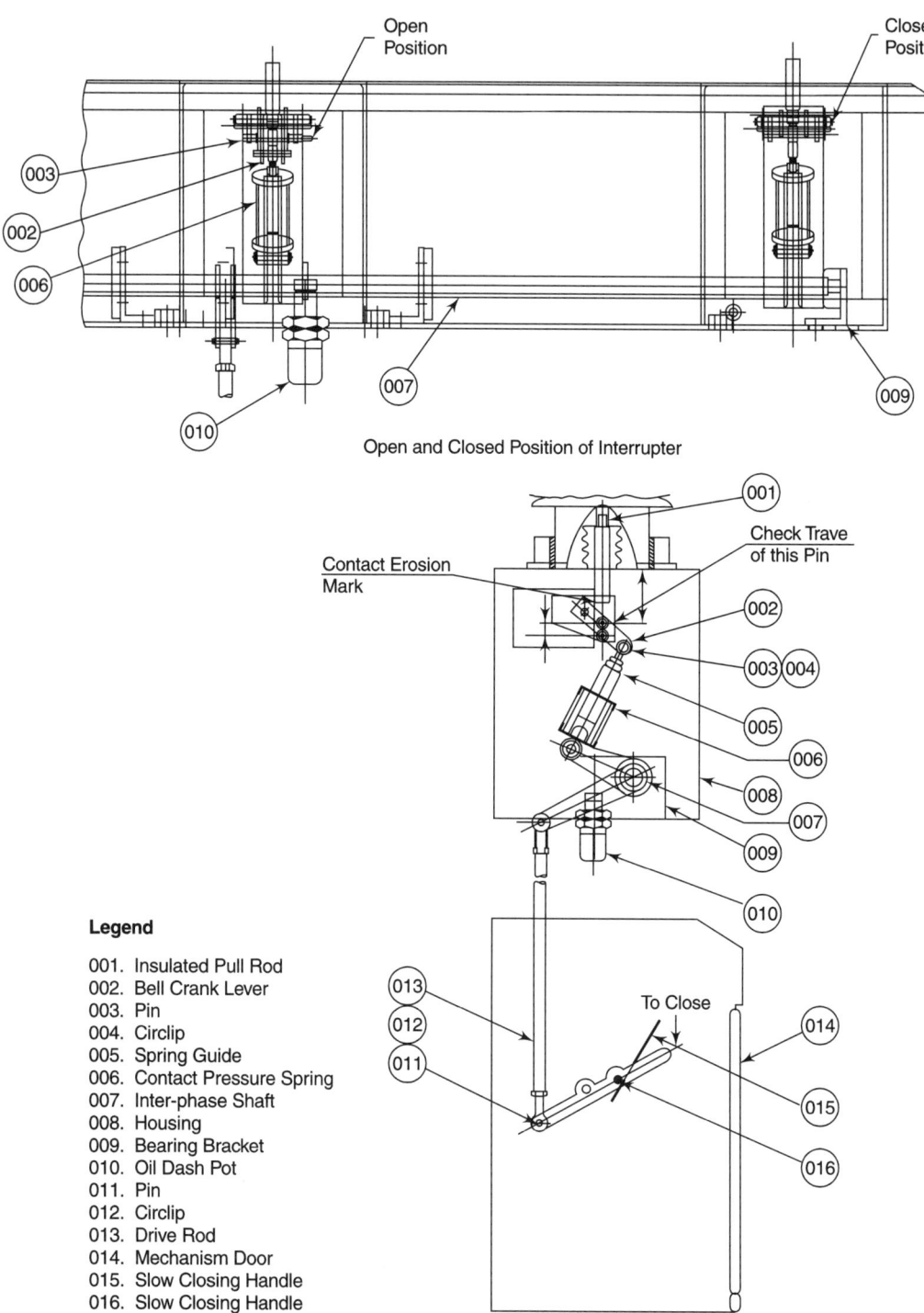

Legend

001. Insulated Pull Rod
002. Bell Crank Lever
003. Pin
004. Circlip
005. Spring Guide
006. Contact Pressure Spring
007. Inter-phase Shaft
008. Housing
009. Bearing Bracket
010. Oil Dash Pot
011. Pin
012. Circlip
013. Drive Rod
014. Mechanism Door
015. Slow Closing Handle
016. Slow Closing Handle

Fig. 17.5: Inter-phase assembly

17.6 ASSEMBLY

Procedure/precaution to be followed for assembly of switchgear are described below.

17.6.1 Outdoor Switchgears

Refer to instructions furnished in the instruction manual. Switchgear/VCB sub-assemblies and devices supplied loose have to be assembled, and the pole unit assembly should be connected properly with the operating mechanism. It is advisable to slow operate the breaker a few times to check the alignment of the moving parts of the mechanism and verify contact stroke (travel), contact pressure spring compression, before operating the breaker/interrupter through the spring mechanism. The spring mechanism should not be operated until it is connected to pole assembly as otherwise it is liable to damage the mechanism parts and hurt the operator.

The assembly of the cable gland should be done in the following manner:

- The undrilled cable gland plate is provided at the bottom of the operating mechanism. Remove this plate and drill holes to suit the cable glands supplied loose with each breaker. The number of holes for the cable glands should be based on the actual site requirement.
- The switchgear should be connected properly by the jumper connection to overhead lines through an isolator. While connecting the terminals of switchgear, care should be taken to avoid undue strain on the insulators. The terminal connector should be mounted on the incoming and outgoing terminal using hardware. The incoming/outgoing conductor should be terminated on this terminal connector.
- The mechanism is normally supplied in an open position with springs discharged. The contacts of the vacuum bottle, when not connected to the mechanism and/or pulled out, will be in a closed condition but without any external contact pressure acting on them. Therefore, do not try to check continuity between the terminal pads of poles not connected to the mechanism and closed under force of the closing spring, otherwise the observations will be misleading and erroneous.

17.6.2 Indoor Switchgear

A loose busbar is sent alongwith the hardwares with each packing case. The manufacturer's instructions and drawings should be referred to for making busbar joints. Connections between two units should be made and tightened properly. Necessary air clearance should be ensured. In case of busbars, due regard should be given to expansion, which is normally taken care of by the manufacturer. Busbars are normally silver-plated or tin-plated at the joints as per the specified requirement or it bare should have a small coating of any conducting grease to ensure good contact. Bi-metallic connectors are used between two different metals such as copper and aluminium. Before making the joints, the faces of the connectors must be cleaned by a scratch brush/emery paper. Petroleum jelly should be applied and wiped clean with a lint-free cloth. The assembly of the busbar joints should then be completed and the hardwares of the busbar joints tightened properly at the specified torque with a torque spanner. Depending upon the design, busbar joints are insulated/shrouded.

The preparation of power cable for connection to the switchgear is a specialised function and an experienced cable jointer should be employed for this purpose. The cable has to be carefully laid to avoid sharp bends, and the length to the cable lug accurately measured to avoid stresses on the cable or on the terminal to which it is attached. Checks should be carried out to ensure proper phase sequence.

17.7 SMALL WIRING AND CONNECTION OF LT CABLES

Small wirings are auxiliary circuits such as tripping, closing, indicating, metering and protection CT/PT circuits. The CT circuit should be made to form a closed circuit as dangerous voltages may develop under open circuit conditions and may even puncture the CT insulation. It must be ensured that the ratings of CTs with meters, relays, etc. are matching. The auxiliary wiring should be neatly supported, and loose wires should be avoided as they may interfere with other devices.

Refer to the schematic and wiring diagram and connect the required cables of auxiliary supply controls, metering and protection. Always use proper glands and lugs.

It is necessary to complete the small wiring connections of the panels to the adjacent cubicle and to connect external multi-core cables to the terminal blocks.

17.8 FILLING WITH INSULATION MEDIUM

Depending upon the design of the switchgear, oil or SF6 gas is to be filled as an insulating medium. After drying out of the switchgear, proper oil level or proper pressure of SF6 gas needs to be maintained as per the manufacturer's recommendations.

In many designs, poles are filled with nitrogen gas above atmospheric pressure (1.5/2 times of the atmospheric pressure) to avoid breathing of moist air.

During the process of ensuring that the joints are oil-tight for the oil circuit breakers, the joint faces, gaskets and the operator's hands should be clean before and during the assembly. Jointing material should be applied as directed and the bolts tightened up evenly not more than half-a-turn at a time and finally pulled up as tight as possible. Similarly for air blast switchgear or switchgear having pneumatically operated mechanism, due care and attention should be given to joints in piping with air at pressure. Tests should be conducted to check for air leakage after installation in order to ensure correctness of such joints. A thin soap solution can be used for checking air leakage.

17.9 EARTHING

Earthing points provided in switchgear are to be connected with the station earth bar/earth grid. Measurements of earth impedance are made to ensure that no dangerous voltages can arise due to fault currents flowing in the earth conductors. All equipment should be solidly and effectively earthed to the earth connection of the switch-boards.

17.10 FINAL INSPECTION

After the switchgear erection has been completed, a final inspection is made covering various aspects of both indoor and outdoor switchgears. The details of this inspection are discussed below.

17.10.1 Indoor Switchgear

The following steps should be taken to inspect the indoor switchgears:
- Check all chambers for complete cleanliness and the absence of foreign material, including tools used in the erection operation.
- Check the tightness of hardwares.
- Check that labels are fitted and visible, wherever required.
- Check that all lubricating points are lubricated properly.
- Check that specified important settings are locked and intact.
- Make a final check for the continuity of the earthing.
- Check the operation of the breaker manually, covering the following:
  — Rack-in and rack-out operation, if provided;
  — Shutter operations;
  — Mechanical interlock operations and satisfactory operation of the interlock schemes;
  — Manual charging of the closing mechanism;
  — Travel of the moving contact;
  — Proper contact wipe in isolating and fixed contact; and
  — Healthiness of the fuse and links.

17.10.2 Common for Both Indoor and Outdoor Switchgear

The slow closing and slow opening operations should be checked as per directions given in Box 1 and Box 2 respectively.

Box 1
Slow Closing Operation

System conditions at the beginning:

Circuit breaker/Interrupter	: **Open**
Closing spring	: **Discharged**
Motor supply	: **Disconnected**
Tripping spring	: **Discharged**

Open the cover of the operating mechanism and place the slow closing handle between the mechanism shaft and pin extension (Fig. 17.5). Follow the instruction of manufacturer for slow closing after the mechanism is latched in closed position. The mechanical position indicator will show CLOSE.

> **Box 2**
> **Slow Opening Operation**
>
> System conditions at the beginning:
>
> | | |
> |---|---|
> | Circuit Breaker/Interrupter | : **Closed** |
> | Closing spring | : **Discharged** |
> | Motor supply | : **Disconnected** |
> | Tripping spring and Contact Pressure spring | : **Compressed** |

For manual charging of the closing spring, insert manual charging handle at the location provided inside the mechanism.

17.11 TESTING AND COMMISSIONING

Commissioning tests are conducted at site after installation of the equipment in order to ensure that:

1. The equipment will perform its duties in service;
2. Inter-connection with other apparatus is correct; and
3. The test data are provided as records for future maintenance and service work.

The basic commissioning checks which are conducted at site are discussed below.

17.11.1 Visual Inspection Check

The following steps should be taken for visual inspection:

(i) Check that the serial number of the base frame and mechanism housing are the same.
(ii) Check that the nuts of foundation bolts are fully tightened and that the spring washers are used.
(iii) Check the tightness of the entire hardware of poles, structure, mechanism, etc.
(iv) Check that porcelains, etc. are free from any damage.
(v) Check the levelling of the base frame and poles.
(vi) Check that all settings marked with a red paint line are undisturbed and properly locked with nuts and spring washer.
(vii) Check that all pins are locked with circlips.

Erection and Commissioning of Switchgears **17.13**

(viii) Ensure that the gaskets for doors and covers are in good condition and pasted in proper places.
(ix) Check that cabling and earthing connections are properly made.
(x) Ensure that porcelains are clean and the mechanism parts duly lubricated.
(xi) Check that the breaker does not operate manually or electrically if the key of the mechanical inter-lock (if any) is removed.

17.11.2 Mechanical Check

The following steps should be taken as part of the mechanical check:
(i) Check the free movement of linkages by operating the slow closing handle and carry out slow closing/slow opening operation a few times.
(ii) Check that the movement of gears is free. Verify that manual charging of the closing spring is effortless for the first few revolutions of crank and no abnormality except higher load is observed during the next few revolutions of the crank.
(iii) Check that the spring charged position indicator, ON OFF indicator and operation counter are operating correctly.
(iv) Check that the travel of moving contact is as specified by the manufacturer. Noting the difference in the position of pin attached to the pull rod breaker closed and open positions can be used to check this. A variation of 1 to 2 mm in travel is possible under electrical checks.
(v) Check the compression of the contact pressure spring.
(vi) Check that the plungers of the oil dash pot (Fig. 17.5) in the operating mechanism are pressed down and rest on the respective casings at the end of the opening operation. Do not disturb this setting. Verify that the plungers lift up during the closing operation and that the damping action is OK while pressing down the plungers manually.

Do not operate the breaker/interrupter if the oil dash pot (Fig. 17.5) is jammed and damping action is absent. The manufacturer should be contacted under such a condition do not disturb the setting of the dash pot as the mechanism is set with reference to the plunger top in the breaker open position.
(vii) Check that the trip and closing coil plunger movements are free and that the coil casing is properly tightened to frame.
(viii) Check breaker operation through manual knobs.

17.11.3 Electrical Check

The following steps should be taken as part of the electrical check:
(i) Check the wiring with the relevant scheme and wiring diagrams. Check the insulation resistance with a 500 V megger.
(ii) Check that the control, indication and auxiliary leads are connected to the respective terminal blocks.

(iii) Ensure the availability of AC and DC supplies.

(iv) Charge the closing spring by motor. The charging time should not exceed 15 seconds. If the timing is more than 15 seconds, check the voltage. If voltage is OK, then check the linkage and gear movement for free movement. At the end of the spring charging operation, the motor supply is automatically cut off by the limit switch.

(v) Electrically close/open the circuit breaker through push buttons/control switch from the local panel.

(vi) Check the contact resistance between the upper and lower terminal pads with breaker/interrupter closed under spring force. The value should be less than the manufacturer's recommendation.

(vii) Check the insulation resistance across breaker open terminals and between lower terminal to earth with 2.5/5 kV megger. The value should not be less than 10,000 mega-ohms or as recommended by the manufacturer.

(viii) Check the opening and closing of breaker/interrupter with remote control.

17.11.4 Operational Check

The mechanism should be operated slowly to check that the contacts are in correct alignment for checking simultaneity. Manufacturers usually explain the slow closing and slow opening operation in the instruction manual, which should be followed.

All switchgears are supposed to operate at the minimum and maximum specified voltage levels to ensure correct tripping and latching at these values. Similarly for the pneumatic mechanism, the maximum and minimum pressure should be used, and the breaker should operate satisfactorily under any of these conditions.

Closing and tripping of the circuit breaker by local control and remote control should be checked including tripping by protective relays.

17.11.5 Insulation Resistance Check

The following insulation tests should be conducted:

(a) *Insulation of main circuit:* The insulation of the main circuit is associated with the main connection. The insulation resistance to earth of each phase and between phases should be measured. The measured insulation resistance of the switchgear should be more than 1,000 mega-ohm for rated voltage above 1,000 V and up to and including 33 kV and 10,000 mega-ohms for rated voltages above 33 kV. For any reading that shows considerably lower values than these, investigation should be carried out to trace the source of the low readings. The drying out method of improving insulation resistance may be followed. Irrespective of whichever method is employed for drying out, the general principle is to apply heat continuously for a considerable time so as to drive out any moisture, which may have become entrapped.

(b) *Insulation of the auxiliary and control circuits:* Insulation resistance to earth of all secondary wiring circuits should be tested using a 500 V megger. Readings obtained for a single circuit or combinations of circuits should not be less than 1 mega-ohm.

17.11.6 High Voltage Check

The following high voltage tests should be conducted:

(a) The application of high voltage to switchgear is not universal but may be conducted. It is preferable that such tests be carried out with ac supply.

(b) Voltage should be applied in accordance with the values specified in the relevant standards for a particular system voltage. The frames and secondary windings of all current transformers and potential transformers included in the test should be earthed.

(c) These tests should be carried out on each phase in tandem with the remaining phases earthed and all units racked in service position and the breaker closed.

(d) The insulation resistance test should be carried out before as well as after the high voltage test to ensure that there has been no reduction in value. If the second reading is lower, each separate component should be tested to find out where the lower figure has occurred.

The following precautions should be observed when conducting the high voltage test:

(a) During the test, one terminal of the testing transformer is connected to earth and to the frame of the circuit breaker.

(b) The test should be commenced at a voltage of more than one-third the test voltage.

(c) The voltage should be increased from its initial value as rapidly as is consistent with its value being indicated by the measuring instrument. The full test voltage should then be maintained for the appropriate duration as the case may be and then reduced to its initial value and switched off.

(d) While conducting the high voltage test, the manufacturer's instructions should be followed.

The HV test on auxiliary and control circuit should be carried out by looping together all the wires and applying 2000 V ac for one minute. All current transformer and voltage transformer earth connections should be removed for this test, and should be replaced immediately thereafter.

17.12 FINAL COMMISSIONING/LOAD TESTING

Once all the checks on the equipment have been carried out, the operational tests are found to be satisfactory, the voltage test passed, and all the functional tests on the relays, etc. completed as per the specifications, standards and schemes, the equipment can be energised. When the circuits are on load, all the instruments and indicators can be checked for correct readings.

If the test blocks are available or the relays have test plug connections, it may be worthwhile to carry out a final check to see that the CT polarity is correct and that the instrumentation is giving correct reading. A typical pre-commissioning checklist is shown below for an outdoor circuit breaker.

17.16 Handbook of Switchgears

Pre-commissioning Checklist

Site Location: Date of Commissioning:

(A) Visual Inspection

Observation   Remarks

1. Breaker S. No.
   (a) S. No. of pole unit
   (b) S. No. of mechanism house
2. Check the tightness of hardware.
   (a) Between foundation frame and support structure
   (b) Between support structure and cabinet assy
   (c) Between cabinet assy and linkage assy house
   (d) Between linkage assy house and pole unit assy
3. Check leveling (Vertical column of support structure to be perpendicular to the foundation)
4. Check porcelains. (To be free from crack and surfaces clean)
5. Check Gasket of
   (a) Doors
   (b) Covers

(B) Mechanical Operation Check

Observation   Remarks

— Counter reading
— Slow closing   — Smooth/Hard
— Manual charging of gear box   — Smooth/Hard
— Closing damper   — OK/Jammed
— Closing spring manual charging   — OK/Not OK
— Manual closing of breaker   — OK/Not OK
— Manual tripping   — OK/Not OK
— Contact travel check   — 22–24 mm
— Compression of contact   — (>/10 mm)
— Pressure spring   — (×)

(C) Electrical Operation Check

Observation   Remarks

— Wiring check as per Drg   —
— Electrical charging of closing spring at 110 V dc
— Spring charging time   — \<15 sec
— Selector switch on local
   *Electrical closing   — OK/Not OK
   *Electrical tripping   — OK/Not OK
— Selector switch on remote
   *Electrical closing   — OK/Not OK
   *Electrical tripping remote
— Main contact resistance (max. 80 micro-ohm)
— Insulation resistance after both terminals are disconnected
   Check by 2.5/.5 kV megger

(Contd.)

(Contd.)

	Observation	Remarks
(a) Between open 2,000 Mega ohm Terminals	— 2,000 Mega ohm	
(b) Between lower 2,000 Mega ohm terminals to earth	— 2,000 Mega ohm	
— Insulation resistance of control/and circuit with 500 V megger >/2M	— /2 Mega ohm	

17.13 DOS & DONTS

Dos and Donts for Site Work	
Do	: Plan Civil, Storage, Erection, Testing, Commissioning Activities
Do Not	: Do Unplanned Site Work
Do	: Train Site Personnel
Do not	: Assign Work to Untrained Personnel
Do	: Prepare Field Quality Documents
Do not	: Neglect Documentation and Records
Do	: Take Safety Precautions
Do not	: Neglect Safety
Do	: Receive and Check Packages on Arrival and Take to Foundation If Ready
Do not	: Neglect Inspection for Transit Damage
Do	: Inspect and Store if Site is Not Ready
Do not	: Keep Packages and Breakers in Field
Do	: Erect the Structure Vertically
Do not	: Erect Structure in Inclined Position
Do	: Place Breaker Assembly on Structure
Do not	: Interchange Breaker and Structure
Do	: Assemble the Operating Rod by Using Slow Closing Handle
Do not	: Disturb any Setting
Do	: Check Mechanical Slow Open-Close Operations
Do not	: Operate Fast Open-Close to Begin with
Do	: Carry out all Pre-commissioning Tests

Do not	: Energise Till Completion of all Tests
Do	: Complete Commissioning Tests Before Energising
Do not	: Forget Safety and Work Permits

Chapter 18

OPERATION AND MAINTENANCE OF SWITCHGEAR

S. Kumar
Neelam Bhogal

Switchgear is designed to ensure the desired operation and to facilitate maintenance of its parts in order to ensure reliability. Normally switchgear is installed away from the equipment to which power is being supplied. Depending upon the requirement, switchgear may be stand-alone as in the case of an outdoor switchyard or in board formation as in case of a power plant. In most of the switchgears, the facility for local as well as remote operation is provided. For convenience, each switchgear is operated from remote. In this case, all switchgear operations like close and trip commands along with relevant indications are provided on the remote control panel.

During the operation of the electrical equipment, the safety of man and machine is of prime importance. It must be ensured that any operation does not energise a part of the system that is not meant to be energised. This is achieved by the proper designing of both mechanical interlocks and electrical interlocks in the control scheme.

No one should work on the switchgear or other equipment without obtaining permission from the person responsible for operating it in the form of a work permit. Once work has started, it should not be possible to connect supply to the system without the permission of the proper authority. Various interlocks and padlocking features are provided on the switchgear to achieve this safety requirements.

In the metal-enclosed switchgear design, all live busbars are placed behind permanently fixed partitions in locked doors. Various padlocking or castle type interlocking keys that are provided on the switchgear should be kept in safe custody and should be issued to the authorised person for maintenance only by a responsible person. Full record of such issues should be maintained. On completion of the work, the original position of interlocking/padlocking should be restored for proper functioning.

The operating personnel should read and understand thoroughly the operation and maintenance manual provided along with the equipment. On the basis of this manual, a written set of rules governing operating safety principles should be drawn and displayed. This written set should preferably be bilingual, with one regional language to be used for enabling local persons to understand the instructions well.

18.1 OPERATION OF SWITCHGEAR

The major functions of switchgear are to protect, control and maintain the electrical network, including the switchgear itself. Safe operation means ensuring that any operation does not energise a part of the system or plant that is not intended to be energised. The system is to be operated in such a way that the safety of both human beings and the plants is ensured.

The switchgear operating conditions are classified as:

(a) Normal Operating Condition

In normal operating conditions, the switchgear is operated (i.e. close/open operation is performed) for load shifting or for normal operational checking. These operations can be performed manually using controls on the mechanism or electrically from local or remote control panel.

(b) Abnormal Operating Condition

The abnormal operating conditions occur when the circuit breaker (CB) operates to clear various system faults after receiving a tripping command from the protective equipment.

The two major functions of switchgear are:

— Closing operation for switching on downstream side load; and

— Tripping operation in normal/abnormal condition for cutting off the power supply to the downstream side load.

These two operations are performed by the circuit breaker operating mechanism. The operating mechanism may be spring-operated, hydraulic, pneumatic or of solenoid type depending upon the design of switchgear. In spring-operated, hydraulic or pneumatic type mechanisms, the operating energy is stored in the mechanism and a latching device is provided for releasing the energy for the closing/tripping operation. In a solenoid type mechanism, the closing operation is performed by a heavy closing solenoid, which directly pushes the moving contact of the CB without using any spring and latches. However, the tripping operation is normally performed by a spring.

The amount of energy to be stored depends upon several factors, viz., the type of mechanism, arc-quenching medium and most importantly the voltage/current ratings of the switchgear. As the system voltage and current increase, the operating energy requirement also increases. For both the operations, energy is stored by using a latching device. As soon as the latch is released manually or electrically, the desired operation is performed by the circuit breaker.

Switchgear should be designed and operated in such a way that the following potentially unsafe mal-operations are avoided: closing onto fault (allowed in case of auto-reclosure), switching on an earthed busbar, inadvertent paralleling of two sources, operation of a no-load isolator without switching the CB off, earthing a live busbar/cable, closing the CB when connected load is undergoing maintenance, etc.

Proper mechanical and electrical interlocks should be used to avoid the above-mentioned mal-operations. The interlocking requirements should be decided at the design stage of switchgear. During operation also, the interlocking requirements should be strictly adhered to.

18.1.1 Dos and Donts for Switchgear Operation

For proper handling, conduction of tests and operation of switchgear panels at site, the site personnel must observe certain basic operating principles. These are enumerated below as a set of Dos and Donts.

Dos and Donts for Switchgear Site Works	
Do	: Train site personnel
Do not	: Assign work to untrained personnel
Do	: Prepare field quality documents
Do not	: Neglect documentation and records
Do	: Take safety precautions
Do not	: Neglect safety
Do	: Check for tightness of all hardwares
Do not	: Ignore loose hardwares
Do	: Check mechanical slow open/close operation of switchgear
Do not	: Operate fast open/close operation to begin with
Do	: Carry out all tests required before energising
Do not	: Energise before completion of all tests

18.2 SWITCHGEAR MAINTENANCE

Like any other major component of the electrical network, switchgear also needs maintenance schedules for trouble-free operation during its lifecycle.

No one should work on the system itself or on any plant controlled by it, without obtaining authorisation/work permit from the responsible person. Once such work has started, then it should not be possible for someone to re-connect the supply accidentally to the associated system or plant. Re-connection of supply should be done only after the work permit has been taken back by the concerned authority. In order to ensure that this safety precaution is implemented, interlock and padlocking features are provided on the switching devices.

18.2.1 Maintenance Requirements

Maintenance covers a wide range of activities, all of which are required to keep the switchgear in ready condition at all times to enable it perform its functions satisfactorily. The parts subjected to normal wear and aging need to be serviced for ensuring full reliability of the operations. These parts may be mechanical components or electrical components. The major mechanical components comprise the mechanism and pole unit where mechanical wear and aging are caused by frequent operations. The electrical components are worn out due to arc interruption and heating. If oil or SF6 is used as an arc-quenching medium, then these also need replacement after a certain number of operations.

The interval at which the maintenance should be carried out depends upon the following factors:
- The number of short-circuit interruptions.
- The switching frequency.
- The actual service time.

The actual conditions of use in terms of the frequency of operations, environment and severity of switching duty vary so much that it is impossible for the manufacturer to give detailed guidelines to the user concerning the frequency of maintenance. A distinction needs to be made between the basic switching device with its moving parts, and the rest of the equipment which performs the supporting role of connection. The latter components are likely to need periodic inspection and servicing only, to ensure that the equipment is free from dirt, dampness and deterioration, such as corrosion of metal work and contamination of insulation. The maintenance of the basic switching unit should be the first priority. For vacuum interrupters, however, no such maintenance is needed.

Applicable standards draw attention to the importance of the methods for determining when various maintenance procedures need to be carried out. As an example, with vacuum circuit breakers the most likely criterion that needs to be monitored occasionally is contact wear, and the timing of the opening operation. The O&M manual of the manufacturer normally describes how to perform these operations and the acceptable levels of performance before putting the switchgear back into service.

The O&M manual of the manufacturer also gives special instructions for the design of switchgear from maintenance point of view, apart from general switchgear practices. For example, the proper method to release any stored energy in the closing mechanism before beginning an examination is spelt out.

18.2.2 Inspection Schedule

The following guidelines must be included while formulating inspection schedules for switchgear:
- Once a year, a general visual inspection should be carried out and insulators wiped clean. More frequent checks are necessary if the breakers are exposed to a dust-laden atmosphere.
- The operating mechanism should be lubricated after two years or on completion of 2,000 make-break operation at lubrication points shown in Fig. 18.1 for a typical mechanism. After a specified number of operations, the mechanism should be overhauled. The number of

operations may vary with the type of breaker. Hence instructions given in the O&M manual of the switchgear should be followed in this regard.

- Constructions of the interruption unit are different for oil, SF6 or vacuum switchgear. Under normal service conditions, vacuum interrupters need no maintenance but in the case of other arc-interrupting media, special care needs to be taken as per the manufacturer's instructions.
- Vacuum interrupters must be replaced after the stipulated number of mechanical operations or when the contact erosion has reached its limit, whichever is earlier.

An inspection should be carried out after an examination or overhaul to ensure that the overall condition of the equipment has been restored after completion of the work. An operation check, or preferably a series of operation checks, must be carried out to establish that the correct performance has been achieved and that the equipment may safely be restored to service. In the case of fully sealed vacuum circuit breakers, manufacturers do not normally recommend examination or overhaul of the pole unit during the economic life of the installation. Moreover, the latest trend is to use gas insulated switchgear (GIS), which is sealed for life. In such designs, carrying out of any maintenance by the staff of the user is not recommended. In the unlikely event of GIS needing an overhaul, this task should be entrusted to the manufacturer. Maintenance should be carried out by trained staff and guidelines provided in the O&M manual should be followed. Earthing before maintenance is very important.

18.2.3 Preparation Before Maintenance

Safety features need to be planned before switchgear units are ordered. The requirement of locking off parts of the system (for carrying out maintenance work on the associated plant) should be finalised. Proper interlocking arrangements should be provided for this purpose. All metal-enclosed switchgears are designed so that all live conductors are placed behind either metal enclosures or locked/bolted doors.

In order to ensure safe operation of the plant, a written set of rules governing these operating safety principles should be available. It is the responsibility of the plant-in-charge to frame such rules. Further, the rules must be understood by all who work in the environment, which necessitates a positive training programme.

The use of switchgear to gain access to the system for earthing conductors, for testing cables and other equipment, along with maintaining the switchgear itself is covered below.

The general safety rules mentioned above must also be followed in maintenance of the switchgear. The person responsible for authorising the action must be identified so that all those concerned know the co-ordinator for each procedure. For example, when the system needs to be earthed at a point, all possible supply routes to that point should be locked off. The earthing should not open inadvertently till work on the earthed equipment is being carried out.

18.2.4 Maintenance Procedure

There are many standards on the subject pertaining to different voltage levels of switchgear, which normally defines four separate aspects of maintenance, with each new stage based on the preceding one, e.g. inspection, servicing, examination and overhaul. These are dealt with in detail below.

18.2.4.1 Inspection

This is a maintenance action, which calls for a careful scrutiny of a switchgear component. The inspection is carried out without dismantling the component from its assembly. The inspection should be carried out by using all the available diagnostic methods for detecting potential malfunction/failure. It may include an operational check.

18.2.4.2 Servicing

Servicing implies work that is carried out for ensuring that the equipment is kept in an acceptable condition. It does not involve any dismantling, and is typically limited to cleaning, lubrication and adjustments as specified in the operation and maintenance (O&M) instructions of the switchgear.

18.2.4.3 Examination

This is an inspection carried out with partial dismantling, as required, supplemented by means such as measurement and non-destructive tests in order to arrive at a reliable conclusion about the condition of a particular switchgear component.

18.2.4.4 Overhaul

Work done with the objective of repairing or replacing parts, which are found to be below standard by examination, so as to restore a particular component or the entire equipment to an acceptable condition, is referred to as overhauling.

A study of above definitions indicates the possibility of an interference in the form of maintenance. Inspection may lead to the conclusion that servicing is needed. The maintenance engineer may sense a problem and call for a detailed examination. After examination, an overhaul may be needed. One of the major reasons for employing new switchgear technologies, such as vacuum, is to eliminate the examination and overhaul procedures from the switchgear maintenance programme as vacuum interrupters are sealed for life. However, proper spares must be available at site for safe operation of switchgear units to their full rating.

18.2.5 Switchgear Spares

Switchgear spares are needed to change the maintainable parts after their useful life is over. Sometimes replacement is required if a fault develops in the parts. This is to ensure a reduction in the downtime and enhancement of reliability. A switchgear operating engineer always likes to keep all necessary spares in his stock but a balance needs to be maintained between inventory cost and chances of failure of the component. The spares needed for a switchgear site are covered in the following broad categories:

(1) *Commissioning spares*—These include items that are consumed during pre-commissioning tests. Typically, these spares include indicating lamps and fuses.

(2) *Recommended spares* (For 2/4/5 years of normal operation)—These include items which are recommended by the manufacturer to be kept at site. The quantities of these items vary with the time period.

(3) *Mandatory spares* (For 2/4/5 years of normal operation)—These include items needed by the user irrespective of any recommendation of the manufacturer for the specified time of operation. Typically, such spares include CTs, VTs, protective relays, switches etc. which are not recommended by the manufacturer.

Table 18.1 gives a typical list of the recommended spares for vacuum switchgear. This list is, however, not exhaustive. In all cases, the recommendations of the switchgear manufacturer should be followed.

S. No.	Item Description
1.	Spare breaker (VCB truck) (a) 2500 A (b) 2000 A (c) 1600 A
2.	Spring charging motor
3.	Closing coil
4.	Shunt trip coil
5.	Bus support insulator
6.	Set of control fuses (30 fuses)
7.	Indicating lamps with fittings
8.	Spring charging limit switch
9.	Breaker control switch
10.	Auxiliary switch
11.	Push button
12.	Anti-pumping relay
13.	Meter of each type: Ammeter Voltmeter
14.	Vacuum interrupter
15.	CT and PT of each ratio (one each)
16.	Set of protection relays: Set of relay (one of each type)
17.	Limit switch (1NO+1NC)
18.	Panel operating switches Switchgear selector switch Auto/Individual/Manual selector switch Trip selector switch Synchronisation selector switch On/Off switch for ac supply On/Off switch for dc supply

(Contd.)

18.8 Handbook of Switchgears

(Contd.)

S. No.	Item description
19.	Set of fixed isolating contacts (1 set comprising 6 numbers) 2500 A 2000 A 1600 A
20.	Set of gaskets
21.	Panel door keys

Table 18.1 List of Recommended Squares for Vacuum Switchgears

18.2.6 Lubrication of Operating Mechanism

Timely lubrication with recommended lubricating oil ensures trouble-free operation of the components of a switchgear mechanism. For a typical vacuum circuit breaker, the lubricating points are marked in Fig. 18.1.

The cover must be removed to lubricate the mechanism parts. All appropriate points need to be lubricated, starting at the top left and working through the mechanism systematically. The parts that are not rigidly fixed (e.g. articulated joints) should be moved slightly to and fro to let the oil penetrate. Articulated joints and bearings that cannot be dismantled should be cleaned with a cleaning agent prior to being oiled.

After lubrication, the circuit breaker needs to be operated several times to test it for smooth functioning. Table 18.2 gives a list of the lubricating points and the corresponding recommended lubricating oil for a typical vacuum circuit breaker.

Lubricating part/point	Manufacturer
Bearings, sliding surfaces and levers of operating mechanism	Hindustan Petroleum (HP)
Connecting joints and rollers of operating mechanisms and bearings of auxiliary switch	Indian Oil Corporation (IOC)
Contact faces for connections and terminal faces of vacuum interrupter	Indian Oil Corporation (IOC)

Table 18.2 Lubricating Points and Corresponding Lubricating Oil Needed

18.2.7 Troubleshooting Guide

Tables 18.3 to 18.6 give troubleshooting guide for a typical vacuum circuit breaker. Such guidelines should be provided by the manufacturer of the switchgear for ready reference in case of any trouble with the operation of the switchgear. The various components mentioned in the tables can be seen in the relevant photographs.

S.No.	Trouble	Reasons	Remedial Action
1.	Pole Bursting	(a) Failure of porcelain insulator due to hair cracks developed during transportation	Replace porcelain insulator.
		(b) Failure of operating rod due to ingress of moisture	Replace operating rod and gasket. Ensure positive pressure inside the pole unit assembly.
		(c) Failure of vacuum interrupter due to vacuum loss or ingress of moisture	Replace vacuum interrupter.
		(d) Failure of outdoor CT which damages the porcelain insulator of breaker.	Investigate CT failure and replace it.
2.	Flashover in Top/ Bottom Porcelain	(a) Contamination of insulator	Clean porcelain insulator. Frequent cleaning may be required depending upon the local environment condition. Check IR value, if found to be less than 10,000 mega-ohm, then replace insulator.
		(b) Dropping of metallic items such as wires by human or birds, climbing of animals and creeping of reptiles	Care should be taken to (i) remove left out metallic wire pieces after maintenance. (ii) provide anti-climbing device; and (iii) remove nests built by birds immediately.
3.	Excessive Heating of Terminal Connectors	(a) Loose connection	(a) Tighten terminal connectors properly.
		(b) Load more than the specified rating.	(b) Reduce load suitably.

Table 18.3 Trouble Shooting Guide for Outdoor Porcelain-clad Vacuum Circuit Breaker

S.No.	Trouble	Reasons	Remedial Action
1.	Flashover in insulating parts	(a) Support insulator failure	Clean the insulator with clean cotton cloth, take IR value using 5 kV megger between busbar and panel. If value obtained is less than 10,000 mega-ohms, replace the insulator.

			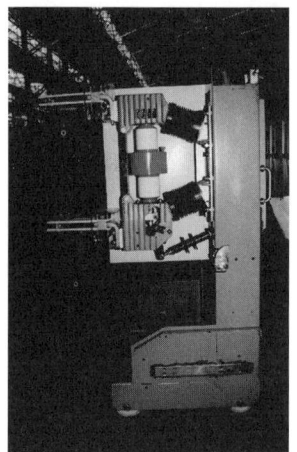
		(b) Inter-panel insulating barrier failure	Clean the barrier with clean cloth, take IR value using 5 kV, megger between busbar and panel earth. If value obtained is less than 10,000 mega-ohms, replace the barrier.
			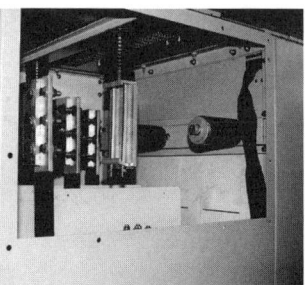

(Contd.)

(Contd.)

S.No.	Trouble	Reasons	Remedial Action
		(c) Insulating rod failure	Take IR value using kV, megger between inserts of operating rod. If value obtained is less than 10,000 mega-ohms, replace the operating rod. 
		(d) Insulating shutter surface tracking	Replace the shutter. 
		(e) Insulation failure due to reptiles	Replace the damaged material. Care should be taken to ensure that doors and covers of the breaker panels are fully closed. The openings provided for earth bus, LT and HT cables should be sealed properly to avoid entry of reptiles.

(Contd.)

(Contd.)

S.No.	Trouble	Reasons	Remedial Action
2.	Vacuum Interrupter Failure	Vacuum loss	Vacuum loss can be checked by applying HV for one minute across terminals of vacuum interrupter after withdrawing breaker truck outside the panel. If it fails to withstand and flashover occurs, then vacuum interrupter should be replaced by new vacuum interrupter.
3.	Current transformer/ potential transformer failure	(a) External flashover due to insulation failure (b) CT saturation	If cracks or burning marks are observed, then replace current transformer/potential transformer. This should be checked by giving secondary current through auto-transformer and measure primary current. If meter reading does not increase with the increase of current in secondary, it indicates CT saturation. Then replace CT.
4. above	Excessive	Loose connections Heating of Joints	Tighten all the joints properly 40–45 Nm torque.

Table 18.4 Trouble Shooting Guide for Indoor Metal-clad Vacuum Switchgear

S.No.	Trouble	Reasons	Remedial Action
1.	Breaker not closing	(a) Motor not charging spring due to supply voltage less than 85%	Ensure supply voltage above 85% of the rated voltage.
		(b) Motor not charging spring due to improper wiring	Check motor circuit wiring as per schematic drawing. If loose connections are found, tighten them.
		(c) Motor not charging spring due to jamming of gear box	Replace gear box.
		(d) Motor not charging spring due to motor shaft being broken or motor key being broken	Replace motor.

(Contd.)

(Contd.)

S.No.	Trouble	Reasons	Remedial Action
		(e) Closing coil burnt due to plunger stuck in closing coil	Replace closing coil. 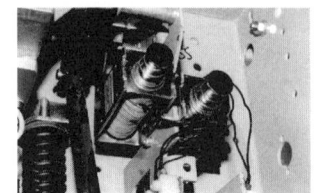
		(f) Closing coil not developing sufficient force	Check the resistance of coil at respective voltage, if not adequate, then replace the coil.
		(g) Mechanism linkage setting disturbed	Setting should be done as explained in the operation and maintenance manual of the switchgear. 
		(h) Auxiliary switch continuity missing due to contamination of contacts	Open the auxiliary switch packets and clean the contacts. 
		(i) Problem in electrical circuit	Check the circuitry and replace the faulty components. 

(Contd.)

(Contd.)

S.No.	Trouble	Reasons	Remedial Action
2.	Breaker not opening	(a) Tripping coil not developing sufficient force	Check the resistance of coil at respective voltage, if not adequate, then replace the coil.
		(b) Tripping coil burnt due to plunger stuck in tripping coil.	Replace tripping coil.
		(c) Problem in tripping circuit.	Check the circuitry and replace the faulty components.
		(d) Mechanism linkage setting disturbed	Setting should be done as explained in the operation and maintenance manual of the switchgear.

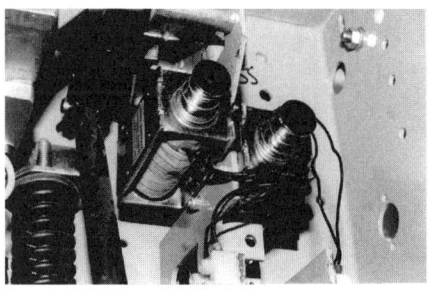

Table 18.5 Trouble Shooting Guide for Spring-charged Mechanism

18.16 Handbook of Switchgears

S.No.	Trouble	Reasons	Remedial Action
1.	Auto-recloser not closing	(a) Closing defeat feature provided in case voltage is less than 80%	Ensure supply voltage above 80% by loadshedding.
		(b) Closing coil, rectifier and contactor burnt due to plunger stuck in closing coil	Replace closing coil, contactor and rectifier.
		(c) Closing coil not developing sufficient force	Check the resistance of coil. If not OK, replace it.
		(d) Closing coil not latching due to setting disturbed	Auto-recloser setting should be done as explained in the manual.
		(e) Relay failure due to electronic circuitry card failure	Replace the card.
2.	Auto-recloser not tripping	(a) Supply voltage through power pack less than 70%	Ensure supply voltage above 70% of the rated voltage.
		(b) Tripping coil not having sufficient torque	Replace coil.
		(c) Tripping coil burnt due to stucking of plunger in the pawl bottom plate	Replace tripping coil. Ensure a 4 mm gap between plunger and pawl bottom plate.
		(d) Relay failure due to electronic circuitry card failure	Replace the card.
3.	Flashover	(a) Contamination of bushing	Clean bushing using a clean cloth. Check IR value, if found to be less than 100 mega-ohm, replace bushing. Care should be taken to:
		(d) Dropping of metallic items such as wires by humans or birds, climbing of animals and creeping of reptiles	(i) Remove left out metallic wire pieces after maintenance; (ii) Provide a anti-climbing devices; and (iii) Remove nests built by birds immediately.
4.	Excessive Heating of Terminal Connectors	(a) Loose connection	Tighten terminal connectors properly.

Table 18.6 Trouble Shooting for Outdoor Auto-recloser

18.2.8 Service Feedback

For maintaining a proper record of failures and for securing timely help from the manufacturer of the switchgear, it is necessary to keep a service feedback record of failures. The following is a typical form for maintaining such a record.

			Before repair	After repair
1.0	Serial No.			
2.0	Type			
3.0	Customer			
4.0	Name of site			
5.0	Date commissioned			
6.0	Date of failure			
7.0	Date inspected/Repaired			
8.0	Nature of complaint			
9.0	<u>Observations</u>		<u>Before repair</u>	<u>After repair</u>
9.1	Counter reading			
9.2	Slow closing		Smooth/Hard	
9.3	Manual charging of gear box		Smooth/Hard	
9.4	Oil damper in operating mechanism		OK/Jammed	
9.5	(i) Electrical charging of closing spring		Motor OK/Motor Burnt/Rectifier Burnt	
	(ii) Spring charging time		_____ sec	_____ sec
9.6	Electrical closing		OK/Not OK	
9.7	Electrical tripping		OK/ Not OK	
9.8	Tightness of hardwares		OK/ Not OK	
10.0	Details of defects/problems			
10.1	Mechanical (settings disturbed)			Yes/No
10.2	Gear box hard/defective			Yes/No
10.3	Motor defective			Yes/No
10.4	Rectifier defective			Yes/No
10.5	Vacuum interrupter R/Y/B/ damaged			Yes/No
10.6	Top porcelain R/Y/B damaged			Yes/No
10.7	Bottom porcelain R/Y/B damaged			Yes/No
10.8	Insulated pull rod R/Y/B damaged			Yes/No
10.9	Dash pot			Yes/No
10.10	Low IR			Yes/No
10.11	Transit damage			Yes/No
10.12	Any other (give details)			

Table 18.7 Service Feedbak Form

Chapter 19

TESTING OF SWITCHGEAR

S. Kumar
Divya Bhatt

The satisfactory and reliable operation of any power system depends upon the reliable operation of switchgear. In a system, a switchgear has to withstand voltage in normal and abnormal conditions to carry varying load currents during its lifetime, to clear faults occurring due to various unavoidable reasons and to carry fault current in case of through-faults. The switchgear should therefore protect systems from severe damages and should also be safe for the operating personnel and neighbouring equipment in the event of any fault occurring in the switchgear panel itself.

The performance of a switchgear under normal/abnormal conditions is verified by subjecting it to different types of tests whose guidelines are laid down in National/International standards with a view to cover as many practical cases as possible. The testing of equipment starts from the concept stage wherein the manufacturer tests all possible parameters to check the performance of the equipment for its lifetime. Normally in these standards, all possible testing requirements are given. However, some of the utilities have different requirements based on their needs and applications. It is very difficult to specify all the requirements at one place for all utilities. Utility now has the choice to select different types of equipment for its voltage levels with different specifications, namely IEC 62271-100/IEC 60298/IEC 60694, etc. But the utility has to see the site condition for which test requirement will be varying for best performance of equipment.

On the basis of the application of the equipment, the following characteristics of a circuit breaker, including those of its operating devices and auxiliary equipment, have to be furnished by the manufacturer:

(a) Rated voltage;
(b) Rated insulation level;
(c) Rated frequency;

(d) Rated normal current;
(e) Rated short-time withstand current;
(f) Rated peak withstand current;
(g) Rated duration of short-circuit;
(h) Rated supply voltage of closing and opening devices and of auxiliary circuits;
(i) Rated supply frequency of closing and opening devices and of auxiliary circuits;
(j) Rated pressures of compressed gas supply and/or of hydraulic supply for operation, interruption and insulation, as applicable;
(k) Rated short-circuit breaking current;
(l) Transient recovery voltage related to the rated short-circuit breaking current;
(m) Rated short-circuit making current;
(n) Rated operating sequence; and
(o) Rated time quantities.

All the stipulated parameters can be measured during the testing of the equipment. Those tests which are performed to verify that each sample of a product is in line with the type tested representative, are called routine tests. These tests help to detect deficiencies in the material or in construction. The test which is carried out on a representative functional unit to verify the design of the product in accordance with the standards, is called a type test. There are some special tests too, which are either specified by the user or necessitated by some application of the product. All these tests can be split into the following main categories:

1. Routine tests;
2. Type tests; and
3. Tests for special applications.

19.1 ROUTINE TESTS

The various routine tests are discussed in detail below.

19.1.1 Dielectric Test on the Main Circuit

This is described under Type Test (19.2.1.1)

19.1.2 Dielectric Test on Auxiliary and Control Circuits

Auxiliary and control circuit shall be subjected to short duration power frequency voltage withstand test. Test is performed in following conditions—

(a) Between auxiliary and control circuits connected together as a whole and the frame of the switching device.

(b) It practicable between each part of the auxiliary and control circuits which in normal use may be insulated from the parts and the other parts connected together to the frame.

Test voltage shall be 2000 V. The test shall be performed according to IEC 1180-1 with a duration of 1 min. The auxiliary circuits of the switchgear and controlgear shall be considered to have passed the test. If no disruitive discharge occur during the test.

Normally the test voltage of motors and other devices used in auxiliary and control circuit shall be same as the test voltage of those circuits. If such apparatus has already been tested in accordance with appropriate specification, it may be disconnected for those tests. But the test voltage may be higher with the agreement between manufacturer and the user.

19.1.3 Measurement of the Resistance of the Main Circuit

The measurements are taken as a dc voltage drop in circuit. The applied current will be minimum 100 A and maximum up to the rated current. The maximum measured value can be 1.2 times of the value obtained at the temperature rise test.

19.1.4 Tightness Test

This test is predominantly carried out for switchgear which is of sealed types, i.e. GIS. In this case, the leakage rate has to be measured. It will decide the life span of the switchgear at site and facilitate a decision regarding its maintenance.

19.1.5 Design and Visual Checks

The circuit breaker should be checked to verify its compliance with the order specification.

The following items should be checked as applicable:
— Language and data on the nameplates;
— Identification of any auxiliary equipment;
— Colour and quality of paint and corrosion protection of metallic surfaces; and
— Values of the resistors and capacitors connected to the main circuit.

19.1.6 Mechanical Operation Tests

Mechanical operation tests include the following:
(a) At specified maximum supply voltage of the operating devices and of auxiliary and control circuits and maximum pressure for operation (if applicable): five closing operations, five opening operations.
(b) At specified minimum supply voltage of operating devices and of auxiliary and control circuits and minimum functional pressure for operation (if applicable): five closing operations, five opening operations.
(c) At rated supply voltage of operating devices and of auxiliary and control circuits and rated pressure for operation (if applicable): five close-open operating cycles with the tripping

mechanism energised by the closing of the main contacts; moreover, for circuit breakers intended for rapid auto-reclosing, five open close operating cycles O-t-C where t shall be not more than the time interval specified for the rated operating sequence.

Mechanical operation tests should be made on the complete circuit breaker.

However, when circuit breakers are assembled and shipped as separate units, routine tests may be performed on the components. In such cases, the manufacturer should produce a programme of commissioning tests for use at site to confirm the compatibility of such separate units and components when assembled as a circuit breaker. For all the required operating sequences, the following tests should be performed and records made of the closing and opening operations: measurement of operating times, where applicable, and measurement of fluid compression during operations, e.g. pressure difference.

Proof should be given that the mechanical behaviour conforms to that of the test specimen used for type testing. For example, a no-load operating cycle can be performed to record the no-load travel curves at the end of the routine tests. Where this is done, the curve should be within the prescribed envelope of the reference mechanical travel characteristic, as defined during the type test, from the instant of contact separation to the end of the contact travel. Where the mechanical routine tests are performed on sub-assemblies, the reference mechanical travel characteristics should be confirmed to be correct, as above, at the end of the commissioning tests on site. If the measurement is performed on site, the manufacturer should state the preferred measuring procedure. If other procedures are used, the results may be different and the comparison of the instantaneous contact stroke may be impossible to achieve. The mechanical travel characteristics can be recorded directly, using a travel transducer or similar device on the circuit breaker contact system or at other convenient locations on the drive to the contact system where there is a direct connection, and a representative image of the contact stroke can be achieved. The mechanical travel characteristics should preferably be in the form of a continuous curve. Where the measurements are taken on site, other methods may be applied which record points of travel during the operating period. In these circumstances, the number of points recorded should be sufficient to derive the time to, and contact speed at, contact touch and contact separation, together with the total travel time. After completion of the required operating sequences, the following tests and inspections should be performed (if applicable):

— Connections should be checked;

— The control and/or auxiliary switches should correctly indicate the open and closed positions of the circuit breaker; and

— All auxiliary equipment should operate correctly within the limits of supply voltage of operation.

19.2 TYPE TESTS

All type tests should be carried out by using the number of test samples specified in clause No. 6.1.1 of IEC 60694 and 6.102.2 of IEC 62271-100.

19.2.1 Dielectric Tests

The power system experiences occasional temporary power frequency over-voltages, which arises during load throw, wrong OLTC operation, insufficient shunt compensation, resonance, etc.

These tests are performed to verify the rated insulation strength of the switchgear to ascertain that when a breaker is put into service, its design is capable of withstanding over-voltages occurring due to above reasons and due to lightning, switching operations, etc. This is verified in compliance with the Standards. The tests which are performed under this category are discussed below.

19.2.1.1 One Minute Dry Power Frequency Voltage Withstand Test

This test is carried out to verify the capability of the equipment to withstand the power frequency test voltage for one minute in dry condition. Power frequency test voltage values with respect to the system voltage have been identified in the standards.

19.2.1.2 One Minute Wet Power Frequency Voltage Withstand Test

This is the same as the one minute dry power frequency voltage test, but is conducted with equipment in wet condition. These are applicable for outdoor installations only.

19.2.1.3 Lightning Impulse Voltage Dry Withstand Test

This test is conducted to verify whether the switchgear is able to withstand over-voltage due to the peak value of standard impulse (1.2/50 micro-sec) during lightning.

19.2.1.4 Switching Impulse Voltage Test (Optional)

This test is conducted to verify whether the switchgear is able to withstand over-voltages due to switching surges. These tests are significant for system voltages above 300 kV. Switching surges occur during opening and closing of the unloaded EHV ac lines, breaking inductive/capacitive loads. Switching surges are comparatively of a longer duration (2500 micro-sec), lower rate of rise and are represented by standard switching impulse test wave of 250/2500 micro-sec.

19.2.1.5 Partial Discharge Test (Optional)

This is a component test and is not recommended to be carried out on complete switchgear, whererin the design of the switchgear consists of a combination of conventional components (e.g. CT and PT, etc.) which can be tested in accordance with their respective Standards. But in the case of switchgear wherein organic insulating material is used, then this test is recommended, such as in the case of integrated switchgear design, specially GIS, where live parts and connections are embedded in solid insulation.

19.2.1.6 Artificial Pollution Tests

These tests are applicable for outdoor installations only and are carried out on the basis of an agreement between the user and the manufacturer. The voltage values for the above tests are specified in standards against the system voltage.

19.2.1.7 Design Aspects

While designing a switchgear, the parts located at different electric potential are to be separated by insulation in order to ensure the safety of personnel working on it and the reliability of its operations, to prevent phase-to-earth, or phase-to-phase flashover.

The insulation in switchgear serves three main purposes. It provides insulation:

1. Between current-carrying live parts and earth;
2. In contact gap during 'breaker open' condition; and
3. Between current-carrying live parts of different phases.

A wide range of materials is used for the purpose of achieving the above stated insulation. These materials can be solid, liquid or gaseous. They are used to fill spaces, support live parts and to extinguish arc depending upon their physical properties and form.

(a) Gaseous Air is the most commonly used gaseous material, which is composed of 80 per cent nitrogen and 20 per cent oxygen. Its properties are therefore close to nitrogen. Air clearances between phases and between phases to earth against the system voltage are as mentioned in: IEC 60694 and IEC 62271-100.

A recent addition to the list of gases used to fill the switchgear enclosure is SF6 which has a dielectric strength of approximately three times that of air. It serves the dual purpose of acting as both an insulation medium as well as an arc-quenching medium. Among gases, air is the only insulating material which can be used effectively at atmospheric pressure.

(b) Fluids A range of fluids has been used for insulation in switchgear. Hydrogen carbon oil often referred to as 'transformer oil' was being used in bulk oil CB. It has the advantage of fulfilling the dual role of both an insulation as well as an arc-extinguishing medium. Its dielectric strength is three times that of SF6 at atmospheric pressure. However, it is now coming under increasing scrutiny from the safety point of view because of its inflammable property.

(c) Vacuum Another medium used for insulation is vacuum. This is very expensive, that's why its use in distribution switchgear is restricted only to circuit interruption devices.

(d) Solid Materials Solid insulating materials include many of natural origin such as mica, asbestos, slate or those derived from natural material such as porcelain and shellac varnish. The principal use of solid materials in switchgear is to insulate conductors where they pass through walls of metal enclosures apart from supporting the conductors. Most of the support insulators are made by resin cast technique. One major advantage of the resin casting technique is the almost limitless range of shapes that it offers. The resin is usually epoxied.

The reasons for the failure of various insulation media in switchgear are:

(a) Tracking of Solid Insulation This takes place due to degradation of the surface glaze by sparking, presence of conductor film on surface due to moisture, etc. In order to prevent this eventuality, the insulation should be clean and undamaged.

(b) Breakdown by Leakage in Solid Insulating Material Surface discharge occurs when the insulator is subjected to higher electric stress, which is greater than the permitted value. Cavities, poor design, presence of moisture, etc. also cause discharge within the insulator. In order to avoid this during the manufacturing process of the insulator, care should be taken to make it void-free with maximum tracking index of that material.

(c) Thermal Breakdown Heat is generated in the electrically stressed insulating parts due to dielectric loss. Heat is imparted to the insulation by neighbouring current-carrying parts, which is lost by it through conduction, convection and radiation. The resistivity of some dielectrics reduces with an increase in temperature. Thermal equilibrium implies a stable temperature at which heat lost = heat gained. If thermal stability is lost due to inadequate cooling, temperature rises indefinitely leading to the insulation failure. In order to avoid this, appropriate cross-section of the current-carrying path should be selected so that the temperature rise in the switchgear is limited within the specified limit by forced/natural cooling.

(d) Failure of Chemical Insulation Moisture causes deterioration of transformer oil and other chemical insulating materials. Moisture gets condensed on the insulator surface and causes flash-over. This can be prevented if ingress of the dust particles and moisture, is restricted.

(e) Effects of Oxygen and Humidity Some organic/inorganic materials oxidise in the presence of oxygen when exposed to light. Rubber oxidises and cracks when exposed to light. Epoxy insulators are not suitable for outdoor applications. Some materials absorb moisture and lose electrical and mechanical strength under humid conditions due to hydrolysis.

(f) Incompatibility of Dielectric Material Some dielectric materials are not suitable for use in certain assemblies due to their incompatibility with the surrounding substance. For example, all the materials are not suitable with SF6 media and care should be taken to select compatible insulating material.

(g) Electrochemical Deterioration In some insulating materials, the impurities get dissociated under electric stress during ionisation of the material and consequently the material deteriorates.

(h) Presence of Arc The insulating gas/oil gets decomposed in the presence of arc. Although the products of decomposition re-combine after arc extinction, some remainders are there, which is why the insulating properties of dielectric get affected.

19.2.2 Radio Interference Test

These tests apply only to switchgear and controlgear having a rated voltage of 123 kV and above, and should be conducted when specified in the relevant standard.

The test voltage shall be applied as per IEC 60694 under following conditions:

(a) In closed position between one terminal and the earthed frame; and
(b) In open position between one terminal and the other terminal connected to the earth frame and then with the connection reversed if the switching device is not symmetrical.

The case, tank base and other normally earthed parts should be connected to earth. Care should be taken to avoid influencing the measurement by earthed or unearthed objects near the switchgear and controlgear and to the test measuring circuit. The switchgear and controlgear should be dry and clean at approximately the same temperature of the room in which the test is carried out. It should not be subjected to any other dielectric test within a period of two hours prior to the present test. The test connections and their ends should not be a source of radio interference voltage of higher values than those indicated in the standard. The measuring circuit should preferably be tuned to a frequency within 10 per cent and 0.5 MHz but other frequencies in the range of 0.5 to 2 MHz may be used. The result should be expressed in microvolts.

19.2.3 Measurement of the Resistance of Main Circuit

The function of switchgear is to carry its rated current through the current-carrying conductor. They have many joints to meet the design parameter. Each joint will have joining resistance. Care should be taken to minimise the total resistance of the circuit, i.e. the total watt loss should be minimum for a low rise in temperature. The measured value will set the limiting parameter during the routine test. The measurements are taken as a dc voltage drop in circuit. The minimum applied current will be 100 A and the maximum up to the rated current.

19.2.4 Temperature Rise Test

This test determines the highest normal current that the switchgear can carry without exceeding the maximum permitted temperature at any point. Since the current carried affects the temperature rise, the current rating is dependent upon the maximum ambient temperature. Sometimes one temperature limit prevails in determining the current rating, and different countries have different limits for similar materials. For example, the ANSI standard states that any part that can be handled by an operator in the normal course of his duties must not exceed 500°C which could become a limitation if solar heating is taken into account. No such limits are quoted in the British or IEC standards, though common sense also plays a part in switchgear design, but even lower temperature limits are given for handles in the BS for LV Switchgear. Also the IEC standard gives 1000°C as the highest permitted temperature for metallic parts in contact with oil, whereas the ANSI document limits this to 90°C.

This test is to verify the normal rated current-carrying capacity of equipment. The rated ac current at rated frequency is passed through the CB in closed condition continuously and temperature at different locations is measured by using thermocouples till steady temperature is achieved. At steady temperature, the temperature measured at different points should be within the limits as specified in the standards. When ac current is passed through a breaker, heat is generated due to I^2Rt loss. The generated heat is dissipated by conduction, convection and radiation. In conduction, heat travels through the solid body, convection uses liquid media and radiation implies transmission through gases.

In order to maintain the temperature rise within specified limits, $I^2 Rt$ loss should be reduced by increasing the conductor cross-section using suitable low resistivity material and by improving heat transfer through conduction, convection and radiation of heat.

Design Aspects

Two materials, namely copper and aluminium, are the most suitable materials for making current-carrying conductors. Copper has lower resistivity and also resists oxidation better than aluminium, particularly at temperatures below about 80°C. Aluminium has a much smaller specific gravity and though larger cross-sections of conductor are needed to give the same resistance as the equivalent of copper, the weight and therefore the cost of aluminium is found advantageous. Following table shows differences between relative properties of copper and aluminium

Properties	Copper	Aluminium
Conductivity for equal areas	1.0	0.50
Electrical	1.0	0.61
Thermal	1.0	0.56
Tensile strength (hard cloured)	1.0	0.40
Hardness (hard drawn)	1.0	0.44
Modulus of elasticity	1.0	0.55
Coefficient and thermal expansion	1.0	1.39
Melting point	1.0	0.61

Table 19.1 Differences between Relative Properties of Copper and Aluminium

For equal conductivity, aluminium is lighter in weight than copper. In many switchboard applications however space considerations are more important than weight and copper is generally preferred over aluminium.

The most important characteristic of the materials used as conductors in switchgear is that they should have a low specific resistance. Excessive temperature rise due to high resistance conductors in a small enclosure will lead to a reduction in the life of the associated insulation materials. The next important characteristic is that the material must resist corrosion. Any corrosion in the joint area leads to an increase in the resistance and higher resistance leads to higher temperature, which, in turn, leads to more corrosion.

The material with least resistance is silver but this is ruled out in switchgear applications because of its cost. However, it does find extensive application in switchgear equipment as a thin coating and in small, light current units in the form of small contact inserts. The coatings are always applied by electroplating and their typical thicknesses are in the range of 0.02–0.05 mm. Coating of the greater thickness is used when sliding motion takes place between the plated surfaces. The use of silver coating permits the conductor to be used for higher currents with the same temperature rise as the base material. It also prevents oxide formation, which imparts longer life to the conductor.

Usually the grade of copper used in switchgear is of higher conductivity. Copper is available in many forms as rods, bars, etc. Aluminium in its pure state is a very soft material. Its use in switchgear is usually restricted to conductors, typically busbars. Aluminium oxide is a good insulator. It is therefore important to follow good jointing practices when aluminium conductors are being bolted together. When bolt pressure is applied, the resistance initially reduces substantially as the oxide filter is crushed and the area of contact around each bolt increases. Then a limit is reached beyond which further tightening causes very little reduction in resistance so a properly designed joint does not need excessive tightening.

Problem arises from the use of aluminium conductor in association with copper, as the electrochemical potentials of these two elements are widely separated. If copper to aluminium joint is allowed to become moist then electrochemical corrosion will begin to take place. In order to avoid this phenomenon, it is customary to coat such bi-metal junctions with a suitable material to avoid moisture. Alternately, jointing methods, which avoid bolted faces of dissimilar material, can be used. It is now possible to electroplate aluminium with either copper or silver by using a special process.

In vacuum interrupters, contact shape is selected to meet the required rated current. For rated current, the temperature rise should be within limits. For this purpose, the control stem should have a sufficiently large diameter to dissipate the heat by conduction to the external heat sinks.

The resistance of the bolted joint depends upon the current distribution in the joint face and live resistance of the joint face itself. The current distribution in the joint is not uniform and the current transfer tends to concentrate towards the ends of the overlap. The joint resistance depends upon surface condition and the joint pressure.

19.2.5 Short Time Current Withstand Test

This test verifies the capability of a switchgear/circuit breaker to carry the specified short-circuit current for a rated duration of 1 or 3 second(s).

Under a particular arrangement of switching, the short-circuit current at the time of the system fault flows through one or more switchgears in series in a group feeder scheme where only one breaker is arranged to open automatically. Those breakers which remain closed during that period must therefore be capable of carrying the through current until fault is cleared elsewhere.

When short-circuit current flows through the circuit breaker, different current-carrying parts are subjected to mechanical and thermal stresses. The insulators in the vicinity of the conductor are severely thermally stressed.

For test, the rated short-circuit, current is passed through the breakers in closed condition for a rated time of 1 or 3 seconds and the oscillogram is taken. The short-circuit current test verifies that due to these stresses, parts of the CB should not get damaged. The circuit breaker should not emit any flame or smoke. The contacts should not get welded and the circuit breaker should be capable of opening and closing after the short-circuit test. Visual inspection and no load operation immediately after the short circuit test are usually carried out.

The electrodynamic forces between adjacent phase conductors are proportional to phase spacing. During short time current, the insulator support experiences impact cantilever force due to electrodynamic forces. The contacts experience temperature and stresses proportional to I^2RT. The resistance R depends upon contact pressure and surface condition.

The bus bars are generally selected on the basis of mechanical considerations of strength and supporting arrangement. When spacing in busbars is small, mechanical forces become significant. During a short-circuit test, at a temperature of about 160°C, aluminium becomes softer and loses its mechanical strength. This sets a limit on the permissible temperature rise during a short-circuit condition.

The dynamic force occurs at the peak of first major loop on short-circuit. Perpendicular conductors tend to straighten out due to electromagnetic forces. The insulators are selected by considering mechanical bending load occurring at that instant of the peak short-circuit current. During a short-circuit fault in the system, short-circuit current flows through the busbars. The insulator supporting the busbars experiences the bending force. The insulator should thus have enough cantilever strength to withstand dynamic force occurring during a short-circuit.

When current flows in two parallel conductors, there is mechanical force of attraction between the conductors when currents are in the same direction, and of repulsion, when currents are in opposite direction. While carrying normal current, these forces are negligible but under short-circuit conditions, they may be appreciable and it is necessary to ensure that the busbar supports can withstand these forces even though they are of momentary duration. These forces may damage the insulators.

It is extremely desirable that the support insulators should be arranged that they are not subjected to lateral forces. The busbar configuration should be designed in such a fashion that minimum forces are generated during a short-circuit.

When circuit breakers are subjected to the short time current test, the contacts may have a tendency to open at the peak current. Therefore adequate contact pressure should be applied to avoid the undesirable opening during the test. After the test, a no-load operation of the breakers should be done. After the test, the breaker should be capable of making and breaking rated fault current. However, this test may not be actually carried out and it is usually found adequate to check the condition of contact on no-load operation. The circuit breaker should also not show any sign of distress.

19.2.6 Verification of the Degree of Protection Test

The design of switchgear depends upon application and installation. Degree of protection tests are conducted accordingly, so that, designed enclosures are safe from the action of environment and wrong application.

19.2.6.1 Verification of IP Coding

This test is carried out to verify the design wherein the performance of mounted equipment is not impaired due to environmental conditions as per IEC 60694.

19.2.6.2 Tightness Test

The tests and specific guidelines have been given in the IEC to check the tightness of vacuum and SF6 circuit breakers. The aim of these guidelines is to show that the total leakage from the system does not exceed the specified leakage rate. For the purpose of identification, three categories have been constituted. These are detailed below.

19.2.6.2.1 Controlled Pressure System In this system the air or gas is automatically refilled from an external or internal source, e.g. air blast circuit breaker and SF6 double pressure circuit breaker and pneumatic operating mechanism. The leakage rate should be measured by the pressure drop over a period and expressed as a percentage of the rate pressure per day or per year. In case of air blast circuit breakers, the test should be carried out in breaker open and close positions.

19.2.6.2.2 Closed Pressure System In this system the gas is refilled periodically by manual external source (SF6 CBS). The leakage shall be measured by the pressure drop over a period and expressed as a percentage. This is normally used in SF6 circuit breakers.

19.2.6.2.3 Sealed System This is sealed for life and no further processing is expected. This can be further divided into sealed pressure system and sealed vacuum system as described below:

(i) If switchgear uses gas then close pressure system/SF6 sealed systems should be used. Any one method can be selected to find out the leakage rate and the time between successive refilling. In these type of systems, since the leakage rates are comparatively smaller, the pressure drop measurements are not possible. Therefore special type of leak meters should be used to detect the leak rate. In case, it is difficult to measure the leak rate of the combined unit, then the leak rates of different parts can be measured separately and the sum total should remain within the specified leakage rate for the complete system. The leak meter should also be calibrated regularly to mention the accuracy.

(ii) Vacuum sealed system (vacuum interrupters): The internal pressure of the vacuum interrupter is measured by the principle of cold cathode ionization process. However the relationship between the vacuum level and electrical parameters will have to be established for each type of vacuum interrupter at the manufacturing stage only. The leakage rate evaluated in this way will be such that the vacuum level will not reach its threshold value (given by the manufacturer) during its lifetime.

Similarly the tightness criteria are defined for liquid tightness also. The leak rate can be evaluated by pressure drop method.

19.2.7 Electromagnetic Compatibility (EMC) Test

If the functional characteristics of electrical or electronic equipment are not disturbed due to the presence of an EMI, the equipment is said to have EMC. This is applicable for the secondary system of the main equipment. The sources of EMI are:

1. Electronic systems like data acquisition system (DAS), distributed digital control system, and PLCs, etc.
2. Electrical systems like thyristor control drives, generators, relays, arc welding, etc.
3. Broadcast systems like TV/radio transmitters, ration, raiders, walkie/talkie; and
4. Lightning, nuclear explosion.

19.2.8 Mechanical Operation Tests at Ambient Temperature

The mechanism of circuit breaker plays a vital role in the switchgear. The reliability of a switchgear depends entirely upon the successful operations of its mechanism. Hence, mechanical endurance test is performed on switchgears to check its reliability. In order to check the mechanical operations, the tests are also carried out at the lower limits of supply voltage, wherein the energy input is sufficient to perform all short-circuit tests without any problem, while the upper limits demonstrate that the energy input is not damaging the parts of the circuit breaker. Apart from usual type tests, it is recommended to check the consumption of operating oils, gas densities, gas pressure, resistance of main circuit and tightness of gas and vacuum. During the test, the circuit breaker should be capable of carrying its rated normal current, making and breaking rated short-circuit currents and withstanding the over voltage level as per the rated voltage of the system.

These operations are performed at no-load as per the procedure mentioned in Standards. The values for speed, travel, opening, and closing time are measured before and after the endurance test, and are compared. These values should be within the limits specified in the Standards.

Design Aspect

In the mechanical endurance test, the mechanism of the switchgear plays the most important role. Thus while designing the switchgear, proper care should be taken to minimise the stresses on the components and adequate factor of safety should be taken for performing a minimum of 30,000 or even more operations in its lifetime. Nowadays sealed type breakers incorporating the concept of use and throw are entering the market. Here the reliability of the mechanism becomes the most important aspect for determining the performance of the switchgear where maintenance is not required for its entire lifetime.

19.2.9 Short-circuit Duties Test

This test verifies that the circuit breaker will reliably interrupt complete range of fault currents up to their maximum short-circuit rating. The different short-circuit or terminal fault test duties simulate variety of fault situations during its service.

The new IEC has revised its recommendation for carrying out terminal fault test duties in view of correctly stressing the circuit breaker having short arcing times. Further, the concept and standard values of initial transient recovery voltage (ITRV) have also been included. The initial period of transient recovery voltage (TRV) may be of importance for some types of circuit breakers and are caused by the initial oscillation of a small magnitude due to reflection from the first major discontinuity along the busbar.

Although the amplitude of recovery voltages are smaller, they are associated with higher rise time in the order of one micro-second. The wave shape of the TRV during this period is referred to as the initial TRV. The phenomena is similar to transient recovery voltage appearing during short line fault tests.

It has been observed that during the testing of circuit breakers of short arcing times, there may be great variations in the actual severity in the same circuit depending upon the incident of contact separation on the current wave. The maximum severity condition during the test can be achieved in the testing procedure. A number of tests required to achieve valid operation may thus be more than prescribed.

Transient Recovery Voltage (TRV) The actual voltage appearing between the circuit breaker after final arc interruption is called TRV which oscillates at high frequency due to the set circuit parameter as called in respective testing Standards. The TRV oscillation usually contains many frequencies because of the complicated network i.e. combination of inductance and capacitance. Under the short circuit condition, faulty circuit is inductive and at current zero, the voltage is maximum. The distributed capacitance of the faulty circuit gets charged to the peak value of the voltage. The interruption of the current between the contact gap resulting to high frequency oscillation with voltage peak approximately twice the circuit voltage, known as transient recovery voltage (TRV).

A number of methods have been given in the IEC to determine the perspective TRV of the circuits of the testing station and power systems. Due consideration should be given to the exponential components superimposed on the power frequency component of the recovery voltage along with the single or multi-frequency component appearing during the transient period. The exponential components have a depressing effect on the peak TRV. The various methods for evaluating the prospective TRV are:

— Direct short-circuit breaking;
— Power frequency current injection;

— Capacitive current injection;
— Model network;
— Calculation from circuit parameters; and
— No load switching of test circuits including transformers.

19.2.9.1 Test Duty—T10

This duty simulates the conditions when the breaker is connected to the primary circuit of a transformer and a fault occurs on the secondary side. The fault current is 10 per cent of the rated short-circuit breaking current. Under these conditions, the rate of rise of recovery voltage and recovery peak voltage is high.

19.2.9.2 Test Duty—T30

This duty represents a situation wherein the faults occur with only the local source in an interconnected system. The fault current is 30 per cent of the rated short-circuit breaking current of the breaker.

19.2.9.3 Test Duty—T60

This duty is carried out to prove the capability of the breaker to withstand a higher rate of rise of TRV at reduced current as compared to 100 per cent short-circuit current. The fault current is 60 per cent of the rated short-circuit breaking current of the breaker.

19.2.9.4 Test Duty—T100s

This duty demonstrates the following:

— The ability of the breaker to interrupt the rated symmetrical short-circuit breaking current at full TRV;
— Reclosing capability to the fault; and
— Ability to perform the rated operating cycles.

The fault current is 100 per cent of the rated short circuit breaking current of the breaker.

19.2.9.5 Test Duty—T100a

During testing of the circuit breakers with short arcing time, it has been noticed that the severity on the circuit breakers will differ if the phase in which the specified dc component exists, when it clears the fault after a minor or major loop. In actual fact, the contact separation in the event of faults may take place at a moment close to the minor loop so that extinction takes place only after arcing during the full subsequent major loop. The arc energy seen by the breaker in this case will be much larger than that compared to the case wherein extinction takes place only after arcing during a minor loop. The sequence of test duties has been devised to arrive at a series of valid tests, which represent the maximum stressed condition with respect to the arc energy input and transient recovery voltage. Some circuit breakers will not clear fault at the end of a major loop. Arcing then continues during the subsequent minor loop and this phase becomes a last phase to clear. However, this test may be considered valid if during a subsequent test it is proven that the longest possible arc duration has been achieved. This duty is not particularly severe for the breaker from the TRV point of view, since due to asymmetry, the current zero arc has shifted from its place, and the condition of voltage peak at current zero is not achieved. However, the maximum severity can occur when the phase with the required dc component is the first phase to clear. This is due to the fact that a higher amount of arc energy in the arc, increases the temperature of the arc column, thereby making the interruption, first phase to clear when combined with the transient recovery voltage. If the interruption in the phase with the required dc component is not the first pole, to clear the test operation will be counted valid, only if, this phase clear arcing during extended major loop. The single-phase test with dc component is performed to test the circuit breaker under maximum arc energy condition then first operation is valid if the interruption occurs after a major loop. The contact operation for this purpose shall be achieved in the preceding minor loop or even earlier. The breaker proves the capability for:

1. Sufficient operating energy with mechanism;
2. Interrupting capability after maximum arc energy; and
3. Exact voltage co-ordination between the main and arcing contact.

A high dc component exists close to the generator centre, hence the generator breaker must have more than an 80 per cent dc component.

19.2.9.6 Single-phase Short-circuit

This duty simulates such a condition and verifies that the unbalanced forces produced due to single-phasing do not adversely affect the operation of the circuit breaker. This test is applicable only to three pole breakers intended for use on an effectively earthed system, with the three poles coupled mechanically and fitted with a common opening release.

19.2.10 Capacitive Current Switching Tests

These include the line-charging current breaking tests ($U_r > 72.5$ kV) and the cable-charging current breaking tests ($U_r < 52$ kV).

The main objective of the capacitive switching test is to demonstrate the ability of the breaker to withstand excessive over-voltage that is produced during the capacitive switching.

The capacitive current switching tests include:
— Rated line charging breaking current test;
— Rated cable charging breaking current test;
— Rated single capacitor bank breaking current test;
— Rated back-to-back capacitor breaking current test; and
— Rated capacitor bank inrush making current test.

The line charging current switching test is not applicable for the circuit breaker having the system voltage less than 72.5 kV. Similarly cable charging current breaking test is considered unnecessary for circuit breakers having system voltage equal to or less then 24 kV. But this test can be conducted on request.

In circumstances when any switching device has to interrupt the capacitance current, the current-breaking duty is calculated and the impedance that will limit inrush currents on closing are also estimated. Particularly important situation is one where, there is more than one capacitor banks to be connected in the system. When one bank is already connected and second bank is going to be switched on in the system which is known as back to back switching and if length of the conductor between one bank and the other is relatively short, then it is very important to ensure that the inrush current when the second bank is connected does not exceed the rated making current of the switchgear.

The reliability of the circuit breakers and over-voltage generation are of probabilistic in nature, hence the performance of the circuit breaker during this test is divided into two parts, including:

1. Class C1 circuit breaker with low probability of re-strike; and
2. Class C2 circuit breaker with very low probability of re-strike.

19.2.11 Environmental Test

(A) Normal Service Conditions The international standard IEC 60694 establishes two sets of normal conditions under which 'standard' Switchgear is deemed to be installed, one for indoor Switchgear and the other for outdoor.

(a) For indoor switchgear the parameters are:

1. Ambient temperature not to exceed 40°C and the average over a 24-hour period not to exceed 35°C;
2. Two classes for minimum temperature, one of –50°C and the other, for more severe environments, down to –25°C;
3. Maximum altitude 1000 m (3,300 ft);
4. The average value of the humidity not to exceed 90 per cent over a month for 95 per cent over any 24-hour period; and
5. Vibrations external to the switchgear, or earth tremors, being negligible.

   The American standards do not specify any condition for humidity and make no distinction between indoor and outdoor switchgear so far as the minimum temperature is concerned, which in the ANSI standard C37.04 is –30°C.

(b) For outdoor circuit breaker the parameters are:

International standard service conditions for outdoor equipment are the same as those for indoor switchgear as far as maximum temperatures, altitude and vibration requirements are concerned. There are again two classes for the minimum temperatures, i.e. –25°C and –40°C.

The additional conditions for outdoor circuit breaker are:

1. Three classes of permissible ice coating are recognised, viz. those that are 1 mm, 10 mm and 20 mm thick respectively;
2. Wind pressure not to exceed 700 Pa (equivalent to a wind speed of 34 m/s); and
3. Account should be taken of rain, condensation, rapid temperature changes, solar radiation and ambient air pollution.

(B) Abnormal Service Conditions

(a) *Outdoor Sites*: Air pollution in the form of ordinary dust, pollens, etc., can be expected not to cause a problem as periodic rain usually washes the insulators free of this and in any case, the design allows for a moderate degree of inert dust pollution. It is when this pollution is of an industrial nature and may be corrosive or abrasive, or the site is near the sea and subject to salt spray, that such circumstances should be brought to the attention of the potential supplier of the switchgear. In the case of outdoor switchgear, regular insulator washing may be prescribed in order to avoid problems.

If the site is subject to earthquakes, then again special precautions need to be taken, and the manufacturer will need to produce evidence of seismic tests.

(b) *Indoor Installations*: For any potential indoor sub-station site consideration needs to be given to whether any of the environmental conditions is more onerous than those envisaged in the specification. Then, attention of potential suppliers should be drawn to the condition(s) or action should be taken to make the environment within the sub-station comply with the 'standard' conditions.

(c) *High Humidity*: In a design of switchgear using porcelain as a main insulation material, relatively frequent condensation may not be a problem to the life of the insulation, but if another material were to be used, the manufacturer would probably fit electric heaters in each panel close to any vulnerable insulation to prevent condensation. In this connection, it is interesting to note that though IEC 60694, considers that with a permitted maximum level of 95 per cent relative humidity, condensation may occasionally occur, IEC 60298 considers that it does not normally occur. Experts in this country believe this to be wrong, as any environment where temperature and humidity are high by day must inevitably be prone to condensation when the temperature falls at night, since the dew point must also be high. This is where the heaters are used to control the humid state. As an alternative, switching on the heaters, at the start of the 'rainy season', and then leaving them 'ON' permanently until the risk is over, is also practised.

(d) *High Temperature*: If both the maximum temperature and humidity are more onerous than the standard, air conditioning of the sub-station could be considered, and may prove to be more economical than specifying special switchgear. Ambient temperatures higher than 40°C are common in many tropical areas of the world. It is usual to ask the manufacturer to compensate for this, which he does by de-rating the switchgear.

Although heat generation is proportional to the square of the current, the heat dissipation is not directly proportional to heat generation. Thus the de-rating factor used will be found to be proportional to current to the power 1.8 or something similar.

(e) *Altitude*: With reduction in air density as altitude increases, the electrical strength of air gaps decreases and the effectiveness of air as a cooling medium also decreases. Therefore, the voltage rating and current ratings both need to be re-appraised, if the altitude is significantly above sea level. The ANSI specification C37.04 indicates reduction factors for maximum rated voltage, BIL and rated normal current at different altitudes. At 1500 m, the voltage ratings are reduced to 95 per cent and the current to 99 per cent. At 3000 m, the corresponding figures are 80 per cent and 96 per cent.

(f) *Excessive Dust or Vermin*: If the site is subjected to heavy dust deposits, which may be corrosive, or conducting, e.g. coal dust or carbon black, then special steps need to be taken to protect the switchgear. Depending upon other factors, fitting effective filters to ventilation openings and arranging air lock doors may be sufficient, or air conditioning may be a wise precaution. Many tropical sites are subject to pollution by vermin, including insects, snakes and other reptiles. It is important to pay attention to the degree of protection, and to operational and maintenance procedures to ensure that the protection is retained when a task is finished, in order to avoid trouble from this cause. One point to be noted is that a reptile may get through a small opening when it is young, and then feed on small insects, until it is large enough to be dangerous!

(g) *Vibration*: On an industrial site, vibration can arise from nearby machinery and affect the sensitive parts of circuit breaker mechanism, causing it to operate inadvertently. Trip latches are usually balanced dynamically to make them insensitive to such vibration. The worst form of vibration is that associated with earth tremors. The need for switchgear to be able to withstand seismic forces is increasing with the high safety standards being applied on nuclear power stations. It is common in the UK for a nuclear site contractor to present the switchgear manufacturer with the seismic profile of the sub-station building and for tests to be performed on a seismic test table to prove that the cubicles, etc. can withstand the resultant forces.

The frequency of the vibrations is usually low, in the order of 1–20 Hz, and these can often excite resonance in the framework, or particularly in components such as instrument panels. Ideally, it is more economical to establish minimum seismic withstand conditions for the switchgear, and then to adjust the sub-station seismic profile to match the latter, thus putting seismic conditions in the same category as temperature and humidity. This matter is under discussion in several places.

On the basis of the application, the following tests are recommended and detailed procedures are given in the respective standards:

- Extended mechanical endurance tests on circuit breakers for special service conditions;
- Low and high temperature tests;
- Humidity tests;
- Static terminal load tests; and
- Test of prove operation under severe ice conditions.

19.2.12 Critical Current Test

This test is applicable for those breakers whose arcing time is significantly higher at test duty T10 with respect to test duty T30, T60. Normally this phenomena comes with an MOCB breaker or EHV SF6 circuit breaker.

19.2.13 Short Line Fault Test

From the short-circuit current point of view, a fault at the terminal of the circuit breaker is considered more severe. However, for certain conditions it is not so as far as transient recovery voltages are concerned. Under short line fault (fault occurring at a few kilometers from the circuit breaker), the line side recovery voltage is a saw tooth wave resulting from a series of travelling waves reflected back. And along the line between the circuit breaker and the fault, this causes a sharp increase in the rate of rise of recovery voltage in the initial stages.

A short line fault test is carried out only on the circuit breakers which are designed for direct connection to transmission line and service voltages of 52 kV and above with the short-circuit current exceeding 12.5 kA. The testing is done in single-phase for an effectively earthed system only since this is the condition which produces maximum strain on the circuit breaker while clearing a short line fault, under single-phase-to-earth fault condition. While performing the test due to limitations of the testing station to provide ITRV, it is permissible to increase the voltage to the first peak of the line side component to compensate for the ITRV deficit at source.

19.2.14 Out-of-Phase Making and Breaking Tests

It is unusual for out-of-phase switching operations to arise on distribution systems, and it is rare to find a specification asking for proof of this feature. It is perhaps possible on large industrial sites, which have local generation and where there is connection to a local utility as well. If this is the case, it is important to check whether parallel switching conditions can be prevented, or if there is a risk that phase shifts between the system voltages on the two sides of any circuit breaker can occur, then the manufacturer must be made aware of this special requirement. A circuit breaker of higher voltage rating may well be required to meet this duty.

19.3 SPECIAL TESTS

For certain applications user may ask following special tests to be performed on switchgear:

(a) Special Electrical Endurance Test Based on the application of the breaker, the user (e.g. m/s Railways) may demand additional electrical type test. If the breaker is used only for load breaking, the breaker is subjected to 500 load-breaking/making electrical endurance without any maintenance.

Similarly, if the breaker is being used for short-circuit breaking/making application, the test requirement is different. The breacler is tested for 200 electrical close/open operation of 6.25 kA with duty cycle close-open-15 secs-close-open. This has to be performed without any maintenance.

(b) Vibration Test This test is based on the application of the equipment. If equipment is being used near the vibrating machine e.g. rolling stock application, then this test has to be performed. The test values are given below.

- Maximum vertical acceleration = 1.0 g
- "Longitudinal" = 3.0 g
- "Transverse" = 3.0 g

(c) Higher BIL Requirement Although applicable standards specify the BIL level, however, some of the users do specify a higher BIL requirement. This BIL level is purely based on application and safety. For railway application, the BIL level is 250 kvp for 25 kV AC system.

# Chapter 20

REFERENCE STANDARDS

Sanjay Dhavle

Reference standards are necessary documents referred for the designing, manufacturing, installing, operation and maintenance of any product. This chapter deals with major standard organisations and various applicable standards for switchgear products.

Standards are evolved considering the interest of manufacturers, as well as users. They fulfil the need of the economy in the sense that they provide a level platform to users, for evaluating one product against another of the same category or type, vis-à-vis standard requirements. At the same time, the manufacturers also get a benchmark to evaluate the design of their own products. This ensures a level playing field for all manufacturers.

A standard is a document setting forth requirements normally dictated by customary practices in industry, science and technology. Such documents may include, and may standardise, terms, definitions, symbols, methods of measurement, tests of parameters or performance of devices, apparatus, systems or phenomena characteristic performance and safety requirements, dimensions and ratings.

On the based of these standards, the users or purchasers formulate documents, called specifications, mentioning their own specific requirements.

A specification is a standard document that specifies all the characteristics and conditions to be met by a product or service to be supplied to the purchaser. Such a document may refer to other standards, selecting from among the specific allowable options. A specification is intended to be a complete purchasing document.

There are various organisations for formulating and publishing standards. The major standards organisations and various applicable standards for switchgear products are covered below.

20.1 STANDARD ORGANISATIONS

20.1.1 Indian Standards

The Bureau of Indian Standards (BIS) is responsible and authorised to issue standard specifications for this country. BIS is a statutory institution established under the Bureau of Indian Standards Act, 1986 to promote harmonious development of the activities of standardisation, marking and quality certification of goods and attending to connected matters in the country.

The BIS is located at Manak Bhawan, 9 Bahadur Shah Zafar Marg, New Delhi, India.

20.1.2 International Standards

In a globally competitive market, it is necessary to have an international agreement on engineering standards. For export-oriented industries especially, it is essential to ensure that their products conform to international standards.

The International Electrotechnical Commission (IEC) is the world organisation that prepares and publishes International Standards for all the electrical, electronic and related technologies. The IEC's principal activity is to develop and publish International Standards and technical reports. The International Standards serve as a basis for national standardisation and as a reference during the drafting of international tenders and contracts. IEC publications are bilingual, in English and French.

The IEC is located at 1 Rue de Varembe, Geneva, Switzerland.

20.1.3 National Standards of Other Countries

Like the Bureau of Indian Standards (BIS), each country has its own standards organisation. Some of them are listed below.

(a)	ANSI	American National Standards Institute, USA.
(b)	BSI	British Standards Institution, UK.
(c)	DIN	Deutsches Institut fuer Normung e.V., Germany.
(d)	JSA	Japanese Standards Association, Japan.

20.2 STANDARDS PERTAINING TO SWITCHGEAR

An indicative list of Indian and international standards applicable to the field of switchgears and circuit breakers is given below.

20.2.1 Indian Standards for Switchgear

These are delineated in Table 20.1.

S.No.	BIS Number	Title
1.	IS:3427	AC metal-enclosed switchgear and control gear for rated voltages above 1 kV up to and including 52 kV
2.	IS:5561	Electric power connectors
3.	IS:7567	Automatic circuit reclosers
4.	IS:9046	ac contactors of voltage above 1000 V up to and including 11,000 V
5.	IS:9135	Guide for testing of circuit breakers with respect to out-of-phase switching
6.	IS:9920	Switches and switch isolators for voltages above 1000 V
7.	IS:9921	Alternating current disconnectors (isolators) and earthing switches for voltages above 1000 V
8.	IS:10118	Code of practice for selection, installation and maintenance of switchgear and control gear
9.	IS:10601	Dimensions of terminals of high voltage switchgear and control gear
10.	IS:11353	Guide for uniform system of marking and identification of conductors and apparatus terminals
11.	IS:12661 Part-1	High voltage motor starter: Direct-on-line (full voltage) ac starters
12.	IS:12729	General requirements for switchgear and control gear for voltages exceeding 1000 V
13.	IS:13118	General requirements for circuit breakers for voltages above 1000 V
14.	IS:13516	Methods of synthetic testing of high voltage ac circuit breakers
15.	IS:13947	Low voltage switchgear and control gear
16.	IS:14657	Gas insulated metal-enclosed switchgear for rated voltage 72.5 kV and above-requirements for switching of bus charging current by disconnectors.
17.	IS:14658	High voltage ac circuit breakers— Guide for short-circuit and switching test procedures for metal-enclosed and dead tank circuit breakers
18.	IS:14659	Alternating current insulation-enclosed switchgear and control gear for rated voltage above 1 kV and up to and including 38 kV
19.	IS:14666	High voltage alternating current switch fuse combinations
20.	IS:14674	High voltage alternating current circuit breakers—Guide for seismic qualifications of high voltage alternating current circuit breakers
21.	IS:14786	High voltage/low voltage prefabricated sub-stations
22.	IS/IEC 60517:1990	Gas insulated metal-enclosed switchgear for rated voltage 72.5 kV and above

Table 20.1 Indian Standards for Switchgear

20.2.2 International Standards for Switchgear

The IEC has published a comprehensive standard IEC 62271:2001, which covers almost all aspects of ac circuit breakers designed for indoor or outdoor installation and for operation at frequencies of 50 Hz and 60 Hz systems having voltages above 1000 V.

IEC 62271:2001 standard is divided into several parts, as listed in Table 20.2.

S.No.	IEC Number	Title
1.	IEC 62271-1:2001	Common specifications
2.	IEC 62271-100:2001	High voltage alternating current circuit breakers
3.	IEC 62271-101:2001	Synthetic testing
4.	IEC 62271-102:2001	High voltage alternating current disconnectors and earthing switches
5.	IEC 62271-103:2001	High voltage switches for rated voltages above 1 kV and less than 52 kV
6.	IEC 62271-104:2001	High voltage switches for rated voltages of 52 kV and above
7.	IEC 62271-105:2001	High voltage alternating current switch-fuse combinations
8.	IEC 62271-106:2001	High voltage alternating current contactors and contactor-based motor starters
9.	IEC 62271-200:2001	Metal-enclosed switchgear and control gear for rated voltage up to and including 38 kV
10.	IEC 62271-201:2001	Insulation enclosed switchgear and control gear for rated voltage up to and including 52 kV
11.	IEC 62271-202:2001	High voltage/low voltage prefabricated sub-stations
12.	IEC 62271-203:2001	Gas insulated metal-enclosed switchgear for rated voltages above 52 kV
13.	IEC 62271-204:2001	High voltage gas insulated transmission lines for rated voltages of 72.5 kV and above
14.	IEC 62271-300:2001	Guide for seismic qualification
15.	IEC 62271-301:2001	Guide for inductive load switching
16.	IEC 62271-302:2001	Guide for short-circuit and switching test procedures for metal enclosed and dead tank circuit breakers
17.	IEC 62271-303:2001	Use and handling of sulphur hexafluoride (SF6) in high voltage switchgear and control gear
18.	IEC 62271-304:2001	Additional requirements for enclosed switchgear and control gear from 1 kV to 72.5 kV to be used in severe climatic conditions
19.	IEC 62271-305:2001	Cable connections for gas insulated metal-enclosed switchgear for rated voltages above 52 kV
20.	IEC 62271-306:2001	Direct connection between power transformers and gas-insulated metal-enclosed switchgear for rated voltages above 52 kV
21.	IEC 62271-307:2001	The use of electronic and associated technologies in auxiliary equipment of switchgear and control gear
22.	IEC 62271-308:2001	Guide for asymmetrical short-circuit breaking test duty T100a
23.	IEC 61233	Switching of inductive loads

Table 20.2 Contents of IEC 62271: 2001 Standard

Some conditions, which are not covered in the above standards, are covered in other standards mentioned below.

20.3 STANDARDS PERTAINING TO ASSOCIATED EQUIPMENT, MATERIAL, PROCESSES AND GUIDELINES REQUIRED FOR SWITCHGEAR

Along with the circuit breaker, a lot of other equipment, materials and processes are required to make a complete switchgear. An indicative list of these items along with BIS standard Numbers applicable for them, is given in Table 20.3.

S.No.	Equipment	BIS Standard No.
1.	Current transformers	IS:2705
2.	Voltage transformers	IS:3156
3.	Electrical relays for power system protection	IS:3231
4.	Static protective relays	IS:8686
5.	Electrical protective relays for use in seismic areas	IS:8714
6.	Indicating Instruments	IS:1248
7.	Integrating meters	IS:722
8.	Control switches for voltages up to and including 1000 Vac* and 1200 Vdc*	IS:13947 IS:6875
9.	Indicating lamps and push buttons	IS:13947 & IS:6875
10.	Miniature air break circuit breakers for ac circuits for voltages not exceeding 1000 V	IS:8828
11.	Contactors for voltages not exceeding 1000 V ac or 1200 V dc	IS:13947
12.	Specification for LV fuses	IS:13703
13.	Application guide for high voltage fuses	IS:9385
14.	Interconnecting busbars for ac voltages above 1 kV up to and including 36 kV	IS:8084
15.	Copper strips for electrical purpose	IS:1897
16.	Copper rods and bars for electrical purpose	IS:613
17.	Wrought aluminium and aluminium alloys, bars, rods, tubes and sections for electrical purpose	IS:5082
18.	Cold rolled carbon steel sheets and strips	IS:513
19.	Structural steel (standard quality)	IS:2062
20.	Code of practice for phosphating iron and steel	IS:6005
21.	Recommended practice for hot dip galvanising of iron and steel	IS:2629
22.	Methods of testing uniformity of zinc coating	IS:2633
23.	Bushings for alternating voltages above 1000 V	IS:2099
24.	Porcelain post insulators for systems with nominal voltages greater than 1000 V	IS:2544
25.	PVC cables up to 1100 V	IS:694
26.	Guide for marking of insulated conductors	IS:5578
27.	Cable sealing kits	IS:13573
28.	Cable glands	IS:2148
29.	Cable lugs	IS:8337
30.	Colours of ready mix paints and enamels	IS:5

*As per IS/IEC, this should be written as ac and dc.

Table 20.3 Standard for Associated Equipment of Switchgears

Chapter 21

STATUTORY REQUIREMENTS AND ELECTRICITY RULES

Neelam Bhogal

The general safety requirement is the most important requirement in the design of equipment. Every country has its own general safety requirements and standards for design, manufacturing and installation of equipment.

It is the prime responsibility of the manufacturer to manufacture equipment that meets the general safety requirements as laid down by the Indian Government in the Indian Electricity Rules, 1956, publication 1992 in addition to the national/international relevant standards.

It is necessary to explain the various parameters which should be considered as per the relevant clauses of the Indian Electricity Rules, 1956 for designing, manufacturing and installation. The relevant clauses are explained below.

21.1 INDIAN ELECTRICITY RULES, 1956 (CLAUSE NO. 29)

Construction, installation, protection, operation and maintenance of electric supply lines and apparatus:

1. All electric supply lines and apparatus shall be of sufficient ratings for power, insulation and estimated fault current and of sufficient mechanical strength, for the duty which they may be required to perform under the environmental conditions of installation, and shall be constructed, installed, protected, worked and maintained in such a manner as to ensure safety of personnel and property.

2. The material and apparatus used shall conform to the relevant specifications of the Indian Standard Institution (including National Electrical Code) where such specifications have already been laid down.

Switchgears and breakers shall be designed, manufactured and tested for rated values like current, voltage, fault current and insulation level. The rated current of switchgear is ensured by conducting a temperature rise test. Voltage and insulation level are confirmed by the dielectric test. Breaking current suitability of switchgear is confirmed by the short-circuit duties test. Mechanical strength and electrical strength are accessed by arc fault test and short time current test. Additional tests of Wet p.f., siesmic shall be conducted for accessing suitability for outdoor application. Safety of personnel shall be ensured by conducting degree of protection and arc fault test. All the above tests shall be conducted as per their National/International Standards IS 13118, IS 3427, IEC 62271.

21.2 INDIAN ELECTRICITY RULES, 1956 (CLAUSE NO. 35)

Danger Notices The owner of very medium, high and extra-high voltage installation shall affix permanently in a conspicuous position a danger notice in Hindi or English and the local language of the district, with a sign of skull and bones as per the relevant ISS No. 2551 on various electrical installations.

The switchgears should always have the danger notices or danger labels on the covers of busbar chamber and feeder chamber as after opening the covers, live busbar and connections are accessible in switchgear.

21.3 INDIAN ELECTRICITY RULES, 1956 (CLAUSE NO. 48)

Precautions Against Leakage Before Connection (1) The supplier shall not connect with his works the installation or apparatus on the premises of any applicant for supply unless he is reasonably satisfied that the connection will not, at time of making the connection, cause a leakage from the installation or apparatus of a magnitude detrimental to safety. Compliance with this rule shall be checked by measuring the insulation resistance as provided below:

1. **High Voltage Equipments Installations**
    (a) High voltage equipments shall have the insulation resistance value as stipulated in the relevant Indian Standard.
    (b) At a pressure of 1000 V applied between each live conductor and earth for a period of one minute, the insulation resistance of HV installations shall be at least 1 mega-ohm or as specified by the Indian Standards Institution from time to time.
2. **Medium and Low Voltage Installations** At a pressure of 500 V applied between each live conductor and earth for a period of one minute, the insulation resistance of medium and low voltage installations shall be at least 1 mega-ohm or as specified by the Indian Standards Institution from time to time.

IR values as per IS 10118 part III 1982 clause No. 2.3.10.2 shall be 100 mega-ohms for 1 kV to 33 kV switchgears and 1000 mega-ohms for 33 kV and above switchgears.

Switchgears shall be manufactured such that the insulation provided is suitable to withstand voltages as per relevant IS/IEC. Further, the insulation resistance between phase-to-earth measured should always be more than 100 mega-ohms for switchgears rated 1 kV to 33 kV and 1000 mega-ohms for switchgears above 33 kV, as per IS 10118 part 3, clause No. 2.3.10.2.

If during the test, IR is found to be less, then HV (high voltage) test should be done to check the health of the equipment.

Before connecting the equipment to line, its IR value should always be checked, by using a meggar.

21.4 INDIAN ELECTRICITY RULES, 1956 (CLAUSE NO. 51)

Provisions applicable to medium, high or extra high voltage installations:

The following provisions shall be observed where energy at medium, high or extra high voltage is supplied, converted, transformed or used:

(a) All conductors (other than those of overhead lines) shall be completely enclosed in mechanically strong metal casing or metallic covering which is electrically and mechanically continuous and adequately protected against mechanical damage unless the said conductors are accessible only to an authorised person or are installed and protected to the satisfaction of the Inspector so as to prevent danger:

Provided that non-metallic conduits conforming to the relevant Indian Standard Specifications may be used for medium voltage installations, subject to such conditions as the Inspector or Officer appointed to assist an Inspector may think fit to impose.

All metal works, enclosing, supporting or associated with the installation, other than that designed to serve as a conductor shall be connected with an earthing system as per standards laid down in the Indian Standards in this regard

(b) Every switchboard shall comply with the following provisions, namely:

   (i) A clear space of more than 1 m in width shall be provided in front of the switchboard;

   (ii) If there are any attachments or bare connection at the back of the switchboard, the space (if any) behind the switchboard shall be either less than 20 cm or more than 75 cm in width, measured from the furthest outstanding part of any attachment or conductor;

   (iii) If the space behind the switchboard exceeds 75 cm in width, there shall be a passageway from either end of the switchboard clear to a height of 1.8 m.

(c) In case of installations provided in premises where inflammable materials including gases and/or chemicals are produced, handled or stored, the electrical installations, equipment and apparatus shall comply with the requirements of flame-proof, dust-tight totally enclosed of any other suitable type of electrical fittings depending upon the hazardous zones as per the relevant Indian Standard Specifications.

In view of above the Safety Requirements for Switchgear can be summarised as below:

1. All busbars and connections should be enclosed in an enclosure which should be continuous electrically and connected to the earthing terminal. This enclosure should be mechanically strong to avoid any danger. Non-metallic conduits can also be provided as per IS/IEC.

2. Any switchboard provided should have clear space of 1000 mm in front of panels for withdrawing movable trucks from panels. Behind the switchboard, either 200 mm space should be provided or more than 750 mm. The passage behind the switchboard should be clear up to a height of 1800 mm for easy movement of the operator.
3. Switchgears required for special application like chemically hazardous and fire-prone areas, should be flame-proof and totally dust-proof as per relevant IS/IEC.

21.5 INDIAN ELECTRICITY RULES, 1956 (CLAUSE NO. 64)

Use of Energy at High and Extra High Voltage The following provisions shall be observed where energy at high or extra high voltage is supplied, converted, transformed or used:

(i) Clearance as per the Indian Standard Code shall be provided for electrical apparatus so that sufficient space is available for easy operation and maintenance without any hazard to the operating and maintenance personnel working near the equipment and for ensuring adequate ventilation; and

(ii) The minimum clearances shall be maintained for bare conductors or live parts or any apparatus in outdoor sub-stations, excluding overhead lines, of HV and EHV installations. These are given in Table 21.1.

Voltage Class	Ground Clearance (Meters)	Sectional Clearance (Meters)
Not exceeding 11 kV	2.75	2.6
Not exceeding 33 kV	3.7	2.8
Not exceeding 66 kV	4.0	3.0
Not exceeding 132 kV	4.6	3.5
Not exceeding 220 kV	5.5	4.3
Not exceeding 400 kV	8.0	6.5

Table 21.1 Minimum Clearances for Apparatus in Outdoor Sub-stations

The windings of motors or other apparatus within reach from any position in which a person may require to work shall be suitably protected so as to prevent danger;

Where transformer or transformers are used, suitable provision shall be made, either by connecting with the earth point of the circuit at the lower voltage or otherwise, to guard against danger by reason of the said circuit becoming accidentally charged above its normal voltage by leakage from or contact with the circuit at the higher voltage.

However, the minimum sectional clearance required to be maintained as per IS 10118 is 3000 mm in place of 2800 mm. Hence for outdoor breakers, sectional clearances shall be maintained as 3000 mm (min) as per IS 10118, part 3.

21.6 INDIAN ELECTRICITY RULES, 1956 (CLAUSE NO. 64A)

The following additional provisions shall be observed where energy at high or extra high voltage is supplied, converted, transferred or used:

21.6.1 Interlocks

Suitable interlocks shall be provided in the following cases:

(a) Isolators and the controlling circuit breakers shall be interlocked so that the isolators cannot be operated unless the corresponding breaker is in open position;

For outdoor breakers (non-withdrawable type), mechanical interlocks shall be provided in the mechanism such that interlock key gets free from circuit breaker mechanical interlock only when breaker is in open condition. This key, when inserted in mechanical interlock provided in the isolator, allows isolator to operate.

In metal-clad switchgear the circuit breaker is isolated by withdrawing breaker truck to test position or outside the metal-clad house. Necessary safety interlocks shall be provided such that breaker truck is withdrawable only when the circuit breaker is open.

(b) Isolators and the corresponding earthing switches shall be interlocked such that earthing switch cannot be closed until the corresponding isolator is in open position.

Separate feeder and busbar earthing trucks shall be provided. These can be inserted only when the feeder breaker truck is withdrawn and circuit is disconnected. Incase integral feeder earthing isolators are provided in indoor metal-clad switchgears then suitable mechanical interlock shall be provided such that isolator will close only when breaker is open. For outdoor breakers suitable mechanical interlocks shall be provided between isolator and earthing switch such that the earthing switch can be closed only when isolator is open.

(c) Where two or more supplies are not intended to be operated in parallel, the respective circuit breakers or linked switches controlling the supplies shall be interlocked to prevent the possibility of any inadvertent paralleling of supplies;

Non-paralleling of source can be achieved electrically or mechanically. Electrically this can be achieved by putting the auxiliary switch contact (NC contact) of one breaker into the closing circuit of the other breaker and vice versa. Mechanically this can be achieved by providing a mechanical interlock on both the supplies breaker and one key. Hence the supply of one breaker only can be put on after the key is inserted in the mechanical interlock provided on that breaker.

(d) When two or more transformers are operated in parallel, the system shall be so arranged as to trip the secondary breaker of a transformer in case the primary breaker of that transformer trips;

This is generally achieved by providing high speed tripping relay contact of upstream breaker in the tripping the circuit of downstream breaker and providing contact in closing circuit for blocking closing.

(e) All gates or doors which give access to live parts of an installation shall be interlocked in such a way that these cannot be opened unless the live parts are made dead. Proper discharging and earthing of these parts should be ensured before any person comes in close proximity of such parts; and

(f) Where two or more generators operate in parallel and neutral switching is adopted, an inter-lock shall be provided to ensure that the generator breaker cannot be closed unless one of the neutrals is connected to the earthing system.

21.6.2 Protection

All systems and circuits shall be so protected as to automatically disconnect the supply under abnormal conditions.

The following protection shall be provided:

(a) Over-current protection to disconnect the supply automatically if the rated current of the equipment, cable or supply line is exceeded for a time which the equipment, cable or supply line is not designed to withstand;

Relays shall be provided which should be set considering system parameters. When the line current exceeds the set value, the relay shall operate and trip the breaker automatically thus disconnecting the circuit.

(b) Earth fault/earth leakage protection to disconnect the supply automatically if the earth fault current exceeds the limit of current for keeping the contact potential within reasonable values.

Earth fault relays provided shall be set considering system parameters. When earth fault occurs, the relay shall operate and disconnect the faulty circuit.

(c) Gas pressure type protection to give alarm and tripping shall be provided on all transformers of ratings 1000 kVA and above;

(d) Transformers of capacity 10 MVA and above shall be protected against incipient faults by differential protection; and

(e) All generators with ratings of 100 kVA and above shall be protected against earth fault leakage. All generators of rating 1000 kVA and above shall be protected against faults within the generator winding using restricted earth fault protection or differential protection or both.

21.7 INDIAN ELECTRICITY RULES, 1956 (CLAUSE NO. 67)

All non-current carrying metal parts associated with HV/EHV installation shall be effectively earthed to a grounding system of mat which will:

(a) Limit the touch and step potential to tolerable values;

(b) Limit the ground potential rise to tolerable values so as to prevent danger due to transfer of potential through ground, earth wires, cable sheath fences, pipelines, etc.;

(c) Maintain the resistance of the earth connection to such a value as to make operation of the protective device effective;

The outdoor breakers shall be provided with earthing stud/bolt of minimum 12 mm diameter as per clause 5.3 of IEC 694. Indoor switchgears shall be provided with earthing strip as per clause 3.5.1 of IS 3427. These shall be connected to sub-station/utility earthing to limit the ground potential and maintain resistance of the earth connection.

In the case of a star-connected system with earthed neutrals or delta connected system with earthed artificial neutral point:

(a) The neutral point of every generator and transformer shall be earthed by connecting it to the earthing system as defined in Rule 61(4) and hereinabove by not less than two separate and distinct connections:

Provided that the neutral point of a generator may be connected to the earthing system through an impedance to limit the fault current to the earth; and

Provided further that in the case of a multi-machine system, neutral switching may be resorted to, for limiting the injurious effect of harmonic current circulation in the system.

(b) In the event of an appreciable harmonic current flowing in the neutral connection so as to cause interference with communication circuits, the generator or transformer neutral shall be earthed through a suitable impedance.

(c) In case of the delta connected system, the neutral point shall be obtained by the insertion of a grounding transformer and current limiting resistance or impedance wherever considered necessary at the commencement of such a system.

(d) Single-phase high or extra high voltage systems shall be earthed in a manner approved by the Inspector.

(e) In the case of a system comprising electric supply lines having concentric cables, the external conductor shall be the one to be connected with earth.

(f) Where a supplier proposes to connect with earth an existing system for use at high or extra high voltage which has not hitherto been so connected with earth, he shall give not less than fourteen days' notice in writing together with particulars to the telegraph authority of the proposed connection with earth.

21.8 INDIAN ELECTRICITY RULES, 1956 (CLAUSE NO. 120)

Switchgear and Terminals Switchgear and all terminals, cable-ends, cable-joints and connections to the apparatus shall be totally enclosed and shall be so constructed, installed and maintained as to comply with the following requirements:

(a) All parts shall be of mechanical strength sufficient to resist rough usage.

(b) All conductors and contact areas shall be of adequate current-carrying capacity and all joints in conductors shall be properly soldered or otherwise efficiently made.

The breaker and switchgears current-carrying parts should have sufficient rating. The current shall not cause temperature rise beyond temperature rise limit specified in IS 13118/IEC 62271 part 100.

(c) The lodgement of any matter likely to diminish the insulation or affect the working of any switchgear, shall be prevented.

(d) All live parts shall be so protected or enclosed as to prevent persons accidentally coming into contact with them and to prevent danger from arcs, short-circuits, fire, water, gas or oil.

Switchgear shall be designed to avoid any accidental touch to live part. The breaker shall be tested for degree of protection as per IEC 62271 part 100/IS 3427 and internal ARC fault test as per IS 3427 1997/IEC 62271-200.

(e) Where there may be risk of igniting gas, coal dust, oil or other inflammable material, all parts shall be so protected as to prevent open sparking; and when circuit breaker operates, the contact separation shall take place in vacuum, oil or SF6 (any insulating media). Further, it should not cause any damage to the adjacent equipment.

(f) Every switch or circuit breaker shall be so constructed as to be capable of opening the circuit it controls and dealing with any short-circuit without danger.

This shall be tested by the short-circuit test and short time current test as per IS 13118 and IEC 62271-100. Further breakers shall be tested for mechanical endurance as per IS/IEC. This will ensure healthiness of mechanism of breaker.

INDEX

A Case Study of Auto-recloser at KPTCL 6.13
A Typical Instrument Transformer Standardisation Model 7.22
Abnormal Service Conditions 19.17
AC Relays 9.4
AC withstand test 8.16
Acceptance of Class PS CT Using Magnetisation Curves 7.18
Accessories 5.16
Accuracy of Meter and Limits of Error 11.3
Adaptive Protection Characteristics 9.16
Additional Functionality 9.21
Air Blast Circuit Breaker 4.4
Air Blast Circuit Breakers (ABCBs) 1.21
Air Circuit Breakers (ACBs) 1.17
Alarm Annunciation Scheme 12.11
Alarm Cancellation Scheme 12.11
Alarm Schemes 12.10
Ambient Considerations 10.3
Applicability of Test Certificates 7.17
Application of Instrument Transformers in Various Protection Schemes 7.12
Application of Surge Protection Devices 8.9
Application of Switchgears in Power System 10.4
Arc Column Characteristics 1.9
Arc Extinction 1.11
Arc Interruption Theories 1.2
Arc Phenomenon: Introduction 1.1
Arcing Process 1.3
Artificial Pollution Tests 19.5
Assembly 17.9
Atmospheric Environment 9.23
Attracted Armature Relays 9.3
Auto-changeover Scheme 12.23
Automatic Bus Transfer Scheme 12.18

Automatic Changeover with Voltage Interruption, under Fault Condition (Fast Changeover followed by Slow Changeover) 12.22
Automatic Sectionaliser 6.9
Automatic Supply Transfer Schemes 12.18
Auto-reclosing Requirements in One-and-a-half Breaker System 15.16
Auto-reclosing 13.11
Auto-synchronising 12.17
Auxiliary Supply Distribution Philosophy 15.6
Auxiliary Supply Requirements 11.3

Back-up Protection 13.14
Balanced Three-phase Short-circuits 3.22
Balanced/Unbalanced Load Meters 11.3
Bay Information 5.3
Blocking of Fast Changeover 12.23
Blocking Scheme Logic 15.12
Breakdown by Leakage in Solid Insulating Material 19.6
Breaker Fixed Portion Auxiliary Switch Interlocks 12.4
Breaker Moving Portion Auxiliary Switch Interlocks 12.4
Breaker Related Protection 15.7
Bulk Oil Circuit Breakers (BOCBs) 1.13
Bus Bar Protection 15.17
Bus Bar 5.7
Bush Fire-prone Areas 6.8

Capacitance Resistance (C-R) Type Surge Suppressor 8.7
Capacitive Current Switching Tests 19.15
Capacitor Protection 13.14
Capacitor Tripping Scheme 12.9
Cathode Region 1.12

Characteristic Time/Current Curves 9.7
Check Synchronising Feature 12.14
Circuit Breaker Closing Requirements 1.32
Circuit Breaker Opening Requirements 1.31
Circuit Breaker Operating Mechanisms 1.29
Circuit Breaker Technology 1.13
Circuit Breaker 5.14
Circuit Breakers for Earthing Applications 10.7
Class 'A' Tripping 14.2
Class 'A1' Tripping 14.3
Class 'A2' Tripping 14.4
Class 'B' Tripping 14.3
Class 'B' Tripping 14.4
Class 'C' Tripping 14.3, 14.5
Classification by External Design 4.3
Classification by Interrupting Media 4.4
Classification by Location 4.3
Classification by Voltage 4.2
Classification of Circuit Breakers 4.2
Classification of Gimes 5.3
Classification of Tripping 14.2
Closed Pressure System 19.11
Common for Both Indoor and Outdoor Switchgear 17.11
Communication Feature 11.4
Communication 9.18
Comparison of Operating Mechanisms 4.13
Component Testing 9.13
Components/Modules 5.7
Conclusion 5.25, 9.23
Conditions for Initiating/Effecting Auto-changeover 12.24
Conditions for Manual Transfer 12.25
Conditions for Specific Blocking of Auto-transfer 12.25
Connectors 5.8
Constructional Aspects of GIS 5.18
Constructional Forms 3.7
Contact Design 3.19
Control and Protection Functions 15.5
Control and Sub-station Monitoring 16.5
Control Panel 5.17
Control Relay Unit for Sectionaliser 6.10
Control Transformer 6.5
Control Unit for Auto-recloser for 12 kV System 6.5
Controlled Pressure System 19.11
Conventional Feeder Protection 2.3
Conventional Incomer 2.1
Conventional Sub-Incomer 2.3
Core Balance CT 7.14
Correction for End-effect 3.21

Cost of Numerical Relays 9.21
Covers and Doors 3.9
C-R Type Surge Suppressor 8.9
Critical Current Test 19.18
Current Balance CT 7.14
Current Reversal Logic 15.14
Current Transfer and Plug-in-Joints 5.22
Current Transformer Recommendations 7.20
Current Transformer 5.15
Current-zero Processes 1.12

Data Acquisition System Testing 9.18
Data Analysis and Archiving Level 16.6
Data Storage 9.16
DC Fail Annunciation 12.13
Dead Tank Switchgears 3.15
Degree of Protection 3.10
De-ionisation, Re-combination and Attachment Processes 1.7
Demands Made on Circuit Breakers 3.27
Derivation of Relaying Quantities 13.2
Design and Construction 3.7
Design and Constructional Aspects of Instrument Transformers 7.8
Design and Visual Checks 19.3
Design Aspect 19.6, 19.8, 19.12
Design Consideration 5.6
Design Considerations of Rogowski Coil and Integrator 7.25
Design Features of Puffer Circuit Breaker 4.8
Design of Instrument Transformers 7.11
Design Parameters 3.15
Design 5.2
Development of Single-Pressure Puffer Breakers 4.6
Development of Static Relays 9.9
Dielectric Test on Auxiliary and Control Circuits 19.2
Dielectric Test on the Main Circuit 19.2
Dielectric Tests 19.5
Differential Protection 7.12, 13.8
Disconnectors (Isolators) 5.13
Disconnectors and Earthing Switches 3.10
Dispatch 17.2
Distance Protection 15.8
Distribution Switchgears 10.6
Dos & Donts 17.17
Dos and Donts for Switchgear Operation 18.3
Double Quantity Relays 9.8
Dynamometer Type Relays 9.5

Earth Switch 5.15
Earthing Truck (Test to Service) Limit Switch 12.4

Earthing 3.10, 17.10
Echo Feature to Detect Remote End Open 15.13
Effects of Oxygen and Humidity 19.7
Electrical Check 17.13
Electrical Environment 9.23
Electrochemical Deterioration 19.7
Electrode Regions 1.5
Electromagnetic Compatibility (EMC) Test 19.12
Electromechanical Relays 9.2
Emerging Trends 2.5
Enclosure 3.8, 5.18
Energisation of Long Transmission Lines 4.18
Environmental Test 19.16
Erection 17.7
Evaluation of Gas Insulated Surge Arrester 8.16
Evaluation of Individual Surge Arrester Block 8.16
Examination 18.6
Expansion Joints 5.20
Experimental Study in Use of Combinations of Surge Protection Devices 8.11
External Faults 13.7

Failure of Chemical Insulation 19.7
Fast Changeover 12.22
Fault Detection 13.2
Fault External to the Transformer 13.7
Faults 13.2
Faults on Transformer Terminals 13.7
Features of Modern Meters 11.2
Feeder Protection 13.9
Field Performance of Instrument Transformers 7.27
Filling with Insulation Medium 17.10
Final Commissioning/Load Testing 17.15
Final Inspection 17.11
First Generation SF6 Circuit Breakers 4.6
Fluids 3.25, 19.6
Future Trends 3.6, 11.9

Gas Circuit 5.20
Gas Insulated Switchgear (GIS) 1.27
Gaseous Materials 3.24
Gaseous 19.6
General Requirements 12.23
Generator Circuit Breakers 10.4
Grid Frequency Management 11.9
Grid Transformer Switchgears 10.5
Grouping of Protections in Class A1, A2, B And C Tripping 14.3

Heat Loss from Arc Plasma 1.11
High Lightning Discharges 6.7

High Voltage Check 17.15
Higher BIL Requirement 19.19
High-pressure Arcs: General Characteristics 1.4
Historical Background 3.1
Housing of Static Relays 9.11
HV and Environment Testing 9.22
Hydraulic Mechanism 1.36, 4.12

Incipient Faults 13.7
Incompatibility of Dielectric Material 19.7
Indian Electricity Rules, 1956 (Clause No. 120) 21.7
Indian Electricity Rules, 1956 (Clause No. 29) 21.1
Indian Electricity Rules, 1956 (Clause No. 35) 21.2
Indian Electricity Rules, 1956 (Clause No. 48) 21.2
Indian Electricity Rules, 1956 (Clause No. 51) 21.3
Indian Electricity Rules, 1956 (Clause No. 64) 21.4
Indian Electricity Rules, 1956 (Clause No. 64A) 21.5
Indian Electricity Rules, 1956 (clause No. 67) 21.6
Indian Standards for Switchgear 20.3
Indian Standards 20.2
Indoor Current Transformers 7.8
Indoor Switchgear 10.2, 17.3, 17.6, 17.7, 17.9, 17.11
Induction Relays 9.5
Inspection on Receipt at Site 17.4
Inspection Schedule 18.4
Inspection Windows 3.9
Inspection 18.6
Installation and Maintenance of GIS 5.23
Installation 17.6
Instrument Transformer Testing and Analysis of Test Results 7.16
Instrumentation 9.17
Insulating Materials and Insulators 5.10
Insulation Aspects of Design 3.24
Insulation Resistance Check 17.14
Integrated Control and Protection 15.18
Interlocks 21.5
Internal Faults 13.7
International Standards for Switchgear 20.3
International Standards 20.2
Interposing CT 7.13
Interrupter Design 4.8
Inter-tripping Schemes 15.8
Inter-tripping 13.9
Ionisation by Electron Collision 1.7
Ionisation Processes 1.5
Islanding Scheme 15.18

Joints 3.19

Large Concentration of Customers 6.8

Leakage current measurement 8.17
Less Severe Fault Conditions 14.6
Lifecycle Cost of GIS 5.24
Lighting/Domestic Load Control 2.4
Lightning Impulse (LI) Over-voltages 8.11
Lightning Impulse Voltage Dry Withst and Test 19.5
Lightning impulse withstand test 8.16
Live Tank Switchgears (Outdoor Porcelain-clad Construction) 3.13
Location Considerations 10.1
Location of Arrester 8.15
Lubrication of Operating Mechanism 18.8

Main and Check Energy Meters 11.9
Main Features of Auto-changeover Scheme 12.24
Maintenance Procedure 18.5
Maintenance Requirements 18.4
Manual Changeover with Voltage Interruption 12.21
Manual Changeover without Voltage Interruption (Involving Brief Parallel Operation) 12.21
Manual Synchronisation 12.14
Measurement of the Leakage Current 8.19
Measurement of the Resistance of Main Circuit 19.3, 19.8
Measurement of the Resistive Component of Leakage Current 8.19
Mechanical Check 17.13
Mechanical Operating Tests 19.3
Mechanical Operation Tests at Ambient Temperature 19.12
Mechanical Relays 9.8
Mechanical Requirements 11.3
Memory Testing 9.18
Metal Oxide Surge Arrester 8.9
Metal Oxide Type Surge Arrester 8.4
Metal-enclosed Switchgears (Constructional Forms) 3.11
Metal-enclosed Switchgears (Indoor) 3.11
Metal-enclosed Switchgears (Outdoor) 3.13
Metal-enclosed Switchgears 3.7
Microprocessors 9.11
Minimum Oil Circuit Breakers (MOCBs) 1.16
Modern Current Sensors (Rogowski Coil) 7.24
Modern Trends in Generator Protection 14.7
Modes of Bus Transfer 12.21
Motor Control 2.4
Motor Operated Relays 9.8
Motor Protection 13.4
Motor-driven Mechanisms 1.38, 4.13
Moving Coil Permanent Magnet Relays (Rotary Type) 9.5

Moving Coil Relays 9.5
Multiple Protection Characteristics 9.16
MV GIS Construction 3.13

National Standards of Other Countries 20.2
Need for Metering 11.1
Need for Auto-recloser 6.7
Normal Current Rating (Thermal Aspects of Design) 3.15
Normal Operating Condtion 12.24
Normal Service Conditions 19.16
Numerical Relays 9.13

Oil Circuit Breaker 4.4
Online Energy Monitoring System 11.4
One Minute Dry Power Frequency Voltage Withstand Test 19.5
One Minute Wet Power Frequency Voltage Withstand Test 19.5
One-and-a-half Breaker System 15.2
Operating Mechanisms for Circuit Breakers 4.10
Operating Principle of Electromechanical Relays 9.2
Operation of Switchgear 18.2
Operational Check 17.14
Other Industrial Load Control 2.4
Other Line Protection Functions 15.15
Out of Phase Switching 4.20
Outdoor Current Transformers 7.9
Outdoor Switchgears 10.2, 17.2, 17.6, 17.7, 17.9
Out-of-Phase Making and Breaking Tests 19.19
Over-current and Earth Fault Protection 13.8, 13.10
Over-current Protection 13.16
Over-current Relays 9.10
Overhaul 18.6
Overloading 13.5
Over-voltage Considerations 10.4
Over-voltage Protection 13.16
Over-voltages in GIS 8.13
Over-voltages in Sub-stations 8.10
Over-Voltages in the Electrical System 8.10

Partial Discharge Test (Optional) 19.5
Partitions and Shutters 3.9
Percentage or Biased Differential Protection 7.13
Performance Evaluation 8.16
Performance Monitoring of Surge Protection Devices 8.19
Permissive Over-reach Scheme Logic 15.11
Permissive Under-reach Scheme Logic 15.10
Phase Comparison Relay 12.22
Philosophy of Protection 13.7

Pilot Wire Differential Protection 15.15
Plasma Region (Arc Column) 1.5
Plasma Region 1.12
Pneumatic Mechanism 1.34, 4.11
Power Frequency Recovery Voltage 4.15
Precautions to be Followed for Indoor Switchgears 17.5
Precautions to be Followed for Outdoor Switchgears 17.5
Preparation Before Maintenance 18.5
Presence of Arc 19.7
Present Status 3.5
Present Trends in Feeder Protection 2.7
Present Trends in Incomer 2.5
Present Trends in Sub-incomer 2.6
Programmable Scheme Logic (PSL) 15.14
Protection and Metering System 16.2
Protection of Parallel Feeders 13.14
Protection Philosophy 14.5
Protection Requirements 10.8
Protection 21.6

Quality Control and Testing 9.12

Radio Interference Test 19.7
Rating Considerations 10.2
Reactor Protection 15.17
Recommendations for Standardisation and Cost Reduction 7.20
Reed Relays 9.4
Relay Circuits Using Analogue Techniques 9.10
Relay Circuits Using Digital Techniques 9.11
Relay Hardware 9.14
Relay Software 9.15
Remote Access 11.8
Remote Site/Difficult Access 6.8
Requirement of Operating Energy for Puffer Breakers 4.10
Residual voltage test 8.17
Resistance of the Main Circuit 3.19
Restricted Earth Fault Protection 7.13, 13.8
Role of Arc in Current Interruption 1.2
Routine Tests 19.2

Safety Schemes Using Position Limit Switches 12.2
Sealed System 19.11
Seals and Gaskets 5.19
Second Generation SF6 Circuit Breakers 4.6
Seismic Considerations 10.3
Selection of Class of Accuracy 7.12
Selection of Primary Current Rating 7.12

Selection of Rated Burden 7.12
Selection of Secondary Current Rating 7.12
Self-check Feature 9.18
Sensitive Earth Fault Protection 13.11
Series Tripping Schemes 12.7
Service Feedback 18.17
Servicing 18.6
Set-point Testing 9.18
SF6 Gas Circuit Breakers 1.22
SF6 Gas Circuit Breakers 4.5
SF6 Gas Dynamic Mechanism 4.12
Short Line Fault Test 19.18
Short Line Faults (SLF) 4.16
Short Time Current Withstand Test 19.10
Short-circuit Current 3.20
Short-circuit Duties Test 19.13
Short-circuit Forces 3.20
Short-circuit Interruptions 3.27
Short-circuit Protection 13.6
Shunt Tripping Scheme 12.6
Single Line Diagram 5.2
Single Quantity Relays 9.6
Single-phase Short-circuit Stresses 3.20
Single-phase Short-circuit 19.15
Single-phasing 13.6
Size of Numerical Relays 9.21
Slow Changeover 12.21
Small Wiring and Connection of LT Cables 17.10
Solenoid and Magnetically Actuated Mechanism 1.38
Solenoid Relays 9.4
Solid Materials 3.25, 19.6
Some Important Design Considerations 7.12
Special Applications 3.29
Special Electrical Endurance Test 19.19
Special Features 15.12
Special Instrument Transformers 7.22
Special Tests 19.19
Spring Charging Mechanisms 1.33
Spring Mechanism 4.11
Stalling of Motors (Locked Rotor) 13.6
Standard Organisations 20.2
Standardisation and Cost Considerations of CTs and VTs 7.19
Standards Pertaining to Associated Equipment, Material, Processes and Guidelines Required for Switchgear 20.5
Standards Pertaining to Switchgear 20.2
Static Relays 9.9
Station Switchgear 10.5, 12.19
Storage and Handling 17.4

Storage and Processing of Meter Readings 11.9
Summation CT 7.14
Support Structure 5.22
Surge Counters 8.19
Surge Protection Devices for Gas Insulated Sub-stations 8.13
Surge Protection Devices 8.2
Switchgear Maintenance 18.3
Switchgear Spares 18.6
Switchgears for Distribution Applications 10.6
Switchgears for Industrial Applications 10.6
Switchgears for Power Plant Auxiliaries 10.5
Switchgears for Rural Applications: (Auto-reclosers and Sectionalisers) 10.7
Switchgears for Transmission Sub-stations 10.6
Switching Functions 3.27
Switching Impulse (SI) Over-voltages 8.11
Switching Impulse Voltage Test (Optional) 19.5
Switching of Low Inductive Currents 3.28
Switching of Unloaded Transmission Lines and Capacitor Banks 4.19
Switching of Very Low Inductive Currents 3.28
Synchronising Schemes 12.13
System Earthing Considerations 10.3
System Features 11.7
System Requirements of Circuit Breakers 4.14
System Topology 11.7

Technical Parameters of Instrument Transformers 7.6
Technical Specifications for an Auto-recloser 6.3
Tee Differential Protection 15.16
Temperature Rise Test 19.8
Temporary Over-voltages 8.10
Terminal Faults 4.14
Test and Service Position Limit Switch Interlocks 12.4
Test and Service Position Limit Switches 12.3
Test Duty-T10 19.14
Test Duty-T100a 19.14
Test Duty-T100s 19.14
Test Duty-T30 19.14
Test Duty-T60 19.14
Testing and Commissioning 17.12
The Current Error (The Ratio Error) 7.4
The Phase Error 7.4, 7.6
The Voltage Error (The Ratio Error) 7.6
Theory of Current Transformers 7.2
Theory of Instrument Transformers 7.2
Theory of Rogowski Coil 7.24
Theory of Voltage Transformers 7.5
Thermal Aspects 3.20
Thermal Breakdown 19.7

Thermal Ionisation 1.6
Thermal Relays 9.8
Third Generation SF6 Circuit Breakers 4.8
Tightness Test 19.3, 19.11
Tracking of Solid Insulation 19.6
Transfer Contacts 3.20
Transformer Magnetizing and Reactor Currents 4.17
Transformer Protection 13.7, 15.17
Transient Recovery Voltage (TRY) 4.15, 19.13
Transmission Line Protection 15.8
Trip Circuit Supervision Schemes 12.9
Tripping Schemes 12.6
Troubleshooting Guide 18.8
Two Main and a Transfer Bus System 15.2
Type Testing of Instrument Transformers 7.16
Type Tests 19.4
Types of Circuit Breaker Mechanisms 1.32
Types of EHV Systems 15.2
Types of Electrical Energy Meters 11.1
Types of Faults 13.4
Types of Instrument Transformers 7.8
Types of Voltage Transformers 7.10
Typical Circuits in an EHV Transmission System 15.2
Typical Groupings of Protections for a Large Unit (500 MW) 14.5

Unbalance Protection 13.16
Under-voltage Protection 13.6, 13.17
Unit Auxiliary Switchgears 10.5
Unit Switchgear 12.19
User's Benefits 6.11
Using Motor Protection Circuit Breaker (MPCB) 12.8
Using Relays 12.7
Using Summation CT 12.7
Using Time Limit Fuses 12.9

Vacuum Arc 1.11
Vacuum as an Interrupting Medium 3.26
Vacuum Circuit Breakers (VCBs) 1.24, 4.5
Vacuum Switchgear 3.25
Vacuum 3.25, 19.6
Various Considerations 10.1
Ventilating Openings 3.9
Verification of IP Coding 19.11
Verification of the Degree of Protection Test 19.11
Vibration Test 19.19
Vibrational Stresses 3.22
Visual Inspection Check 17.12
Voltage Selection Schemes 12.5

Voltage Sensors 7.26
Voltage Transformer Recommendations 7.21

Watchdog Timer 9.18
Weak Infeed Logic Feature (Both End Breakers Closed) 15.13

Why Gas Insulated Metal Oxide Arrester (GISA) for GIS? 8.13
Working of Auto-recloser 6.6

Zero-phase Sequence Systems 13.3
Zone 1 Extension Scheme 15.9